RICHARD S. HYSLOP • LIN WU • SARA A.
California State Polytechnic Univer...

CALIFORNIA
Eclectic

SECOND EDITION

A TOPICAL GEOGRAPHY

Kendall Hunt
publishing company

Kendall Hunt
publishing company

www.kendallhunt.com
Send all inquiries to:
4050 Westmark Drive
Dubuque, IA 52004-1840

Contents

Author Bios   xi

Chapter 1 Introduction and Overview   1

Introduction: California in the Twenty-First Century 1
Travelogue Tales and Romance Musings 2
California through the Eyes of the Spaniards 2
Early American Images 4
Literary Visions 5
A Taste of Today 6
The Lure of Mild Climate 7
The Legacy of Horace Greeley 7
Midwestern Graffiti 7
The Grapes of Wrath 8
Lifestyle Change: Can California Maintain Its Mystique? 8
Is the Mobile Culture Affordable? 9
Is the Poolside Goddess a Thing of the Past? 10
Can Californians Become Energy Spartans? 10
California in Its Locational Context 11
Summary 13
Selected Bibliography 13

Chapter 2 Physical Landscapes of California   15

Introduction 15
The Forces Behind 18
Plate Tectonics 18
Geological History of the State 19
Rock Structure and Distribution 22
Earthquakes in California 22
General Concepts 22
Lessons from the Past 24
Prediction and Future Earthquakes 27
External Forces and Processes 27
Glacial Processes and Landforms 28
Fluvial Processes and Landforms 28
Aeolian Processes and Landforms 29
Coastal Processes and Landforms 29

Geomorphic Regions 31

    Dividing the Regions 31

    Peninsular Ranges. 31

    Transverse Ranges. 32

    Coast Ranges 33

    The Great Basin and the Southeast Deserts. 34

    Sierra Nevada 35

    Klamath Mountains 38

    Modoc Plateau and Southern Cascades 38

    Central Valley 39

Summary. 41

Selected Bibliography 41

Chapter 3 Historical Geography of California   **43**

Introduction. 43

The Original Californians. 44

European Exploration. 46

    Spanish Dominance 46

    Mission Settlement Patterns 47

The Mexican Period. 49

    The End of the Mission 49

    The Romance of the Rancho 49

Foreign Incursions and Early Statehood 50

    Mountain Men, Sailors, Pioneers, and Heroes. 50

    The Gold Rush 50

    The Decline of the California Indian Population. 52

    The Rise of the Beef Industry. 53

Transportation and California's Evolution 54

Dry Farming and Irrigation Colonies. 56

The "Black Gold" Rush: The Rise of the Petroleum Industry 57

The Ascent of the Western Stars: The Making of the Movie Capital. 58

World War II: Enter Defense Plants, Exit Japanese Americans. 59

Postwar California: American Suburbia 61

Patterns for the Present and Future. 62

Summary. 63

Selected Bibliography 64

Chapter 4 Climates of California   **65**

Introduction. 65

Temperature 65

Controls of the Temperature 66

Latitude.. 66

Topography... 67

Water and Land 69

Ocean Currents 70

Precipitation 70

Controls of the Precipitation 72

Pressure Belts and Storm Tracks.. 72

Orographic Process.. 72

Convectional Precipitation.. 72

Fog and Low Clouds 74

Fog 74

Photochemical Smog... 75

June Gloom 77

Regional and Local Winds... 77

Westerly Winds... 77

Santa Ana Winds 77

Sundown Wind of Santa Barbara... 78

Sea and Land Breezes 79

Winds in the Mountains and Valleys....................................... 80

Missing Hurricanes and Rare Visits of Tornadoes 80

Climate Regional Variations 81

Climate Classification... 81

Mediterranean Climate 81

Desert Climate 83

Highland Climate... 84

Climate Change 84

Summary... 85

Selected Bibliography 85

Chapter 5 Contemporary Folkways, Cultural Landscapes   87

Introduction... 87

Cultural Geographic Oddities 88

Isolation or Uniqueness 88

The Perceptual Regions of California 88

Logging Paul Bunyan Style: Redwood Country, Northern Forests, and Plains........ 90

Argonauts and Ghost Towns: Gold Country................................. 91

A State without Wine Is Like a Day without Sunshine: Wine Country 92

Sophistication: The San Francisco Bay Area................................. 94

Nashville West: Bakersfield, the Central Valley, and the Farm Belt.................. 96

Neon Glitter: Southern California and Los Angeles 98

        Beachboys, Boating, and the Body Beautiful: Southern Coastal Playgrounds 101

        Resort Mecca and Commercial Hub: The San Diego Region 103

    Mad Dogs and Californians 104

        Creating Dreams: Disney, Knott, and Others. 104

        Eating Your Way to Nirvana—By Railroad, Bistro, Pub, and Drive-in 105

        Isolating Age: Leisure World and Age Ghettos 107

        The Art of Artsy California Burial 108

        California's Salad Bowl Culture 109

    Recreation or Else 111

        The Resort Mentality: A Place in the Sun 111

        Spectator Spectacles: There's More Than One Coliseum 111

        Active Play: Everybody Is a Star 112

    Summary. 112

    Selected Bibliography 113

Chapter 6 Vegetation of California   115

    Introduction. 115

    Taxonomic Divisions of California Flora 116

        Endemic and Relict Species. 117

        Geographic Divisions of California Flora 118

        Floristic Provinces 119

        Plant Biomes and Plant Communities 119

    Californian Floristic Province 119

    Vancouverian Floristic Province 120

    Great Basin Floristic Province. 121

    Sonoran Floristic Province 121

    Coniferous Forest Biome. 121

        Vertical Cross Section of Elevation 123

    Oak Woodlands Biome. 124

        Anthropogenic Impacts on Oak Woodlands 126

        Sudden Oak Death 126

    Grasslands and Marshes Biome 126

        Vernal Pools 127

    Chaparral and Coastal Sage Scrub Biome 128

        Adaptation to Fire 129

    Desert Scrublands and Woodlands Biome 129

    Habitat Destruction. 131

    Introduced Plants. 131

    Summary. 131

    Selected Bibliography 132

Chapter 7 Lifestyle Choices: From Urban to Rural   133

Introduction.. 133

    Choices of Lifestyles.. 133

    Changing Economic Impacts 134

    Impacts of the "Third Wave"... 134

Rural California 135

    Where Is "Rural California"? 135

    What Is "Rural California"?... 136

Rural Economic Phenomena 137

    General Patterns 137

    Agriculture's Role in the Rural California Environment.......... 137

    Population Trends.. 138

    Challenges and Problems in Rural California...................... 139

Along the Spectrum: Urban, Suburban, and Other Variations.......... 140

    The Context.. 140

    The Three Classic Categories 141

A New Jargon of Living Places... 143

    Megalopolis... 143

    Micropolitan 144

    Penturbia 144

The Geopolitical Landscape of California 145

    Counties... 145

    Cities 146

    School Districts 147

    Special Districts 147

Summary: Patterns for the Future 147

    Population Growth and Challenges 148

    Sprawl and Space 148

    Gridlock... 148

    Budget-Related Issues... 148

Selected Bibliography 149

Chapter 8 California's Water Resources   151

Introduction.. 151

The Hydrologic Cycle 151

California's State Hydrology 154

    Droughts 154

    Floods 157

    Snowpack... 158

    Groundwater... 159

    California's Hydrologic Regions 161

The History of Water Use in California 163

    Native American Period 164

    Spanish and Mexican Period 164

    American Period 165

    The Evolution of Water Rights in California 166

    Riparian Rights 166

    Appropriative Rights 166

    California Doctrine and California Water Code 167

    Groundwater Law 168

Development of California's Water Distribution System 168

    The State Water Project 169

    Central Valley Project 170

    Colorado River Aqueduct and Canals 171

    Los Angeles Aqueduct 171

    Mono Lake Extension of the Los Angeles Aqueduct 172

    Historic Lake Levels 174

    Tuolumne River and Hetch Hetchy Delivery System 174

Impacts and Issues of Water Use 174

Summary: Sustaining California's Water Supply into the Future 177

Selected Bibliography 177

Chapter 9 Economic Geography   179

Introduction 179

Components of the California Economy 179

Population 180

Entertainment Industry 182

Tourism 184

California's Agriculture 189

    Natural Resources 190

    Farms and Farmers 190

    Major Products and Distribution 190

    Livestock and Poultry 191

    Field Crops 191

    Vegetables, Fruits, and Nuts 192

    Opportunities and Challenges 193

California's Housing Market 194

    Coastal versus Inland Markets 195

The Silicon Valley 196

Summary 197

Selected Bibliography 197

Chapter 10 California's Coastal Ocean Region   199

Introduction 199

The Chumash 200

Ocean Economy 201

Environmental Protection 201

The Pacific Ocean and the California Current 202

Nearshore Waters 204

    Kelp Forests 204

    El Niño 205

The Southern California Bight 205

    Municipal Wastewater and Stormwater Runoff 206

Offshore Waters 207

    Sardine Fisheries 207

Marine Mammals 208

Coastal Landforms 209

    Sea Cliffs, Marine Terraces, and Headlands 211

    Beaches and Dunes 211

    Wetlands 213

    Islands 214

Summary 215

Selected Bibliography 215

Chapter 11 Sustaining California and Beyond   217

Introduction 217

Needs of a Growing Population 217

    Food 217

    Shelter 218

    Water 218

    Energy 218

Environmental Impact 219

    Ecological Footprints 219

    Land Degradation 220

    Air Pollution, Water Pollution, and Solid Wastes 220

    Impact of Climate Change 222

Solutions and Challenges 222

    Renewable and Alternative Energy Sources 222

    Better Cars and Fewer Cars 227

    Zero Waste 230

    Alternative Agricultural Practices 231

    Sustainable Home 232

    Protect and Improve Biodiversity 233

Summary 233

Selected Bibliography 234

Chapter 12 California on the Threshold   235

Introduction 235

The Prospects 236

The People 237

        Growth and Population 237

        Immigration, Migration, and Ethnicity 238

        The Social Fabric 239

The Pacific Rim–The Pacific Century 239

The Natural Environment 240

        Federal Regulations 240

        State Regulations 241

The Water Problem 244

The Problem of Movement 245

The Future 248

Summary 249

Selected Bibliography 249

Glossary   251
Index   261
Color Section   275

        Map 1 Selected Geographic Features in California 275

        Map 2 California Population Distribution 276

        Map 3 Geomorphic Regions of California 277

        Map 4 California Climate Classification 278

        Map 5 Precipitation Distribution in California 279

        Map 6 January and July Temperature Distribution 280

        Map 7 Floristic Provinces of California 281

        Map 8 Distribution of the Five Primary Biomes of California Vegetation 282

        Map 9 California Aqueducts 283

        Map 10 California Renewable Energy Resources 284

Richard Hyslop is Emeritus Professor of Geography at California State Polytechnic University, Pomona, where he has taught for many years, earning several awards for outstanding teaching. In addition to his teaching, he has served variously as a practicing attorney, deputy fire chief, department chair, and coauthor of several texts. Areas of particular expertise include environmental law, urban and rural geography, California, and emergency management, including ongoing research on hazards management. He received his JD degree from the University of California, Los Angeles, and his PhD from the University of California, Riverside.

Lin Wu is a Professor of Geography and a faculty member of the Master of Science in Regenerative Studies Program at California State Polytechnic University, Pomona. Her research and teaching interests include urban climatology, physical geography, cartography, geographic information systems, environmental modeling, and environmental sustainability. She has been teaching Geography of California since 1989, and her online version of the course was recognized by the university as one of the best online courses. She received her PhD from the University of California, Los Angeles and currently serves as the chair of the Geography and Anthropology Department. She enjoys traveling and landscape photography.

Sara Garver is a professor of Geography at California State Polytechnic University, Pomona, where she has been a member of the department of Geography and Anthropology since 1997. Her areas of expertise and research include the use of tree rings to examine climate change and drought cycles in the western Unites States, the study of upper ocean ecosystem dynamics in the world's oceans using remotely sensed imagery, and currently, changing vegetation patterns in the Mojave Desert. She teaches classes in the Geography of California, Global Positioning Systems (GPS) and Field Techniques, Remote Sensing, Digital Image Processing, Advanced Physical Geography, Climate Change, and Geographic Information Systems. Dr. Garver did her graduate work at the University of California, Santa Barbara. She spends her summers camping, hiking, and backpacking in the Sierra Nevada.

Introduction and Overview

"Eclectic \e-'klek-tik\ adj: Selecting or made up of what seems best of varied sources"

–Langenscheidt's Pocket Merriam-Webster Dictionary

Introduction: California in the Twenty-First Century

California is unique in many ways, and the variety and complexity of its physical and human geography is unparalleled in comparison with the other 49 states. Although this may seem a bold assertion, the fact remains that California has represented the leading edge of styles, attitudes, fashions, and environmental, political, and social movements for many years. As we more fully discuss in subsequent chapters, the Golden State does seem to be composed of some of the best (and some of the most troubling) of varied elements—a *truly eclectic mix.*

Evidence of California's popularity can be found in the state's continuing population growth, over 38 million people according to 2012 US Census estimates. This is an increase of about 5 million people, around a 15 percent change, since the 2000 census. Traditionally, California has been a destination of choice for both domestic and foreign immigration. In fact, only 54 percent of current Californians were born in the state, 19 percent have migrated from other states, and the remaining 27 percent have come from other countries. By contrast, the United States as a whole has a different pattern of domestic and foreign immigration. Nationwide, only 12.8 percent have come from another country. California also boasts an ethnically diverse population, with the white population no longer the ethnic majority, down from 57 percent to 40.1 percent in comparing 1990 and 2012 census data. In comparison, the Hispanic population has grown from 26 percent to 40 percent during the same time period.

In terms of economic vitality, according to the California Legislative Analyst's Office, if California were an independent nation its economy would rank eighth in the world, after the United States, Japan, Germany, the People's Republic of China, the United Kingdom, France, and Italy! Although California has had years of prosperity, along with bouts of economic insecurity, the diversity and growth of its businesses and industries remain a remarkable mainstay of the state's economy, helping California lead the country in gross state product (2006 data). Important contributing industries include entertainment, tourism, agriculture (including dairy, fruits, vegetables, and wine), aerospace, and computer and high tech. In short, both its diverse physical environment and its varied human and cultural environment ensure that the Golden State will remain an economic, social, environmental, and political leader in the United States and beyond. More detailed demographic information will be provided in Chapter 9.

Travelogue Tales and Romance Musings

There has always been a certain mystique about California. The name itself evokes thoughts of golden sunshine, golden opportunities, and the good life. This is hardly a purely contemporary phenomenon; much of the area's written history abounds with these images. Whether in references to the fabulous Seven Cities of Gold, the glowing reports of the Forty-Niners, the frequently glamorized television commercials, or the Hollywood images that have flashed on the world's screens, California is seldom portrayed as anything less than magical. Somehow, California seems a natural home for "lifestyles of the rich and famous."

How did such a glowing identity emerge? It is astonishing how consistent this enthusiastic boosterism has been throughout the region's written history. Before Europeans ever set foot in California, a tourist tale equal to any travelogue promotion had been concocted on the Iberian Peninsula. During the sixteenth century, romance writers in Spain indulged in many fanciful and exotic literary creations. (Cervantes's tale of Don Quixote was a classic parody of these imaginative stories.) One such popular tale, by Ordóñez de Montalvo, spoke of an idyllic kingdom on an island near the Indies called "California." This mythical (or not so mythical) paradise came complete with many of the characteristics we now readily associate with the Golden State: balmy climate, beautiful women, handsome men, easy wealth, and eternal happiness. The tale was a travel agent's dream, a classic early advertisement for a mythical land that really existed—and one that would try to live up to these early fantasies three or four centuries later.

Speculation also exists about whether this mythical land was known to the Chinese and described as "sang" (literally, "rich mountain") well before the Spanish tales emerged. Although this hypothesis has not been definitively established, at least one respected historian, Charles Edward Chapman, has concluded that the adventurous Chinese did indeed reach California as early as the fifth century. Whether actually true or false, the tale merely lends further gloss to the romanticization of the region's history.

California through the Eyes of the Spaniards

Among the first eyewitness accounts of California were those provided by the Spaniards. Indeed, some of the earliest written historical remembrances of the land are found in the diaries and reports of early Spanish explorers and missionaries. For the most part, they reinforce rather than contradict the Spanish literary myth. Working their way northward from Mexico beginning in the mid-sixteenth century, the explorer-missionaries made some initial superficial observations about California, particularly about its coastal regions, where they discovered both "beautiful valleys and groves" and flat and rough country. As for the native inhabitants, many of the explorers described them in glowing terms as mostly peaceful, gentle, and attractive people. *Gentle*, however, did not mean "weak," and several Spanish soldiers described them as brave and determined when threatened or pushed. Other comments on the climate, water, and fertility of the soil were frequent and effusive. In fact, these explorers and missionaries described a California familiar to most of present-day America: a beautiful, sun-drenched land with friendly natives lolling about, usually grateful recipients of culture from the rest of the world. The expected and familiar result of such enthusiastic reports was the establishment of missions and permanent settlements by Hispanic (Spanish and Mexican) colonists in this remarkable land, initiating what was to become a common theme throughout the state's history.

This is not to say that there were no detractors of the area. Indeed, one crusty Spaniard, Gaspar de Portolà, speculated on whether it would have been better to let the Russians have the country. In all fairness, Portolà's disposition may have been somewhat soured by several weeks of dieting on mules that he had been forced to kill along his route to feed his expedition. His journey of exploration from Baja (now Mexico) into Alta California (now the US state) in 1769 was fraught with perils, and the terrain was often sparse and dry as seen in modern day Brea Canyon where Portolà's presence is documented (Figure 1.1). Moreover, Portolà's dim view of Baja California probably carried over into Alta California; his expedition occurred at the tail end of a long period of exploration, after the Spaniards had finally decided to settle Alta California.

Over the course of approximately 230 years, beginning in 1542 with Juan Rodriguez Cabrillo, a respected sailor and navigator, sea and land expeditions were sent out from Mexico on a sporadic basis to explore and report back. But not until the 1770s did the Spaniards make any significant efforts to colonize Alta California. Perhaps this delay was partly because 2,000 miles separated this area from the Spanish population centers in Mexico—certainly not a casual stroll around the plaza! Once Spain perceived a threat to its territory by other powers, however, it began its three-part (pueblo–presidio–mission) colonization of Alta California; by the close of the eighteenth century it had established the area as

Spanish Alta California. The attitudes and commentaries of the various mission fathers, settlers, and officials during the subsequent period reflected the desires of the Spanish government to encourage and expand the scope of settlement (Figure 1.2). The people who most benefited from the largesse of the government—the Californian rancheros, soldiers, and priests—were most often those who wrote extolling the virtues of the land. Through their eyes, pastoral Spanish California was transformed into a romantic "Spanish Arcadia," an image that has had amazing staying power.

The Spanish settlers of California took little active note of the revolt against Spain taking place in Mexico. When they learned that an independent government had been formed in Mexico, however, they soon adopted the new political loyalty. The predecessor Spanish period lasted until the early 1820s, when it was succeeded by this new Mexican period. Mexican rule in California was quite abbreviated—a quarter of a century notable for three significant changes. The first change was the complete secularization of the missions with resulting economic and social dislocations that fell especially heavily on the indigenous inhabitants of California, whom the Spaniards dubbed "Indians." The second change was the rise of the cattle rancho, replacing the mission as the focus of settlement. The third change was the increase of foreign influence and importance in California, particularly

Figure 1.1 Brea Canyon—legend on monument at site reads "Don Gaspar Portolá with 60 men camped here July 31, 1769, on his first exploring march from San Diego to Monterey. Dedicated June 2, 1932, by Grace Parlor No 242. Native Daughters of the Golden West."
(Photo: Richard Hyslop)

Figure 1.2 Mission San Juan Capistrano.
(Image © Thomas Barrat, 2009. Used under license from Shutterstock, Inc.)

of US interests. The Americans who trickled into California during this brief period sang the praises of the locale and eased the way for acceptance of US rule in 1848. Although the Treaty of Guadalupe Hidalgo made California a part of US territory in 1848, it was the discovery of gold the same year that made statehood a reality in 1850.

Early American Images

Although Spaniards were the first serious white colonizers and "tourists" of California, they were not alone in their efforts. Others, including the Russians and the British, as well as the Americans, were active in their tentative incursions into the area.

Following the American Revolution, the British navy indulged in some exploration, mapping, and surveying of the California coast. In a somewhat stuffy vein, the British explorers frequently described the manner of life they observed as being too lazy and nonproductive. For example, British naval officer George Vancouver remarked on the Spanish Californians' "habitual indolence and want of industry" and concluded that true civilization would probably be a long time in coming.

The British observations were often reinforced by various Yankees who frequented the ports of California for purposes of trade. Primarily focusing on the Spanish and Californian officials with whom they dealt, the US seamen often found little to admire. Californians were frequently described as pompous or cowardly, greedy or lazy, disorganized or frivolous, or all of these things. Indeed, many other Americans seemed to share the opinion of US Navy Lieutenant Charles Wilkes: "Although the Californians are comparatively few in number, yet they have a distinctive character. Descended from the old Spaniards, they are unfortunately found to have all their vices, without a proper share of their virtues."

Other visitors, however, found a people eager to socialize, ready to play, and quick to display their talents in dancing, riding, and other physical activities. Rather than uniformly greedy, pompous, or irritating, the Californians described by commentators, such as the respected trader Alfred Robinson, were gracious, generous, and kind. Although these comments tended to be *by* men *about* men, complimentary remarks about the California women were also common. The presence of many other Americans in California during the Spanish and Mexican periods attests to the fact that the California way of life held strong attractions for at least some Americans.

However mixed their comments about the inhabitants may have been, the descriptions of the area's climate, land, and physical features were hardly discouraging. One of the first American writers, Captain William Shaler, in distinctly noneffusive terms, described the land and climate as "dry and temperate, and remarkably healthy," whereas another, Alfred Robinson, perhaps more typically spoke of the "fairy spots met with so often in California."

Contact of the initial American coastal settlements with California were soon followed by a series of reports, comments, letters, and enthusiastic tall tales from explorers who crossed the continent seeking the golden clime and opportunities that beckoned from the land of the dons. Beginning with mountain man Jedediah Smith, Americans gradually began to reach California by the overland route. After their arduous trek across mountain, plain, and desert, California did indeed appear a paradise to many of these travelers. As men and women began to settle and establish themselves in the area, they composed enthusiastic letters and comments on the wonders of California. For example, John Bidwell described his "Journey to California" to the folks back home and detailed the fabulous character of soil, flora, and fauna in the early 1840s. Seeking to encourage more Yankees to settle in the area and thus increase US influence, these promoters described a fertile, verdant area ripe for settlement—that is, until 1848, when such promotion was no longer necessary. The cry of "Gold!" became the overnight stimulus to the growth of California, beginning an odyssey of fortune seeking that has continued in various forms up to the present.

Literary Visions

Although truth is often stranger than fiction, the "factual" reports of California frequently came close to being outright fantasy. Likewise, fiction writers who used the region as a backdrop in their works may have been no more imaginative or creative than the authors of the "true accounts." In any event, the images of California that emerged from the pens of some of America's most popular literary figures did not vary greatly from their previous (and contemporary) "truthful" colleagues. Continuing in the now-established positive and colorful vein, these literary visions usually perpetuated the myth of the Golden State—and certainly did nothing to discourage new immigration to and interest in the region. Although the number of writers who described California is too great to permit exhaustive treatment here, a stellar few bear mentioning.

Richard Henry Dana is noted not for his promotion of California but for his crusade for the rights of sailors. When his classic work *Two Years before the Mast* was published in 1840, however, it did act as a major influence in attracting American settlers to California. His vivid descriptions of handsome people, attractive land, pleasing climate, and boundless opportunities helped lure ambitious Americans to this economy.

Another well-known figure, Washington Irving, wrote of "the plains of New California a fertile region extending along the coast, with magnificent forests, verdant savannas and prairies that look like stately parks." The great regional authors of the mid-1800s also found eager audiences for their tales of the Far West. Bret Harte, Artemus Ward, and Mark Twain spun romantic tales of the gold rush towns, colorful characters, high Sierra country, and unusual occurrences of the California scene. Twain's story "The Celebrated Jumping Frog of Calaveras County" was only one of many amusing fictions that provided enticing whiffs of the golden land. Even Walt Whitman created a tribute to one aspect of California in "Song of the Redwood Tree"; and although Rebecca of Sunnybrook Farm never visited California, her creator, Kate Douglas Wig-

gins, sang the praises of the state, and particularly of Santa Barbara, which she called her "Paradise on Earth." Even more widely read were the romantic crusades of Helen Hunt Jackson, who in *Ramona* and other works about the old Spanish period both shed further light on the development and historical roots of the state and created more romanticized myths about California.

As means of transportation and communication improved, writings by and about Californians became even more popular with the American reading public. Of major importance to the naturalistic school of American fiction were several California authors now considered important writers in their own right. Frank Norris effectively used California's diverse landscape as a dramatic backdrop for his epic novels, such as *McTeague* (with scenes ranging from glittering San Francisco to a dramatic climax in Death Valley) and *The Octopus* (which deals with wheat production of the San Joaquin Valley and the farmers' battle with railroad interests). Likewise, Jack London derived inspiration and mood from his Bay Area birthplace. His descriptions of fog in the Golden Gate and of the slums and dives of San Francisco, Sausalito, and Oakland and the characters he had known there were particularly vivid. Indeed, he began one of his best-known works, *The Sea Wolf*, with a tremendously evocative depiction of the clammy, blanketing fog of San Francisco Bay—a description instantly recognizable to anyone who has experienced that phenomenon. Other well-known writers, such as Upton Sinclair (*Oil!*), portrayed the scattered and sometimes tawdry development of the state, and Sinclair Lewis exposed San Francisco to the public eye. Many other popular and respected authors, poets, and journalists also paid tribute to the Golden State in their works, either directly or through the settings. For example, mystery writer Raymond Chandler provided one of the most evocative passages yet penned to explain the effect of the famous Santa Ana winds. In his short story "Red Wind," he described the "desert wind blowing that night. It was one of those hot dry Santa Anas that come down through the mountain passes and curl your hair and make your nerves jump and your skin itch. On nights like that every booze party ends in a fight. Meek little wives feel the edge of the carving knife and study their husband's necks. Anything can happen."

In all, California provided a rich source of material for literary endeavors. This literary activity effectively drew attention to the state, as did *Sunset Magazine* and publications by the Southern Pacific Railroad, further enhancing its mystique and nurturing the desire in multitudes of people to see firsthand the areas that had been so vividly described. Thus, both directly and indirectly, the literary visions of three more centuries further expanded the almost hypnotic appeal first created in Ordóñez de Montalvo's sixteenth-century tale of a mythical California.

A Taste of Today

Americans' fascination with California has not substantially subsided even to this day. The region still has an aura of otherworldliness. While temperatures plummet in the East and Midwest, millions of Americans enviously watch the sun shine on scantily clad cheerleaders and spectators at the Rose Parade and Rose Bowl game televised every January 1. Shivering through another winter, many people cannot help but view the balmy surroundings pictured on their television sets or computers and tablets as a slice of paradise.

Modern writers have continued to express this interest in their writings, wherein California's unique position (climatically, geographically, and historically) plays a key role. Certainly the works of John Steinbeck top the list in terms of influence, style, and importance; his great *Grapes of Wrath* described a California able to produce winter crops that were impossible to grow in many other parts of the country (at a time when no one could afford to buy them). In this and other works, Steinbeck exposed the very soul of much of the state, particularly the rich farm areas such as the Salinas Valley, and in the process became a modern California classic himself.

Nor is Steinbeck alone. William Saroyan brought Fresno alive in his works; Joseph Wambaugh brought notoriety to Los Angeles through his widely read police stories; and Eugene Burdick exposed politics, California style, to millions of readers. Other contemporary popular fiction writers like Walter Mosley and Robert Crais have used Los Angeles as their settings. Furthermore, national magazines and newspapers continue to examine the phenomenon that is California, with "special California issues" and with famous writers regularly and vigorously praising (or panning) the state. In print media as diverse as *Popular Science*, *Newsweek*, *Westways*, and *American Photo*, the fascination continues to be expressed. Thus, through these and other current sources, the myth-making process continues. Television shows, movies, magazines, and newspapers constantly extol the virtues and mystique of the golden land, in the process keeping intact the august tradition of travelogue tales and romantic musings about California.

Recently, however, a slightly different tone has begun to creep into some of the iconology of California. It is not uncommon now to find negative commentary interspersed in media depictions of the region. Graphic reports of urban gang violence and freeway shootings sound an ominous note in parts of the state. Official recognition that the South Coast Air Basin, which spans the greater Los Angeles area, contains the worst air quality in the nation certainly removes some romance from the image of glamorous Southern California. Widely broadcast "fictionalized" accounts of the "great Los Angeles earthquake" give pause to possible immigrants. The now-famous 1989 World Series broadcast from Candlestick Park in San Francisco, which was interrupted by the Lorna Prieta earthquake, demonstrated that fact could be as dramatic as fiction—especially as the number of casualties soared into the hundreds. Similarly, the deadly 1994 Northridge quake was widely reported and videoed for the rest of the nation and the world.

Yet, even in the face of these negative depictions, as noted earlier, the population of California has continued to grow. According to 1990 census figures, in the decade from 1980 to 1990, the population increased by over 26 percent, and by the 2000 census, the population numbered just barely under 33 million people. By 2012, that number exceeded 38 million and showed every indication that growth would continue. As the undisputed most populous state in the country, California can continue to claim its place as the destination of choice for millions of newcomers. Although the image may be somewhat tarnished, the state retains its golden aura.

The Lure of Mild Climate

Placed against this mixed backdrop, the traditional lure of the Golden State has relied on an irresistibly mild climate and a romantically embellished image. As detailed earlier, the romance of California has been promoted by many people for many reasons. A persistent element of this mystique, however, has remained the notion of "forever summer" weather. Rightly or wrongly, the climate has provided a consistent measure of value for immigrants to the state.

In reality, although only a small portion of the state has the favorable climate and physical environment for its inhabitants, it is true that the majority of Californians do find a spot in the golden sunshine or in the coastal belt. Any population distribution map shows a highly concentrated population distribution in sunny southern California and in the San Francisco Bay Area, and a thinly scattered population in the northern and eastern areas of the state, where the climate and physical environment are less than perfect. The county-based population varies significantly from around 1,200 people in Alpine County, to almost 10 million in Los Angeles County, with its more than 300 sunshine days per year. The population distribution pattern is also highly confined by the topography and affected by climate variations. It is amazing to see that with California having the highest population among the 50 states, the majority of its vast land is still unpopulated, which has contributed to the various romantic and literary visions detailed earlier.

The Legacy of Horace Greeley

When Horace Greeley issued the now-famous dictum "Go west, young man, go west," he could not have hit on a more appropriate theme for California. (Greeley borrowed the phrase from J. B. L. Soule, who had first published that advice in the *Terre Haute Express* in 1851. As a result of the greater prominence of Greeley, however, his restatements of this advice became better known and credited, especially since he was able to publish it in the *New York Tribune*. Thus, Greeley is more widely credited with this directive than is Soule.) In addition to economics and adventure, however, climate proved to be a dominating motive drawing tourists and immigrants to the state, and especially to the southern portion. To a native of New York or Minnesota, the weather was a source of amazement and delight, producing a sense of wonder that was conveyed in letters, books, advertisements, and lectures. Some went so far as to proclaim marvelous cures for all ailments that would soon accrue to pilgrims if they would but bask in the state's climate.

Thus, by the early 1900s, Southern California had acquired a reputation as a Shangri-La for the infirm, invalid, and afflicted, as well as the able-bodied and the ambitious. As Carey McWilliams recalled, the classic, if somewhat cynical, motto of the times became: "We sold them the climate and threw in the land." The draw of climate and claims of health benefits constituted part of the appeal of California to the rest of the nation. More than health considerations alone drew many immigrants to the state, however.

Midwestern Graffiti

Various scholars and observers have concluded that Southern California generally owes its particular ambience to the dominant influence of displaced midwesterners. Indeed, as early as the 1930s, a popular saying identified Southern California as the western seacoast of Iowa. The saying might well have substituted Illinois, Ohio, Nebraska, or any other midwestern state, because the number of native-born Californians seldom accounted for a very large percentage of the total population. Annual Iowa-day picnics or Ohio-day reunions were once common occurrences in California, providing an opportunity to swap reminiscences of life "back home." Although such reunions became less common after the late 1950s, the immigration patterns certainly continued.

An interesting cultural effect results from this rather constant midwestern migration. The continual flow of new residents into the state has created a layered form of cultural identity wherein the newcomers strive to develop characteristics that may be attributed to the "California personality" but in the process create an overall sense of never-ending growth and development with a midwestern twang. As we discuss more fully elsewhere, this national and regional cultural diffusion is rapidly being augmented and perhaps even transformed by international influences and cultures as well.

The Grapes of Wrath

The role of California as a key agricultural state has contributed to its steady appeal. Frequently, when poor weather or economic conditions created untenable situations in other parts of the country or world, California came to be viewed as an escape—a sunny, Mediterranean-like setting where food could be grown at least most of the year and the climate would be hospitable in the meantime. Having lived through brutal winter blizzards or scorching dust storms, was it any wonder that immigrants to the area saw a chance to improve their lives? The exaggerations of the fecundity of the soil and climate were misleading and often extreme, but the myths contained just enough truth to encourage people to continue to see California as the fulfillment of unrealized dreams. John Steinbeck's *Grapes of Wrath* (movie and book) partially questioned the popular image of California, but for the most part the seekers of dreams persisted in chasing their dreams to the West.

In fact, the trend of American immigration into California had a curious twist. Initially seen as a resort for the rich, increasingly the state came to be considered a land of opportunity for the less affluent as well—an image promoted by land companies, railroads, and others. Throughout the early 1900s, this in-migration occurred at a steady rate, with occasional swells in the level of movement. The Depression years, for example, brought well over 350,000 people across the country in old cars and trucks. As time passed, California increasingly lured working-class people from such diverse places as Ontario, West Texas, New York, and Oklahoma. Much of this flow, however, could still be traced to the small-town folks and farmers who looked to California as an enchanted land where orange and palm trees grew in abundance and even country folk could share in some of the magic. And in modem times, California has also lured many foreign and non-English-speaking poor. As demographic patterns demonstrate, Hispanic and Asian immigrants in particular look to California for a demonstrably better way of life.

Lifestyle Change:
Can California Maintain Its Mystique?

Like most Americans, Californians assumed for many years that the sources of energy were unlimited. Within the past few decades, however, this assumption has undergone serious challenge. As a consumer of one-fifth of the total energy resources of the world (with less than one-fifteenth of the total population), America has been forced to reexamine its future. The United States no longer has a monopoly on petrochemical energy sources. Alternate sources are needed, and alternate lifestyles may be necessary. The question is, can Americans in general, and Californians in particular, meet this challenge?

In many respects, California epitomizes the Western world in its attitude toward and use of nature and energy. The commitment of Western civilization to science and technology has permitted great improvements in living standards for humankind. Much of this progress, however, has been accomplished by following a basic social and religious premise: subdue the earth. Subduing the earth has, in the short term, brought about a lifestyle of unheard-of luxury, nowhere more evident than in California.

Part of the great mystique of California has involved the use of the land for pleasure. No place is inaccessible; if there is no way to get there, build a road. Freeways and roads have thus become an important feature of the state.

Because the beaches and shoreline are beautiful, everyone should be able to enjoy their natural beauty; thus, the beaches and shoreline have been "improved" with vast concrete parking lots, artificial fishing jetties, and sand groins to prevent erosion. To this way of thinking, mountains may be striking evidence of nature's majesty, but ski lifts and lodges, artificial snow, mountain-forest condos, and improved roads make them even better. Similarly, the vast California deserts, which are among the last American frontiers, possess a fragile, balanced ecology, so dirt bikes and off-road and recreational vehicles must be brought in, the better to see this wilderness before it is destroyed. And, of course, in each instance nature is brought to heel for the enjoyment of humans and the "toys"—beer coolers, jet skis, snowmobiles, and so on—they bring along to desert, beach, or mountain as inevitably as trash and garbage. Thus, democracy, religion, technology, capitalism, urbanization, and mobility bring new meaning to the concept "natural environment."

Is the Mobile Culture Affordable?

More than any other single artifact, the automobile is the icon of California, a state that has relied heavily on individualized transportation. Although mass-transit systems exist, these are the exception; for the most part, the typical pattern is single-passenger transport. Even in the face of reduced gasoline supplies, most Californians have been reluctant to give up the convenience of the automobile, as Figure 1.3 demonstrates. The vast numbers of commuters have shifted their approach only to the extent of moving toward smaller, more fuel-efficient cars, or to the increasingly popular hybrids.

Although some attempts have been made to get Californians out of their cars, these efforts have had only limited success. Los Angeles's much-vaunted light rail system suffered setbacks from the beginning, and efforts to establish a "subway" in the City of Angels were plagued with such problems as fire, tunnel collapse, and questionable geologic siting.

Figure 1.3 Typical California freeway scene.
(Image © egd, 2009. Used under license from Shutterstock, Inc.)

The Bay Area's Rapid Transit (BART) system has enjoyed some success, but the total volume of single-occupant automobile commuters remains astronomical. Even San Diego's experiment with a novel "trolley" system has not met the most optimistic expectations. In short, Californians seem blissfully wedded to their automobiles as a means of maintaining their sense of independence and privacy—at least while commuting to work.

In terms of recreation in the state, Californians have seldom been deterred by distance in pursuing their entertainment, either as spectators or participants, be it theater, sports, hiking, or sightseeing. And it is not just a matter of the family van, sedan, or sports utility vehicle (SUV); the proliferation of recreational vehicles, off-road vehicles, motorcycles, and sports cars has not slackened perceptibly into the new century, even though the availability and cost of gasoline have become important factors.

As the new century evolves, the question of mobility will become more vital. Clearly, Californians cannot continue in their present pattern. If mobility is to remain a part of the state's identity, forms of transportation must change because the costs and impracticality of the current mobile culture cannot be borne much longer. This becomes even more evident as gridlock spreads from one metropolitan area to another, as state and local governments struggle to maintain roads, and as air quality continues to deteriorate.

Is the Poolside Goddess a Thing of the Past?

The aquatic image is also a part of the California mystique. Southern California in particular is a vast checkerboard tract filled with backyard swimming pools and Jacuzzis. The lifestyle promoted in fashionable magazines, decorators' drawings, newspaper ads, and "house beautiful" publications includes the swimming pool and spa as vital elements of the good life. The social and sexual revolution in suburbia focused on the aquatic symbol, and common wisdom dictates that a home's value jumps with such an addition. No apartment complex or condominium development could survive without the accompanying water sports playground, and cost traditionally has not been a consideration.

Heating the water for such a playground, however, is not an inexpensive proposition. Rising energy costs have mandated a reexamination of this luxury symbol. The stereotypic golden-tanned poolside goddess relaxing by bubbling, heated water may soon become an anachronism. With both water and energy becoming critical resources in the state, priorities must be considered. Swimming may well continue, but in winter months, at least, Californians may have to develop hardier constitutions.

Can Californians Become Energy Spartans?

The United States traditionally has been one of the world's largest consumers of energy, accounting for about one-fifth of the world's total energy use. Californians generally hold their own in this energy usage in terms of both amount and variety.

Californians, like Americans elsewhere, continue to insist on some relatively marginal uses of energy. The extent of these uses can easily be seen during the Christmas season with the prolific and prodigious displays of lights and electrical decorations. During the rest of the year, Californians insist on widespread outdoor lighting of houses, patios, tennis courts, pools, and driveways. Likewise, luxury and convenience appliances remain big sellers in department and discount stores. Californians seem unable to function without garage door openers, trash compactors, televisions, and electric toothbrushes, can openers, and knives. Yet in the face of rising costs and limited supplies, Californians may have to become reacquainted with their wrists, shoulders, and backs as energy sources.

The California penchant for air conditioning may undergo a severe test. Although hermetically sealed skyscrapers and schools, enclosed shopping malls, and tract houses may not have enough design flexibility to permit natural air circulation, thermostatically controlled indoor climates may soon be an expensive luxury. California power providers' "voluntary" programs to reduce usage during peak seasons are evidence of this growing concern. The assumptions of unlimited energy that have made such design and construction possible in the past can no longer be indulged. Californians must face the harsh reality that their energy uses are often wasteful and that the old assumptions do not serve. Petroleum and water, electrical, and other forms of energy are valuable assets to the state. Simple belief in unlimited existing sources cannot take the place of responsible searches for alternate forms of energy.

Not only must the state look into new sources for its energy supply, but it also must examine the impacts of the energy consumption it generates. Facing the threat of global warming, Californians find themselves at the center of the current national and international environmental crisis and debate. On the one hand, Californians are part of the problem and potential victims of global warming. California emits more greenhouse gases into the air than many small countries combined, and California will be severely affected if the projected global warming trend continues. The potential land loss due to melting snow pack and increased extreme weather fluctuations are just two of the many consequences the state may face. On the other hand, California plays an important role in leading the national and worldwide efforts to reduce greenhouse gas emissions and develop solutions to the problems resulting from the global warming. Such efforts include developing alternative fuel sources, promoting various solar and other green solutions (such as wind power seen in Figure 1.4) to homeowners and business practices, and tightening policies and regulations regarding greenhouse gas emissions.

Figure 1.4 Wind farm, Rio Vista, California.
(Image © Terrance Emerson, 2009. Used under license from Shutterstock, Inc.)

California in Its Locational Context

In the following chapters, we explore various geographic themes throughout the state. Before we start, let us briefly review the topics in the context of the state's geographic location—a key element that distinguishes geography from other disciplines. The major locational factors are illustrated by the locational map (Map 1).

One locational aspect of the state is its location over two major tectonic plates. Because of this duality, the geological processes along this major plate boundary contribute to the formation of the state's dramatic and diverse topography. These geological and geomorphologic processes, and the resulting landforms, are examined in Chapter 2.

Sandwiched between a large ocean and a major mountain system, the state is heavily influenced by both its connections to the outside world via the ocean routes and mountain paths, and its locational isolation because of the oceans and mountains. Major historical events

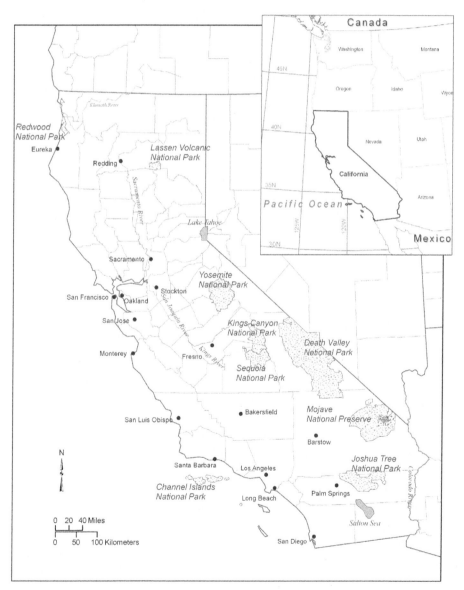

Map 1 Selected geographic features in California.

that shaped the geography of the state today are explored in Chapter 3.

As illustrated by Map 1, the latitudinal extension of the state is approximately 10 degrees, from 32°N to 42°N. Chapter 4 looks at the diverse climates and the formation processes that are partly the result of the state's long latitudinal stretch, as well as its location at the edge of the North American continent and the Pacific Ocean.

Chapter 5 explores the cultural geography of the state, which is heavily influenced by the state's location as the "far west" of the United States, its connections to South American countries via its southern border, and to Asian countries via its Pacific Rim connections.

The state's location as a transitional zone from ocean to mountain, and the isolation created by the mountains, have helped shape the flora and fauna of the state, resulting in very high species diversity and a high degree of endemism (species not found elsewhere). The characteristics and distribution patterns of the diverse plants are surveyed in Chapter 6.

The state's vast area is a perfect stage to display the drastic contrast between rural areas and urban centers, as illustrated by Map 2. Very few US states have both the most-developed and most-populated urban centers in the coastal areas and the smallest rural villages scattered in the isolated inland valleys. Chapter 7 showcases characteristics of the state's urban and rural geography.

Being located next to the largest water body on earth—the Pacific Ocean—does not help the state with its water needs. Its mountains, however, combined with its latitudinal range, have created the conditions for high precipitation in both the mountainous and northern parts of the state, which in turn becomes the major water sources for the state. Chapter 8 presents the dilemma and the opportunity the state is facing regarding its water resources.

Chapter 9 presents samples of diverse economic activities. The economic geography of the state is the culmination of the other geographic characteristics presented in the previous chapters, which reflect the state's locational characteristics.

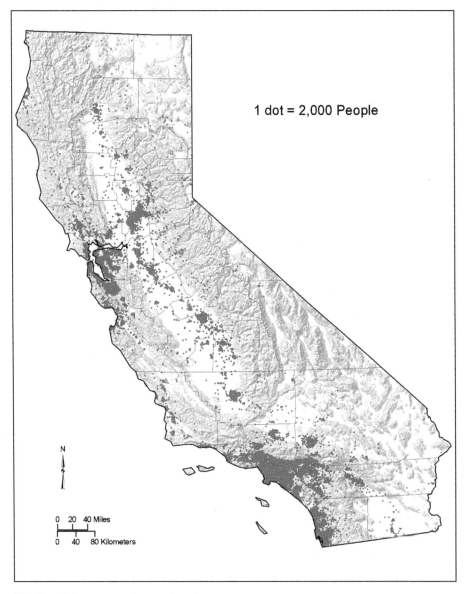

1 dot = 2,000 People

N

0   20   40 Miles

0   40   80 Kilometers

Map 2 California population distribution.
(Sources: US Geological Survey, ESRI Data and Map, US Census Bureau, Lin Wu)

Very few places in the world possess the location of a Mediterranean coast and the rich marine environment that comes with it. Chapter 10 explores California's marine environment and the interaction of the ocean and land in the coastal zone.

The state's sunny southern area made it a perfect place to develop technology and practices to use solar and other renewable energy sources as alternative resources. Chapter 11 explores such opportunities, along with the challenges, and energy alternatives.

Finally, Chapter 12 wraps up all the topics explored in the various chapters and looks into future challenges and opportunities that will continue to evolve and play out in the state of California.

Summary

The variety, extremes, and challenges of California frame the state in monumental terms. From its rich historical development to its increasingly complex cultural mosaic, California presents a wealth of topics for the interested student. As an overview, this chapter presents a quick view of the history of California and its evolving population patterns. Tourism, promotion, recreation, and economic bases are addressed, and challenges for future development are touched upon.

Selected Bibliography

Atherton, G. (1914). *California: An Intimate History*. New York: Harper and Bros.

Bancroft, H. H. (1890). *History of California (1886–1890)*. San Francisco, CA: The History Co.

California Department of Finance website (www.dof.ca.gov).

California State Government website (www.ca.gov).

Caughey, J. W. (1953). *California* (2nd ed.). Englewood Cliffs, NJ: Prentice Hall.

Cleland, R. G. (1951). *The Cattle on a Thousand Hills: Southern California, 1850–1880*. San Marino, CA: Huntington Library.

——. (1944). *From Wilderness to Empire: A History of California, 1542–1900*. New York: Alfred A. Knopf.

Dana, R. H. (1840). *Two Years Before the Mast*. New York: Harper and Brothers.

De Voto, B. (1950). *Year of Decision: 1846*. Boston, MA: Houghton Mifflin.

Federal Writers Project. (1939). *California: A Guide to the Golden State*. St. Clair Shores, MI: Somerset Publishing.

Lee, W. S. (1968). *California: A Literary Chronicle*. New York: Funk and Wagnalls.

McKnight, T. L., and Hess, D. (2008). *Physical Geography: A Landscape Appreciation*. Englewood Cliffs, NJ: Prentice Hall.

McWilliams, C. (1973). *Southern California: An Island on the Land*. Santa Barbara, CA: Peregrine Smith.

Miller, C., and Hyslop, R. (2000). *California: The Geography of Diversity*. Palo Alto, CA: Mayfield Publishing Company.

Trover, E. L. (State Ed.) and Swindler, W. F. (Series Ed.). (1972). *Chronology and Documentary Handbook of the State of California*. Dobbs Ferry, NY: Oceana Publications, Inc.

US Census Bureau website (www.census.gov).

Vidal de la Blache, P. (1918). *Principles of Human Geography* (E. de Martonne, Ed., M. T. Bingham, Trans.). London: Constable Publishers.

West, N. (1939). *The Day of the Locust*. Cutchogue, NY: Buccaneer Books.

Physical Landscapes of California

"Go where you may within the bounds of California, mountains are ever in sight, charming and glorifying every landscape.... But with this general simplicity of features there is great complexity of hidden detail."

—John Muir

Introduction

Landforms, the shapes of the land, are the physical features of the earth's surface—mountains, valleys, plateaus, and so on. Landforms are the foundations that support, confine, and shape other features of the earth. Each landform interacts with climate, vegetation, wild animals, agricultural land, cities, and other geographic phenomena in its unique ways, which results in endless possibilities of regional and local environments. Exploring the nature of each landform and the processes behind it helps us understand the stages and settings where other geographic features coexist.

This chapter explores various landforms and processes through eight unique geomorphic regions in California (Map 3). Figure 2.1 gives a few samples of these diverse landforms. Figure 2.1a is a glimpse of one of the most spectacular landforms in California—the *glacier landforms* in the Sierra Nevada. Called the "backbone" of California, the Sierra Nevada,

characterized by its spiky peaks, sharp ridges, and steep-sided valleys, provides clear evidence of the combined works of plate tectonics and glaciation. To the north of the Sierra Nevada stand two snow-topped mountains: Mount Lassen and Mount Shasta. These two volcanic mountains represent many of the *volcanic landforms* in the state (Figure 2.1b). Descending from the gentle western slopes of the Sierra Nevada, the land flattens into the Central Valley, a sediment-filled depression that hosts one of the most productive agricultural regions in the world. In the northwest region of the state, the oldest mountain system in California—the Klamath Mountains—represents a *fluvial landform* with a complex stream network superimposed on it (Figure 2.1c). Delineating the western border of the state, the Coast Ranges are characterized by both emerging and submerging *coast landforms* (Figure 2.1d). The coastal areas also exhibit *aeolian landforms,* shapes of the land created by wind. The southern end of the Coast Ranges is connected to the Transverse

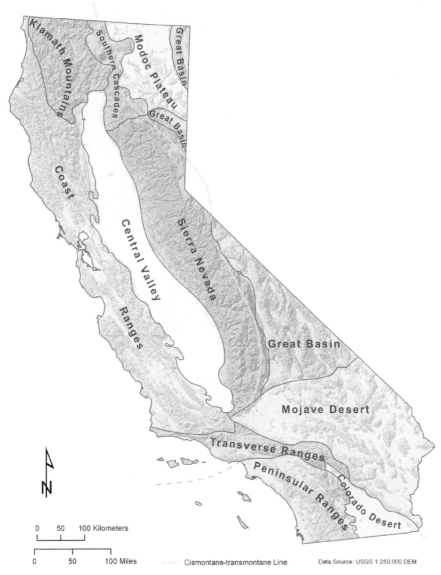

Map 3 Geomorphic regions of California.
(Source: Lin Wu)

Ranges, the only west–east running mountain range in California, which serves as a natural divide between Southern California and the rest of the state. To the south of the Transverse Ranges lies the Peninsular Range, with the Los Angeles Basin on its western side. The Los Angeles Basin, as one of the most extensive flat areas in the coast, hosts the largest population cluster in the state (Figure 2.1e). In the southeast region of the state, many elongated, alternate patterns of hills and valleys give the distinct characteristics of the *basin and range landforms* (Figure 2.1f). What made this group of diverse landforms coexist in such close proximity? The first part of this chapter explores the internal and external forces and processes that created these landforms.

Physical landscapes in California are not always static scenes poised for pictures. The landforms are always in transition, constantly changing

Figure 2.1a Glacial landforms in Sierra Nevada.
(Photo: Lin Wu)

Figure 2.1b Mount Shasta, stratovolcano of the Southern Cascades.
(Photo: Lin Wu)

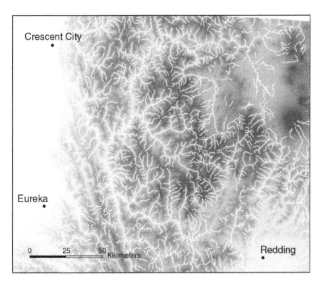

Figure 2.1c Computer rendering of the complex stream systems in the Klamath Mountains.
(Source: Lin Wu, DEM data: USGS)

Figure 2.1d Coastal landforms at Point Reyes National Seashore.
(Source: USGS)

Figure 2.1e L.A. Basin on the Peninsular Range.
(Photo: Lin Wu)

Figure 2.1f Computer rendering of Death Valley, the deepest valley in the Basin and Range region.
(Source: Lin Wu, DEM data: USGS)

and evolving. The changes are sometimes slow and gradual, such as the deepening and widening of a valley by the carving force of running water over thousands or millions of years. Other times, the changes are sudden and violent, such as earthquakes and volcanic eruptions that change the shape of the land in days and minutes. Although earthquakes are part of the natural landform-building process, they can be extremely destructive and remain as a major threat for people living in the state. The second part of the chapter looks into the general concepts, history, predictions (or lack of such), and effects of earthquakes in California.

Whereas the threatening nature of earthquakes, landslides, volcanic eruptions, and other

hazardous land-building processes might be the downside of living in an active land-building zone, the upside is that the diverse landforms and the land-building processes contributed to the abundant natural resources for the state. Petroleum, natural gas, raw materials for construction, gold, silver, and other commodities from the land contribute heavily to the dynamic economy of the state. With the majority of the state located in an arid to semiarid zone, the state would have had much less water if it had no major mountains (we will learn in Chapter 4 that the mountains contributed heavily to the amount and distribution of precipitation in California). The diverse and spectacular landforms also attract millions of visitors each year. This chapter touches on some of the landform resources; these topics are also explored in other related chapters.

The last part of the chapter provides an overview for each landform region delineated on Map 3.

The Forces Behind

Plate Tectonics

California is the meeting ground of two major *lithospheric plates*: the Pacific Plate and the North America Plate. According to the theory of *plate tectonics*, which explains earth's lithospheric structure and mountain-building processes, the earth's surface is divided into several lithospheric plates. These plates have joined and split many times in the earth's history. It is believed that the last time the plates were all joined was more than 250 million years ago in the form of a super continent called *Pangaea*, a Greek word meaning "whole land." Pangaea broke up about 250 million years ago. The separated plates drifted, collided,

and eventually formed today's plate distribution pattern, as illustrated by Figure 2.2.

The sizes of these plates vary from continental scale to thousands of square kilometers. A plate may be composed of *oceanic crusts* with dense, heavy *basaltic rocks*, or it may be composed of *continental crust* with lighter, *granitic rocks*. Most of the plates, however, have continental crusts on one side and oceanic crusts on the other side. These solid lithospheric plates, with thicknesses that vary from 16 kilometers (10 miles) to more than 200 kilometers (124 miles), "float" over the *asthenosphere*, a "soft" layer beneath the solid *lithosphere*. It is believed that rocks under extreme heat and pressure in the asthenosphere are "soft" and move slowly in circular motions. These circular motions are believed to be the driving force underneath the moving lithospheric plates. Although the plate movements measure only centimeters per year, over thousands and millions of years, they can add up to hundreds and thousands of kilometers in distance.

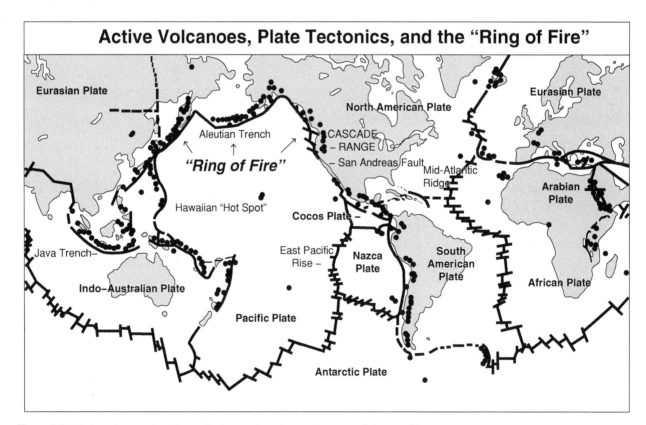

Figure 2.2 Major plates, plate boundaries, and active volcanoes of the world. (Source: USGS)

Based on the ways two plates interact along their contact zone, three types of plate boundaries are identified: convergent, divergent, and transform (or transcurrent) plate boundaries (Figure 2.2). Along a *convergent plate boundary*, where an oceanic plate meets a continental plate, long mountain ranges on land and deep trenches in the ocean form. The formation is a result of the compression and crumpling effect caused by the oceanic plate's subduction underneath the continental plate (Figure 2.3). The submerging plate melts under the extreme heat and pressure inside the earth and turns into magma, causing frequent volcanic eruptions along the subduction zone. Frequent earthquakes are another indication of the collision process. The interaction of the two plates creates enormous stress; when the stress is released, an earthquake occurs. (The initial formations of the Sierra Nevada, the Coast Ranges, and the Central Valley were associated with a convergent plate boundary.) Converging actions may also occur along two oceanic plates or two continental plates. The former creates long volcanic island arcs in the ocean and the latter forms extreme high mountains on the land.

Along a *divergent plate boundary*, where one plate splits into two plates, deep valleys form on the land and mid-ocean ridges develop in the ocean. Hot magma from inside the earth rises along divergent plate boundaries to fill the gap between the two plates. As a result, new crust forms along divergent plate boundaries. This land-building process is often accompanied by volcanic activities and earthquakes. Distribution of the major divergent plate boundaries is shown in Figure 2.2.

When two plates slide laterally past each other, a *transform plate boundary*, also called a transform fault, forms. A transform plate boundary does not destroy old crust, nor does it create new crust; however, the interaction of the two plates does build up stresses, and whenever these stresses are released, earthquakes occur. The San Andreas Fault, a fracture zone that runs almost the entire length of California, is a transform plate boundary that separates two major plates: the Pacific Plate, moving northward, and the North America Plate, moving southward. In addition to the San Andreas Fault, many other faults occur along the plate boundaries (Figure 2.4). These faults are the main contributors to the earthquakes in California.

Geological History of the State

Although the oldest rocks found in California date back more than 2 billion years ago, the most important geological period that shaped today's physical landforms in the state began about 250 million years ago when the supercontinent Pangaea broke apart and a subduction zone began to form along the western edge of the ancient Farallon Plate and the North America Plate. The Farallon Plate, located between the Pacific Plate and the North America Plate, began its subduction beneath the North America plate about 225 million years ago. The subduction created a long mountain where today's Sierra Nevada is located. Volcanic rocks dating back to that period have been found throughout the Sierra Nevada, the Klamath Mountains, and

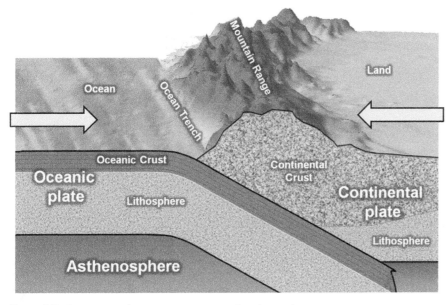

Figure 2.3 Dynamics along a convergent plate boundary.

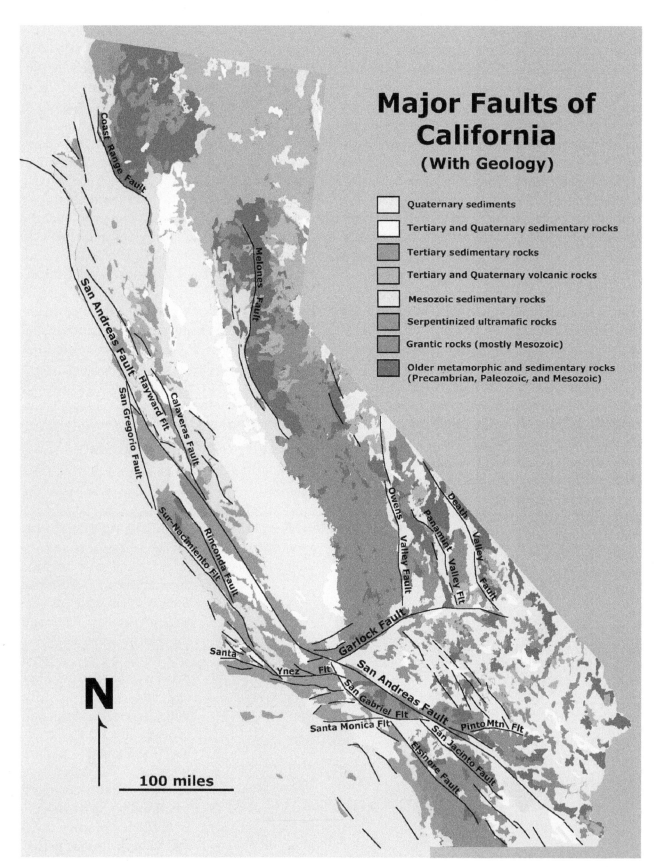

Figure 2.4 Major fault distribution in California.
(Source: USGS)

other regions associated with the subduction zone, indicating frequent volcanic eruptions at the time. To the west of the mountain systems, continuous sediment deposition over the subduction zone created *sedimentary rocks*, which became the foundation of the Central Valley and the Coast Ranges. About 100 million years ago, the subduction zone between the Pacific Plate and the Farallon Plate shifted westward to where today's Coast Ranges are located. The convergent force created ancient Coast Ranges, and the depression between the Coast Ranges and the ancient Sierra Nevada became the base of the Central Valley.

The continuous subduction of the Farallon Plate eventually made it disappear under the subduction zone and resulted in the direct contact between the Pacific Plate and the North America Plate approximately 27 million years ago. The remnants of the Farallon Plate became a few small plates. Among them are the Cocos Plate to the south of California, and the Juan de Fuca Plate to the north of California. Along the direct contact zone between the North America Plate and the Pacific Plate, a transform plate boundary, the San Andreas Fault, was formed. The right lateral fault system (opposite block moving toward the right) is a very complex zone of fractured rocks that is more than 1,300 kilometers (800 miles) long, 16 kilometers (10 miles) deep, and from a few hundred meters to more than 1,000 meters (3,300 feet) wide. Although the fault, because of its large scale, may not always be visible when you walk past it, viewed from a distance, the linear zone is clearly visible (Figure 2.5).

Over the past 20 million years, the Pacific Plate has been slowly sliding along the San Andreas Fault northwestward relative to the North America Plate with an average speed of 5 centimeters (2 inches) per year. This shift is not a smooth, consistent movement through the whole system. The movements usually occur along one segment of the fault zone at a time, and the distance shifted at each movement varies from undetectable to more than 6 meters (20 feet) during one earthquake. Over time, it is believed that the accumulated movement is now

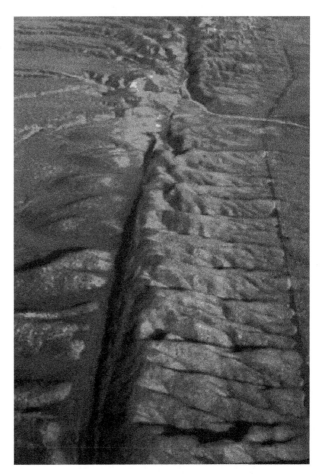

Figure 2.5 An aerial view of the San Andreas Fault. (Source: USGS)

more than 600 kilometers (360 miles) since the formation of the San Andreas Fault. It is believed that this trend of movement will continue in the foreseeable future. It is not too hard to imagine that where Los Angeles is located today was next to Mexico 20 million years ago, and that Los Angeles will eventually leave California to visit Oregon, Washington, and so on in the next 20 million years.

Plate movements not only shaped the landforms in areas near the plate boundaries but also significantly affected the formation of landforms in the interior regions of the state. During the early contact of the Pacific Plate and the North America Plate, the tectonic process created a stretching force over the interior areas; the force created a series of *block faults* that established the foundation for the basin and range topography. A block fault is characterized by a middle block of rock structure dropping down between two normal fault lines, as illustrated in Figure 2.6.

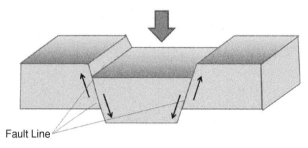

Figure 2.6 Block faulting.

At the same time that basin and range topography was developed, the right lateral motion of the two plates along the San Andreas Fault created the Transverse Ranges. As the name indicates, the Transverse Ranges are the only mountain ranges in California that do not follow the northwest–southeast trend. Geologists believe that the Transverse Ranges broke from the North America Plate during the early history of the San Andreas Fault. While its northern end was still attached to the North America Plate, its southern end was dragged by the Pacific Plate to move northward, resulting in a clockwise rotation that turned the mountain rages to run west–east.

Rock Structure and Distribution

The dynamic land-building processes along plate boundaries created many different rock types and structures in California. In the Klamath Mountains in the northwest, *metamorphic rocks*, rocks deformed from previously existing rocks, are found. Over the Modoc Plateau in the northeast, *extrusive igneous* rocks, a type of volcanic rock, are abundant. *Granite*, an *intrusive igneous rock*, forms the backbone of the Sierra Nevada along the east side of the state. The Coast Ranges are composed of primarily *sedimentary rocks*, rocks formed from solidified sediments. The geological history of the state was revealed for the most part by these rocks in terms of their formation, structure, age, and location.

Earthquakes in California

General Concepts

California belongs to the *Pacific Rim of Fire*, a belt surrounding the Pacific Ocean marked by frequent earthquakes and volcanic eruptions. Although, on average, the North America Plate

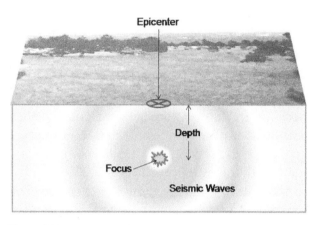

Figure 2.7 Structure of an earthquake.

moves about five centimeters (two inches) per year toward the south relative to the Pacific Plate, in reality, these movements are not steady and smooth along the fault lines. Tension and pressure from the moving plates build up gradually along the fault zones, and when energy associated with the tension and pressure reaches a critical point, the land suddenly moves along a fault line to release the pressure while triggering an earthquake. The potential magnitude of an earthquake is then proportionally related to the amount of energy stored and released along the fault. In an active plate boundary zone, the longer the wait between earthquakes, the more the energy stored and the bigger the earthquake when it happens.

During an earthquake, the underground location where the sudden movement occurs is called the *focus of the earthquake* (Figure 2.7). The energy released during the earthquake travels from the focus outward in the form of seismic waves. When these waves reach the earth's surface, they cause the earth's surface to move and shake, causing destruction along their path. The point on the earth's surface right above the focus is called the *epicenter of the earthquake* (Figure 2.7).

The amount of energy released during an earthquake is called the *magnitude* of an earthquake. It is measured by the Richter scale, a method developed by California geologist Charles Richter in 1935. The Richter scale starts from zero and goes up with increase of magnitude. An increase of a whole number (e.g., from 4 to 5) on the Richter scale indicates an increase in energy level approximately 32 to 33 times higher than the previous number. For example, the energy released is equivalent to 1 million tons of TNT during a Richter scale 6 earthquake and is 32 million tons during a Richter scale 7 earthquake. Although the Richter scale has no upper limit, the highest recorded magnitude of earthquake was around 9. The 1964 Alaska earthquake, with a Richter scale of 9.2, was one of the few such earthquakes in the world. Most of the earthquakes with magnitude less than 3 have no impact on people or structures; with increasing magnitudes, however, they come with increasing injuries and damages. Fortunately, anywhere in the world, the higher the magnitude of an earthquake, the lower the possibility it occurs. In addition, the relationship between an earthquake's magnitude and its impact on people and structures is not a simple linear one. A strong earthquake in a remote area may cause no or few injuries and structural damages; however, a moderate earthquake in a heavily populated and poorly constructed area may cause mortality and massive structural damage. An earthquake's destruction is also related to the geological foundation where it occurs. Earthquakes with the same magnitude may cause more ground shaking in one place and less in another place with the same distance to the epicenter.

The ground motion and resultant impact of an earthquake is measured by the *intensity* of an earthquake. One method often used to measure earthquake intensity is the Mercalli scale, initially developed by Giuseppe Mercalli, an Italian geologist, in 1902 and later modified by others. The Mercalli scale uses the observations of people to estimate earthquake intensity. It ranges from I to XII, with I being almost no impact to people to XII being catastrophic impact (Table 2.1).

During an earthquake, there is only one magnitude measurement: the Richter scale is the same for any place, no matter how close or how far it is from the epicenter. The intensity, on the other hand, varies from location to location during an earthquake. In most cases, the closer a place is to the epicenter, the higher is its Mercalli number. The Mercalli number distribution may vary because of the underlying geological

Mercalli Scale	Description
I	Not felt.
II	Felt only by a few people at rest indoors or in high-rise buildings.
III	Felt indoors, similar to light truck passing.
IV	Clearly felt indoors. May cause light damage.
V	Light shaking. Felt indoors/outdoors, objects swing. Light damage.
VI	Moderate shaking. Felt by all. Trees visibly shaken. Moderate damage.
VII	Strong shaking. Difficult to stand. Felt by drivers. Moderate to heavy damage.
VIII	Very strong shaking. Difficult to drive. Moderate to heavy damage. May trigger landslides, break tree branches.
IX	Violent shaking. Serious damage to structures.
X	Very violent shaking. Most structures damaged or destroyed.
XI	Most buildings collapse. Underground pipes destroyed. Railroad tracks bent.
XII	Catastrophic. Damage nearly total.

Table 2.1 Intensity of Earthquake by Mercalli Scale. (Source: USGS, Federal Emergency Management Agency.)

structure and the variation in population density and structural characteristics.

Earthquakes are a part of the land-building processes, as discussed earlier. During the San Fernando earthquake in 1971, Santa Monica Mountain rose 2 meters (about 6 feet), and during the Northridge earthquake in 1994, the same mountain rose another 15 centimeters (6 inches).

Lessons from the Past

There are many memorable earthquakes in California's recent history. Learning what happened during these earthquakes helps us understand and prepare better for the inevitable future earthquakes.

The earthquake with the highest recorded magnitude in California was the 8.7 quake that occurred on January 9, 1857, along the San Andreas Fault near Fort Tejon. During that earthquake, a long section of the San Andreas Fault ruptured. The ripped line ran from Parker Field to near Wrightwood, a stretch of 350 kilometers (220 miles). The lateral shifting of the land reached 9 meters (28 feet) over the Carrizo Plain. Geologists believe that this earthquake marked the latest major movement along the southern section of the San Andreas Fault. It is now often used as the starting point for estimating the energy stored along this section when predicting the next "Big One" in Southern California. Although 8.7 is the highest magnitude of recorded earthquakes in California and the ground shaking was felt as far away as San Diego and Las Vegas, only one casualty was documented and very few structures were damaged or destroyed. This less-than-massive destruction is not hard to understand because there were very few people or structures over this area in 1857. One of the few reportable stories associated with the earthquake was that fish in Tulare Lake, more than 100 kilometers (60 miles) from the epicenter, were thrown out by water waves triggered by the earthquake and were found kilometers away from the lake.

It was quite a different story a half century later when the northern section of the San Andreas Fault moved. On April 18, 1906, fault movement triggered the most storied earthquake in California: the 1906 San Francisco earthquake (Figure 2.8a). Although the magnitude was 7.8, not as strong as the one happened at Fort Tejon, it still generated a rupture zone more than 300 kilometers (190 miles) long running from San Juan Bautista to Point Arena and into the ocean. The ground shaking lasted one minute. Within that minute, horizontal shifting of land occurred in many places, the largest displacement being a 6.4-meter (20-foot) offset of land near Point Reyes. Damage to structures was widespread, with an epicenter so close to a densely populated area. The heaviest damage associated with the earthquake was not from the direct shaking impact, however, but from the subsequent fire. Fire triggered by gas mains ruptured during the earthquake quickly burned out of control. The rupture of one of the city's main water pipes by the earthquake, along with many other problems, hindered efforts to put out the fire. After it was all over, many building skeletons remained standing, but their wooden parts were burned down (Figure 2.8b). If not for the fire, many of these buildings would have survived the earthquake. It was estimated that over 90 percent of the damage was caused by the fire. An estimated 3,000 people died during the earthquake and subsequent fire. Property loss was estimated more than half a billion (billions in today's value) dollars.

Unlike the southern section of the San Andreas Fault, which has had no major movement since 1857, the northern section has been more active. Less than a century after the 1906 San Francisco earthquake, the same fault triggered another major earthquake on October 17, 1989—the 6.9 magnitude Loma Prieta earthquake. The epicenter was at the Santa Cruz Mountain where the San Andreas Fault runs through. Although the largest displacement caused by this earthquake was less than one meter (three feet), the impact reached the densely populated Bay Area. Most of the damage caused by this earthquake happened in the heavily built San Francisco and Oakland region, 100 kilometers (60 miles) north of the epicenter. Compared to events that happened decades ago, people understood more about earthquakes and were better prepared. The earthquake occurred at the start of the evening rush hour on a day of a major baseball game in the area. Most people remained calm during the earthquake, and the emergency response was effective. The interruption of public services by the earthquake was kept to a minimum. Despite all the efforts, however, numbers still reveal a bleak picture: 63 people lost their lives, 3,757 people were injured, and property damage reached six billion dollars. Among the 63 killed, 41 died because of the collapse of the Cypress viaduct connected to the Bay Bridge (Figure 2.8c). Within the same proximity to the epicenter, how could the Bay Bridge, built in the late 1950s as the first double-decker freeway in the state, fail, while the Golden Gate Bridge, built two decades earlier, stood unharmed? Among the explanations were the difference in ground foundation and the technology and materials used. Although we will never get complete answers to these questions, the more we learn from an earthquake, the better we can be prepared for the next one. The repair, retrofit, and update of the damage sustained by the Bay Bridge during the 1989 earthquake lasted almost a quarter of a century. On September 2, 2013, the opening of a $6.4 million eastern span of the San Francisco–Oakland Bay Bridge, a side-by-side bridge built parallel to the old bridge, marked another new chapter of the Bay Bridge and the continued effort of getting ready for the "Big One".

Figure 2.8a After the 1906 San Francisco earthquake, damage to the region was widespread. The pictured was the Agassiz statue at Stanford University. (Source: W. C. Mendenhall)

Figure 2.8b After the subsequent fire of the 1906 earthquake, many structures were left standing, but their wooden components had burned down. (Source: W. C. Mendenhall)

All three historical earthquakes discussed here were associated with the San Andreas Fault. Although the San Andreas Fault remains the largest fault in the state and poses the biggest threat to Californians, smaller faults with lower earthquake magnitudes are equally capable of generating the "Big One." One lesson we learned was from the 1994 Northridge earthquake in

Figure 2.8c The Loma Prieta earthquake on October 17, 1989, caused the collapse of a segment of the Bay Bridge.
(Source: G. Plafker)

Figure 2.8d A collapsed parking structure on the campus of California State University at Northridge.
(Source: M. Celebi)

Northridge name was already widely used in the media when the epicenter was accurately located, the earthquake was never renamed. The earthquake was triggered by movement along an unknown *blind thrust fault* (a fault line where the rocks primarily shift vertically but the fracture does not reach the surface). The earthquake's vertical movement caused the Santa Monica Mountain to rise 15 centimeters (6 inches). Although little surface rupture was observed after the Northridge earthquake, the associated ground shaking was one of the most violent, with many areas experiencing IX Mercalli scale intensity. The strong ground shaking combined with the epicenter's location in a heavily populated area made this earthquake the most costly one in California history. The death toll reached 60, more than 7,000 were injured, and more than 20,000 were left homeless. With damage estimated at more than twenty billion, it is ranked among the most costly natural disasters in the country. Most structural failures during this earthquake occurred when the steel and concrete support beams of freeway overpasses and parking structures buckled under sideways shaking (Figure 2.8d). One of the most important lessons learned

Southern California. The earthquake occurred January 17, 1994, at 4:30 AM. The epicenter was right in the middle of a heavily populated area in the San Fernando Valley, near Northridge. The epicenter was actually located at Reseda, a city south of Northridge, but because the from the Northridge earthquake was the need to improve structural support for freeway overpasses and parking structures. Scientists and engineers learned that the construction materials and structural designs needed to be enhanced to sustain not only vertical pressure but also strong horizontal movement during strong earthquakes.

Prediction and Future Earthquakes

Despite continuous efforts in finding ways to predict earthquakes, there is still no method for predicting an earthquake so precisely and accurately that the public could be warned of the danger ahead of time. Most predictions today place the potential occurrence of an earthquake in a large area with a long time span (e.g., in Southern California for the next decade or two). Such imprecision serves little to help the public avoid the dangers associated with an earthquake. However, these predictions do help policy makers, government officials, planners, and the public prepare for an earthquake and minimize its associated impact.

It is believed that the San Andreas Fault has been and will continue to be the major earthquake threat in California. The last time an earthquake occurred along the northern section of the San Andreas Fault was the 1989 Loma Prieta earthquake in the Bay Area. Over the southern section, the fault has not moved since the last earthquake on the fault line in 1857 near the Fort Tejon area. Geologists believe that the tension accumulated along the fault line is now capable of generating earthquakes at magnitude 8 levels.

The San Andreas Fault is not the only earthquake threat, however. Many faults are already confirmed throughout California (Figure 2.4), and many more are yet to be discovered. Among the fault lines already discovered, scientists predict that many are capable of producing earthquakes up to Richter scale magnitude 6 or 7 levels, which can be very destructive, as demonstrated by the Northridge and Loma Prieta earthquakes (magnitude 6.7 and 6.9). Unfortunately, most areas with favorable climate and plentiful resources in the state coincide with active fault zones. This fact places most Californians in the moderate to high risk seismic zones. With increasing population, even a moderate earthquake could produce deadly and costly consequences.

Diligent planning and preparation are needed to minimize the impact of an earthquake. In the past decades, significant progress has been made in improving emergency services, tightening building and transportation safety regulations, and educating the public to be prepared.

Despite all these efforts, the unstable ground remains the most destructive natural hazards threat in California and requires everyone's continuous effort to minimize the impact of these unavoidable natural events.

External Forces and Processes

Internal forces, forces generated from inside of the earth, such as earthquakes, fault movements, and volcanic eruptions, created California's high mountains and deep valleys. These landforms were, and continue to be, shaped by external forces, forces from the earth's atmosphere and hydrosphere. Internal forces tend to increase the relief of the earth's surface, whereas the external forces tend to flatten the landforms by weathering and eroding down higher lands and filling up lower lands with the sediments from the erosion. External forces work within the landform structure created by internal forces and shape the land into distinctive landscapes in different locations. The most important external processes in California are glacial (moving ice), fluvial (running water), aeolian (wind), and coastal (wave) processes. Although these processes are distinctively different, they do share something in common: they are either erosional, which weather down the higher portion of the land, or depositional, which fill up the lower part of the land. Connecting the erosional and depositional processes are the transporting processes, which constantly move the eroded materials from one place to another, often from higher lands to lower lands.

Glacial Processes and Landforms

Throughout the earth's history, glaciers have come and gone many times. Most evidence of these extreme cold events, however, was erased by later events, such as warmer periods and more recent glaciers. The glacial landforms still present in today's California were formed during the most recent Ice Age (the Pleistocene epoch) that began 1 to 2 million years ago and may have ended less than 8,000 years ago. During that period, continental ice sheets covered a large part of Canada and the northern United States; mountain glaciers extended farther south to the mountain ranges in the west. Evidence of glaciation was found as far south as the San Bernardino Mountains in Southern California, but it is most significant in the Sierra Nevada.

Glacial landforms can be divided into two distinctive categories: those associated with erosional processes, and those associated with depositional processes. *Erosional glacial landforms* are created by the enormous carving power of moving ice, a force unmatched by any other external forces on earth. Mountain glaciers form in rugged highland areas when snow accumulates over a period of years and compresses into glacial ice. This ice further develops into mountain glaciers as it moves downslope, pushed by snow and ice accumulating behind it and pulled by gravity. When a mountain glacier moves downslope, it scours and smoothers the rugged mountainsides with the great pressure of the ice and the rock debris carried by it. It is so powerful that a steep-sided V-shaped valley could be carved into a widened, U-shaped valley. When a U-shaped valley is truncated by another, deeper valley running sideways, a *hanging valley* is created, which often leads to the formation of a waterfall. Examples of such waterfalls are abundant in Yosemite Valley in the Sierra Nevada (Figure 2.9). Where two or more U-shaped valleys meet, sharp mountain ridges and peaks, called *arêtes* (French for "mountain ridge") and *horns* (sharp peaks), form between them. The place where the original glacier accumulation occurred is often marked by a bowl-shaped depression, called a *cirque* (French for "circus").

Figure 2.9 One of the many waterfalls in Yosemite Valley.
(Photo: Lin Wu)

Many cirques later became lakes filled with fresh mountain water. U-shaped valleys, hanging valleys, arêtes, horns, and cirques are typical erosional glacial landforms found over the highland areas of the Sierra Nevada (Figures 2.1a and 2.9).

Opposite to the process of snow and ice accumulation in the highland areas, *ablation*, the process of glacial ice melting (changing from a solid to a liquid form of water), and *sublimation* (changing from a solid directly to a gas form of water) occur in the lower part of the mountains. Rocks and other debris left by the melting ice form depositional glacial landforms. These landforms are characterized by piles of rocks varying in sizes and shapes that were moved long distances by glaciers. *Moraines*, rolling small hills composed of rocks and debris deposited by melting glaciers, are often found along the foothills of glaciated mountains. Depending on the location of the moraines in relation to the glacier, they are called *terminal moraines*, moraines located at the end of a glacier; lateral moraines, moraines located at the sides of a glacier; or recessional moraines, a sequence of terminal moraines that mark the retreat route of a glacier. These spectacular glacial landforms are further discussed in the section on geomorphic regions.

Fluvial Processes and Landforms

Running water is the most common and consistent external force that keeps working on the landforms. Landforms shaped by running water are called *fluvial landforms*. Fluvial landforms dominate humid regions where precipitation is frequent and abundant as in the North Coast Ranges and the Klamath Mountains. Even in places where rainfall is rare, however, occasional storms may also create strong imprints on the surfaces as in the basin and range topography and the desert areas. In humid areas, water running on the surface forms stream channels.

In the upper reach of a stream, the primary erosional process of the stream water is to deepen the valleys. As a result, deep V-shaped valleys form in the mountains where the bases of the streams are high above sea level. When a stream reaches its outlet, usually close to sea level, the erosional processes change from deepening the valley to widening the valley. At the same time, the depositional process becomes more dominant in the widened valleys. With combined impact from both erosional and depositional processes, the lower reach of a stream is often characterized by *floodplains* with *meandering streams* winding through it (Figure 2.10).

In areas where precipitation is high and the underlying geologic structure is less confining, running water channels into connected streams to form a drainage pattern that resembles tree branches. It is called a dendritic *drainage* pattern (Figure 2.1c). Dendritic drainage pattern is common in humid climates around the world; however, in California it is present only in the northwest regions, in the Klamath Mountains. Most of the streams in California are confined by the underlying geological structures. For example, rivers in the northern Coast Ranges mostly flow northwest–southeast, following the trend of the mountain ranges.

In the desert area, with little precipitation, running water, along with wind, is still the major external force. Because desert land surface is less protected by vegetation, when the occasional storm occurs, the erosional power can be very strong. As a result, deep gullies and rough, rigid terrains form. In the basin and range geomorphic region, when streams from the range reach the valley, water in the streams fans out and deposits sediments on the valley floor, creating fan-shaped landforms called alluvial fans (Figure 2.11). Alluvial fans are prime locations for agriculture and settlements in the humid and semiarid regions.

Aeolian Processes and Landforms

Wind-shaped landforms, called *aeolian landforms*, are often seen in sandy areas, such as sandy deserts and sandy coasts. Aeolian landforms are one of the least stable landforms; they migrate and reshape constantly. Crescent-shaped sand dunes can be found in the Colorado Desert in Imperial County. Along sandy coastal areas, the formation of sand dunes is often a result of natural or planted vegetation that serves as anchors of the dunes (Figure 2.12).

Coastal Processes and Landforms

The long stretch of California coast line is a natural canvas on which coastal waves and stream water craft diverse coastal landforms. Both

Figure 2.10 An aerial view of a flood plain near the Sacramento River. (Photo: Lin Wu)

depositional and erosional landforms are present along the coastal areas (Figure 2.13).

Erosional coast landforms, created by the pounding force of the coastal waves, are characterized by steep cliffs, sea caves, and sea stacks. Sea stacks are isolated rocks along the coast with their underlying geological structures connected to the mainland. Sea stacks form when the surface area connected to the coastal land was eroded away by the ocean waves. Sea stacks may appear as scattered rocks along a coastal area or they could be as large as an island, such as the Channel Islands (Figure 2.14). Most coastal landforms in California are rocky coasts dominated by the erosional processes.

Depositional processes and landforms often coexist with erosional processes and landforms along the coastal area. The most common depositional coastal landforms are beaches, narrow strips of sand deposited along the coasts. Beach sand essentially came from sediments of mountain erosions carried to the ocean by streams. The coastal waves, with their constant onshore and offshore motion, bring the sand back to the

Figure 2.11 A classic example of an alluvial fan, at Copper Canyon Fan in Death Valley National Park.
(Source: USGS)

Figure 2.12 The imperial sand dunes.
(Source: Bureau of Land Management)

Figure 2.13 A coastal area in Santa Barbara—sea cliff, beach, and longshore currents are pictured.
(Photo: Lin Wu)

Figure 2.14 The Channel Islands, off the coast, were protruded mountain ranges from the mainland. Over time, constant movement of the ocean water eroded their connections to the mainland.
(Image © Thomas Barrat, 2009. Used under license from Shutterstock, Inc.)

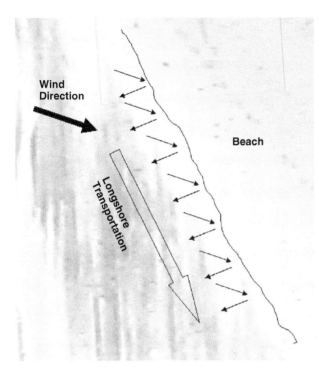

Figure 2.15 Coastal wave motions and processes.

coast (Figure 2.15). In an area with gentle slope and sheltered from strong wind, the sand carried by the waves is deposited to form a beach. Like most depositional landforms, beaches are not permanent land forms; they are constantly moving and changing and often exhibit daily and seasonal patterns.

Geomorphic Regions

Dividing the Regions

One important subject of any scientific discipline is the classification methods developed to describe and study its subjects. To conduct a classification, clear criteria are often developed and the subjects are measured against those criteria and labeled according to the characteristics of the subject. In geography, the most commonly used classification method is to divide areas into regions and subregions based on different criteria. In this section, we explore the geomorphic regions of California, a classification based on the characteristics of the different landforms and the processes that created them. One unique perspective of geographic classification is that we cannot create groups by moving places to

other locations even though they may fit the same criteria. Therefore, geographic regions are often divided on the basis of their locations and the most dominant features that fit the criteria. In the following sections, we explore examples of different geomorphic regions. Landforms in some regions are quite uniform within their regional bounds and others can be quite diverse in the same region, but are bound by their geographic locations and their formation processes.

California's eight geomorphic regions, as illustrated by Map 3, are based on geological and geomorphologic characteristics. They are described in the following sections.

Peninsular Ranges

The Peninsular Ranges is located at the southwest end of California (Map 3). This area is home to more than 4 out of 10 Californians and includes all or part of Los Angeles, Orange, San Diego, Riverside, and San Bernardino counties. The Peninsular Ranges is delineated on the north by the Transverse Ranges, on the east by the Colorado Desert, and on the west by the Pacific Ocean. It includes the southern part of the Channel Islands: Santa Barbara, Santa Catalina, San Nicolas, and San Clemente. Starting from south of the Transverse Ranges, the Peninsular Ranges runs 240 kilometers (150 miles) to the southern border and extends another 1,200 kilometers (750 miles) south of the border. Like most regions in California, ranges in this area have a north–south trend. The ranges are higher in the north, with the San Jacinto Mountain peaking at 3,296 meters (10,804 feet) and lowering toward the south (Figure 2.16). Similar to the Sierra Nevada, the ranges have steeper slopes on the eastern side and gentler slopes on the western side. Most rock structures in the region also resemble that of the Sierra Nevada, with Mesozoic granite rocks dominating the region, especially in the eastern part.

One major feature in this region is the Los Angeles Basin, home to millions in the city of Los Angeles and its surrounding areas. Framed by the mountains and hills in the area, Los Angeles Basin is a sediment-filled depression built on a complex *folding* and *faulting system*. The geological structure of the basin created the ideal condition to form oil reserves, which became the source of the "black gold rush" in the 1890s. To this day, oil is still being produced throughout the Basin. Los Angeles Basin sits on a very complex and active fault zone (Figure 2.4).

The western part of this region is characterized by various coastal landforms. Along the San Diego coast line, marine terraces, natural harbors, and sandy beaches are important resources supporting the economy and lifestyle of the region. Moving northward, the coast line along Orange County is characterized by rocky shorelines, narrow beaches, and offshore islands. Farther north, beaches along the Los Angeles coast draw millions of visitors from around the country and the world.

Transverse Ranges

As its name indicates, the Transverse Ranges runs west–east and is the only mountain range in California to do so. As discussed earlier, during the early history of the San Andreas Fault, the Transverse Ranges broke off from the North America Plate. With its northern end still attached to the North America Plate and its southern end dragged by the Pacific Plate moving northward, it turned clockwise and formed the west–east running mountains and valleys. Major mountains in this region include, from west to east, the Santa Monica Mountains, the San Gabriel Mountains, and the San Bernardino Mountains. The highest peaks in the ranges are San Gorgonio Peak (3,505 meters) in the San Bernardino Mountains, and San Antonio Peak (3,068 meters) in the San Gabriel Mountains. The heavily populated Oxnard Plain and San Fernando Valley are among the lowlands in the region. The northern part of the Channel Islands—Anacapa, Santa Cruz, and Santa Rosa—also belong to this region (Figure 2.16).

Late Mesozoic era and early Cenozoic era sedimentary rocks are present in the Santa Monica Mountains and the western part of the

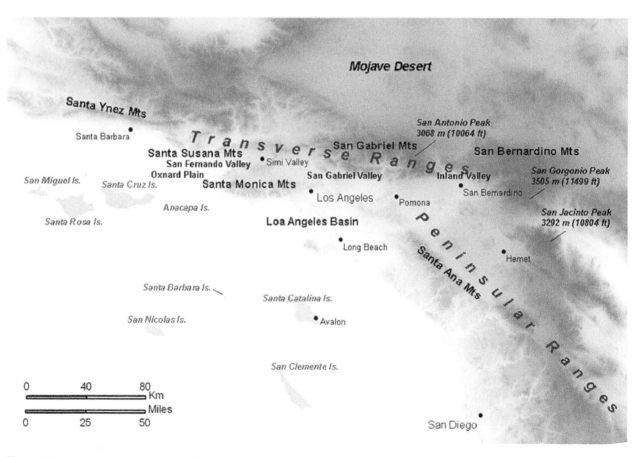

Figure 2.16 Major landforms in the Transverse Ranges and Peninsular Ranges.
(Source: Lin Wu, DEM data: USGS)

region. Toward the eastern side of the ranges, in the San Bernardino Mountains, complicated metamorphic rocks similar to rocks in the Basin and Range topographic region and granite rocks that resemble the Sierra Nevada are found.

The San Andreas Fault, along with many other faults, runs through the region, making this one of the most geologically active regions in the state. In fact, the Transverse Ranges region is one of the most rapidly uplifting areas in the world. During the 1971 San Fernando earthquake, the Santa Susana Mountains rose 2 meters (6 feet) and rose 70 centimeters (28 inches) again during the 1994 Northridge earthquake.

Coast Ranges

The Coast Ranges run from north of the Transverse Ranges all the way to the northern Oregon border. These ranges are 1,000 kilometers (600 miles) long and extend from the coastline to 130 kilometers (80 miles) inland bordered by the Klamath Mountains to the north, the Central Valley to the west, and the Transverse Ranges to the south (Map 3). Although the Coast Ranges do not rise as high as the Sierra Nevada, they nevertheless are impressive mountain ranges that, in some sections, rise sharply from the coast to more than 1,500 meters (5,000 feet) above sea level in a very short distance. The Coast Ranges are divided by San Francisco Bay into northern and southern sections; both are composed of a series of northwest–southeast trending long mountain ranges and valleys (Figure 2.17). The Coast Ranges were formed as part of the subduction plate tectonics process. After the Sierra Nevada was formed, the plate boundary between the North America Plate and the Pacific Plate moved to the west; the subduction of the Pacific Plate caused the uplifting of today's Coast Ranges. Today, the San Andreas Fault runs parallel to the east of the Southern Coast Ranges, crossing at the Bay area and running into the ocean to the west of the Northern Coast Ranges (Figure 2.4).

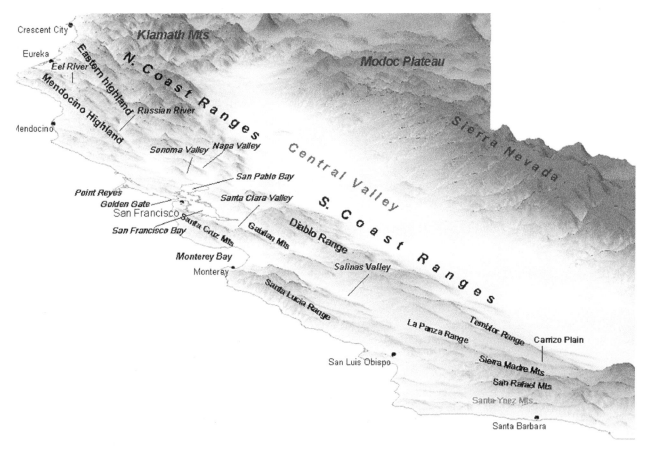

Figure 2.17 Major landforms in the Coast Ranges.
(Source: Lin Wu, DEM data: USGS)

The major ranges and valleys in the southern section of the Southern Coast Ranges include the Santa Lucia Range along the central coast; the long, narrow Salinas Valley; the Gabilan Range to the east of the Valley; and the Diablo Range and Temblor Range that border the Central Valley (Figure 2.17). The Southern Coast Ranges are hilly, with limited lowlands in the valleys and in the southern section of the coastal areas. The Salinas Valley is the largest lowland in the region and hosts one of the major agricultural areas in the state. Salinas, located at the end of the valley with a population more than 150,000, is the only place in this region with a population that exceeds 100,000. Although the region has only a few small coastal communities, it connects the two major population centers in the state: Southern California and the Bay Area. Being part of an active plate boundary with the San Andreas Fault running through it, this region often shakes and is threatened by major earthquakes.

Geologically, the Northern Coast Ranges consist of sedimentary folded rocks, with faulting being less important here than in other regions of the state. The valleys in this region are often made up of soft shale, with hard sandstone ridges surrounding them. Most of the small streams eventually channel into two major rivers: the Eel River, flowing northward, and the Russian River, flowing southward. The river valleys divide the ranges into the two systems: the Mendocino Highlands, which reach the coast, and the Eastern Highlands, which are more inland. Figure 2.18 is a coastal scene near Crescent City, at the northern end of the Coastal Range.

Figure 2.18 A coastal scene near Crescent City.
(Photo: Lin Wu, DEM data: USGS)

The Bay Area separates the Southern and Northern Coast Ranges. A structural depression, the Bay Area was flooded by seawater through the Golden Gate about 10,000 years ago when melting glacial ice caused the sea level to rise. The southern part of the region, San Francisco Bay, is a large estuary, a body of water partly isolated from the ocean. The northern part of the Bay, the San Pablo Bay, connects to the *delta* created jointly by the Sacramento River and the San Joaquin River. Except in a few valley bottoms, flat land in the Bay Area is limited. Most areas are hilly.

The Great Basin and the Southeast Deserts

The digital topographic rendering in Figure 2.19 shows the unique shape of the Basin and Range topographic region in the interior western United States: elongated ranges and basins running parallel northwest–southeast. The lengths of the basins and ranges vary from 75 to 250 kilometers (50 to 150 miles) long. The Basin and Range topography region in California marks the western edge of this topographic region (Map 3). A typical cross section of the topography shows steep mountain sides that rise thousands of meters and flat basins that may extend more than 30 kilometers (20 miles). Transitions between the basins and the ranges are abrupt and are often marked by *alluvial fans* created by stream deposits. When streams from the ranges reach the basin floor, alluvial fans are formed (Figure 2.11). The majority of basin and range topographic regions are characterized by *interior drainage basins*, streams that do not have outlets to the ocean. Over the years, water evaporates from the basins, leaving a layer of accumulated salt deposits behind painting the surface with a light tone. Looking from the distance, travelers not familiar with the region sometimes confuse the scene as snow-covered ground.

Among the major ranges and basins, the most spectacular basin is Death Valley. Created by block faulting, the valley bottom dropped so drastically that a large portion of the valley floor is under sea level, with its lowest point 86 meters (282 feet) below sea level. It is the lowest land spot in the United States. The valley is flanked

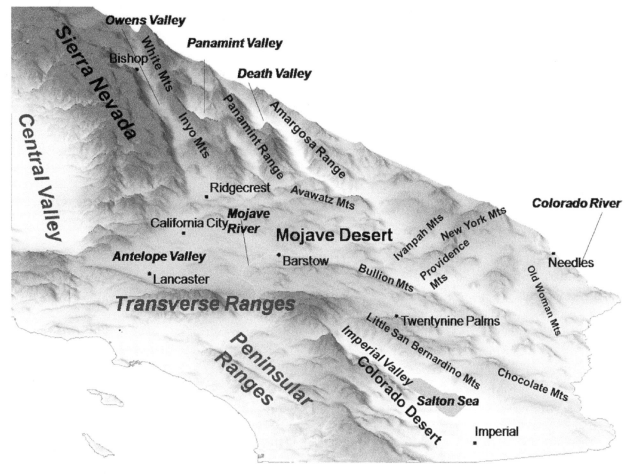

Figure 2.19 Major landforms in the Southeast.
(Source: Lin Wu, DEM data: USGS)

by steep mountain sides with exposed rugged rock surfaces. Running water carved deep into the exposed surfaces, creating a land so rugged that it obtained a special name: badland. Badlands may be bad for developmental use, but they provide spectacular scenes that showcase the carving arts created by natural processes and attract people from all over the world.

Another important valley in this region is Owens Valley. Located just east of the Sierra Nevada, Owens Valley serves as the catch basin for both surface and subsurface water from the Sierras. Although yearly precipitation in the valley is only a few centimeters, Owens Lake is one of the main water sources for metropolitan areas in Southern California.

Sierra Nevada

Sierra Nevada, called the "backbone of California," stretches 700 kilometers (440 miles) along the state's eastern border. The south end of the mountain range runs from east of Bakersfield to its north end south of Lassen Peak (Map 3). The north–south trend of the mountain range runs parallel to the coast lines and the Coast Ranges. The mountain range is 65 to 130 kilometers (40 to 80 miles) wide and rises from near sea level to more than 4,000 meters (13,000 feet). This massive "snowy mountain" serves as a natural barrier for both natural environment and human settlement. The mountain range demonstrates full windward and leeward climatic effects, with the western side of the range getting the highest amount of rain and snow in the state and the eastern side receiving very little precipitation. Following the climate patterns, the mountain range provides a classic example of vertical zonation of vegetation (further explained in Chapter 6) on the western slope and typical dry desert vegetation on the eastern slope. Historically, the mountain range was an immense barrier for those who migrated to California; yet the Sierra Nevada contributed to the first population boom in the state because of the gold discovered at the foothills of the mountain in the mid-1800s.

Water from the precipitation generated on the western slopes of the Sierra Nevada supports the agriculture in the Central Valley as well as the population in the metropolitan areas of Southern California and the Bay Area. Without the orographic precipitation (further explained in Chapter 4) caused by the massive mountain ranges, California would have been a lot drier and the land could not have hosted one of the greatest farmland areas in the world and supported more than 38 million people. In addition to water, minerals, and forestry resources, the region displays some of the most spectacular scenery in the country and attracts tourists from all over the world.

As discussed earlier, the geological origin of the Sierra Nevada is quite complex. Originally, the land was flat with horizontal sedimentary layers. The convergent force associated with tectonic processes caused the uplift of the mountain range, and magma from below formed massive granite batholiths (deep-rooted granite rock structures). Over time, erosion stripped off the overlying sediments, exposing the underlying granite we see today. At the same time the mountain was uplifted, a massive block faulting occurred along the eastern side of the mountain to form the steep eastern slope. The uplifted block tilted toward the west, forming a gentle slope on the western side of the mountain, which resulted in the asymmetrical profile of the range.

Glaciation during the *Pleistocene* further shaped the landforms. In the highland areas, glacial erosion created U-shaped valleys, hanging valleys with waterfalls, cirques, and other glacial erosional landforms. On the western side of the slope, glacial water carried the eroded material to the lowlands where depositional landforms dominate. Glacier melt water carried glacial debris into the Central Valley. As a result of the large amount of water washing on the western side of the mountain and heavy vegetation cover, very few depositional features remain today. On the less glaciated eastern side of the range, the depositional features can still be seen.

The High Sierra is located in the central part of the mountain range extending from Sequoia National Park into Alpine County (Figure 2.20). It is an alpine wilderness with many mountain peaks above 4,000 meters (12,000 feet). Only two of the higher-than-4,000-meters mountain peaks in California, Mount Shasta and White Mountain peak, are located outside of the High Sierra. The area was covered by an alpine glacier during the Pleistocene glaciation and exhibits typical erosional glacial landforms today with sharp mountain ridges and peaks. Mount Whitney, with an elevation of 4,421 meters (14,505 feet), is the highest mountain peak in the contiguous United States. It is only 123 kilometers (76 miles) from Death Valley, the lowest land spot in North America, with an elevation 86 meters (282 feet) below sea level. Most of the High Sierra is above the *tree line*, with scattered alpine vegetation over the barren land.

From the High Sierra, elevation decreases toward the north. The Northern Sierra, with summits around 3,000 meters (10,000 feet), is covered with vegetation. Elevation also decreases from the High Sierra toward the south. The southern prong of the mountain range connects to the Transverse and the Coast Ranges. Two major mountain passes, Tahachapi and Tejon Passes, are both located in the southern area.

To the west of the High Sierra, down the gentle slope, lies the mountain's park belt: Yosemite, Sequoia, and Kings National Parks. Yosemite National Park was established on a U-shaped glacial valley. It is best known for its glacial landforms with many waterfalls (Figure 2.9), steep granite mountainsides, small lakes, and vegetation that exhibits *vertical zonation*. With more than 95 percent of the over 3,000-square-kilometer (1,200-square-mile) park being designated wilderness areas, Yosemite National Park, along with Kings Canyon and Sequoia National Parks, attracts tourists from all corners of the world. The heavy rainfall resulting from the orographic precipitation created many streams on the western slope. The streams carved out deep canyons and drains into the Central Valley.

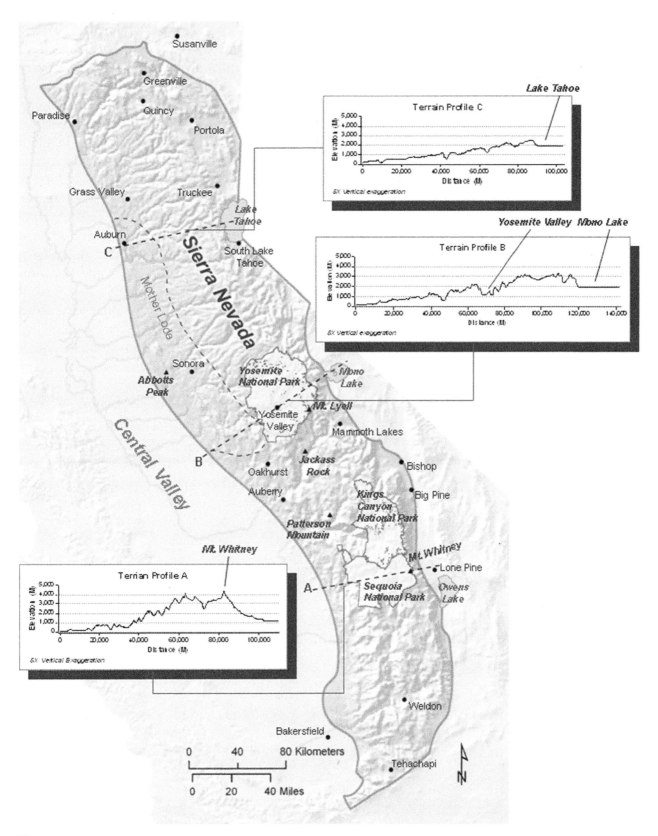

Figure 2.20 Major landforms in Sierra Nevada.

(Source: Lin Wu, DEM data: USGS)

From the park belt farther down the slope, the foothill region of the Sierra Nevada is known for its Mother Lode country, a continuous zone of gold-enriched rocks and quartz veins that runs nearly 200 kilometers (120 miles) long. The discovery of gold in Sutter's Mill located in Coloma at the bank of the American River triggered the gold rush in 1848–1855. Compared to other commodities the state holds, such as construction materials and agricultural products, gold production in the state is always ranked low on the economic scales. Nevertheless, the discovery of gold drew hundreds of thousands to the state and started the state's first population and economic boom. Gold mining, recreational and commercial, remains one of the state's economic activities.

Moving toward the east from the High Sierra is a huge block-faulting scarp that drops abruptly to the valley below. As a result of the rain shadow effect, the area is very dry with exposed barren surfaces and scattered desert vegetation. Figure 2.21 is a view of the eastern Sierra near Inyo National Forest. The arid terrain at the foreground is a stark contrast to the towering clouds rising from the humid western side of the Sierra.

Klamath Mountains

The amorphous Klamath Mountains are the oldest and most complex mountain systems in the state (Map 3). Located in the northwest region,

Figure 2.21 Eastern Sierra Nevada near Inyo National Forest. (Photo: Lin Wu)

the area is characterized by heavy rainfall, thick natural vegetation coverage, and extensive stream networks. These mountains vary lithologically (by the characteristics of the rock) and are rugged with steep slopes and irregular drainage patterns. Here the oldest metamorphic rocks of the state are found. Because of abundant rainfall in the area, valleys were created by the down-cutting power of the streams. Most of the small streams eventually join the Klamath River, which reaches the coast. Many peaks in the Klamath region exceed 2,000 meters (6,500 feet), with some more than 2,500 meters (8,200 feet). Flat land in the region is limited, with two exceptions: the Smith River lowland near Crescent City and the Eel River flood plain near Eureka (Figure 2.22).

Modoc Plateau and Southern Cascades

The Southern Cascades and the Modoc Plateau region are characterized by volcanic landforms. The Southern Cascade stretches from Mount Lassen all the way to the Canadian border. The transition from the Sierra to the Cascades is not always clear on the ground. Feather River Canyon is generally considered to define the boundary between the two ranges (Map 3).

Topped by volcanoes and covered by volcanic debris, the Southern Cascades are irregular and with more rolling mountainous area compared with most of the aligned mountain ranges in California. Two prominent volcanic peaks, Mount Shasta and Mount Lassen, dramatically rise above the land accompanied by a dozen smaller cinder cones scattered in the area. Mount Shasta (Figure 2.1b), is a stratovolcano, or composite volcano— a volcano created by multiple eruptions with overlapping layers. Mount Shasta is the most prominent mountain peak in the area, with an elevation more than 4,300 meters (14,000 feet) above sea level. It is considered a "dormant" volcano, which has never erupted in historic times, but it does give off gases and could come out of its dormancy at any time. Mount Lassen, with an elevation more than 3,000 meters (10,000 feet), erupted sporadically from 1914 to1918. It was considered the only active volcano in the lower 48 states until the 1980 eruption of Mount St. Helens in Washington.

To the east of the Southern Cascades lies the Modoc Plateau, with the basin and range topography farther to the east. The landform exemplifies a typical plateau with a flat surface area and steep edges that drop abruptly more than 300 meters (1,000 feet) along its southern and eastern margins (Figure 2.23). The flat top of the plateau is held up by resistant rock layers created by lava flows that poured from fissures in the ground. Its surfaces are rigid in some areas with exposed lava beds and in other areas covered by pine and sagebrush (Figure 2.24). The Mammoth Crater (Figure 2.25) contributed much of the lava covering the area. The fluid basaltic lava erupted sporadically, with one of the major eruptions occurring about 30,000 years ago. The eruptions created tubes that facilitated lava flow, and today hundreds of these tubes remain as lava caves in the area. Figure 2.26 is an example of a small lava tube, and Figure 2.27 is an example of a large lava cave. To give an idea of the scale of the cave, the rounded boulder near the bottom right of the photo is about the height of a person.

Central Valley

The Central Valley, also called the Great Valley, stretches 650 kilometers (400 miles) from its northern end near Red Butte and its southern end near Bakersfield (Map 3). Averaging 80 kilometers (50 miles) wide, the valley is a depression filled with alluvial sediments (deposited by running water) as deep as 300 meters (1,000 feet). Surrounded by

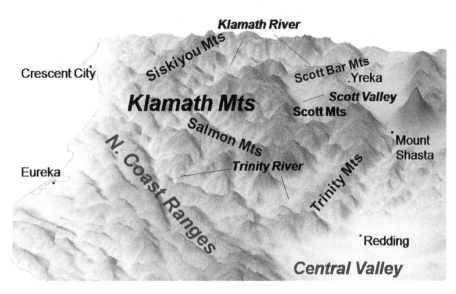

Figure 2.22 Major landforms in the Klamath Mountains.
(Source: Lin Wu, DEM data: USGS)

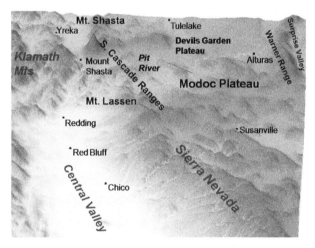

Figure 2.23 Major landforms in the Northeast Region.
(Source: Lin Wu, DEM data: USGS)

Figure 2.24 Modoc Plateau; notice the lava beds at the front of the picture and the volcanic cinder cones at the background of the photo.
(Photo: Lin Wu)

Figure 2.25 Mammoth Crater at the southern part of the Modoc Plateau.
(Photo: Lin Wu)

Figure 2.26 A small lava tube at the Modoc Plateau.
(Photo: Lin Wu)

Figure 2.27 Inside Skull Cave, one of the ice caves in the area; the cave is so deep that the cold winter air sinks to the bottom of the cave, leaving a layer of ice year round.
(Photo: Lin Wu)

spectacular mountain ranges, this seemingly featureless long stretch of flat terrain is one of the most prominent agricultural lands in the world. The valley is also known for its extensive water development projects.

During the plate tectonic period when the ancient Sierras were being constructed, about 225 million years ago, sedimentary rock layers that would later become the foundation of today's Central Valley and the Coast Ranges began to form. About 100 million years ago, the convergent force of the plates created ancient Coast Ranges, and the depression east of the ranges became the base of the Central Valley. Over the next millions of years, sediments from river erosion on the Sierras and the Coast Ranges were deposited on this great depression, forming today's Central Valley.

The thick alluvial deposits in the valley became the foundation of the rich soils evolved over the time. Many streams that contributed to the formation of the Central Valley are still active streams. Two major streams in the valley divided it into northern and southern sections: The Sacramento River Basin in the north covers about one-third of the valley, and the San Joaquin River Basin covers the other two-thirds on the southern end. The Sacramento River flows south and the San Joaquin River flows north, joining the Sacramento River in the delta area before draining into the ocean. In addition to these two major rivers, many smaller river systems flow into the two major river systems or drain into the southern end of the valley. Most of these streams originate from the Sierra Nevada.

Among the few features that interrupt this monotonous flat land are the Sutter Buttes. Called the smallest mountain range in the world, Sutter Buttes is a circular range, remnants of an ancient volcano. The range is about 17 kilometers (10 miles) in diameter, and the highest peak is about 700 meters (2,300 feet) above ground level. Originally named Marysville Buttes, the Sutter Buttes are located near Marysville in the Sacramento Valley (Figure 2.28).

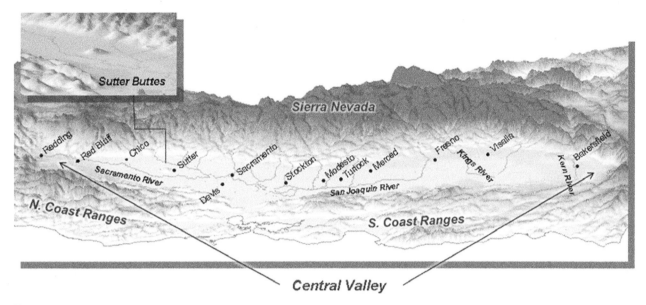

Figure 2.28 Major landforms in the Central Valley.
(Source: Lin Wu, DEM data: USGS)

Summary

This chapter explores the diverse landforms of the state and the forces and processes that created them. The history and the potentiality of earthquakes are focused on because it is important to understand the impact that a modern day land-building process could bring to us. Knowledge of the main characteristics of the eight distinctive geomorphic regions in the state helps connect and understand other geographic themes explored in the following chapters.

Selected Bibliography

California Geological Survey. *California Geologic Provinces—CGS note 36.* Retrieved August 2013 from http://www.conservation.ca.gov/cgs/information/publications/cgs_notes/note_36/Documents/note_36.pdf.

California Geological Survey. *Simplified Geologic Map of California.* Retrieved August 2013 from http://www.conservation.ca.gov/cgs/information/publications/ms/Documents/MS057.pdf.

Harden, D. R. (2004). *California Geology.* Upper Saddle River, NJ: Prentice Hall.

Hess, D., and Tasa, D. (2011). *McKnight's Physical Geography: A Landscape Appreciation,* Upper Saddle River, NJ: Prentice Hall.

Hill, M. (1984). *California Landscape: Origin and Evolution.* Berkeley, CA: University of California Press.

Lantis, D. W., Steiner, R., and Karinen, A. E. (1989). *California, the Pacific Connection.* Chico, CA: Creekside Press.

McKnight, T. (2004). *Regional Geography of the United States and Canada.* Upper Saddle River, NJ: Prentice Hall.

Southern California Earthquake Center (SCEC). Retrieved August 2013, http://www.scec.org/.

United States Geological Survey (USGS). *California Earthquake Information.* Retrieved August 2013 from http://earthquake.usgs.gov/regional/states/?region=California.

Historical Geography of California

*"It was a splendid population—for all the slow, sleepy, sluggish-
brained sloths stayed at home—you never find that sort of people
among pioneers—you cannot build pioneers out of that sort of
material. It was that population that gave to California a name
for getting up astounding enterprises and rushing them through
with a magnificent dash and daring and a recklessness of cost or
consequences, which she bears unto this day—and when she projects
a new surprise the grave world smiles as usual and says, "Well, that is
California all over."*

—Mark Twain, *Roughing It*

Introduction

Scholars have long debated how much influence the physical environment has had on the unique development of cultures throughout the world. According to the theory of *geographic determinism,* the human race, its creations, and its evolution are essentially conditioned by the natural environment. At the other extreme, scholars deny that nature and geography alone are so omnipotent in the unique development of a culture or people.

Perhaps the most balanced view is to recognize the interrelationships between people and the environment. Certainly, many factors influence the evolution of a particular culture, and geography is one of these important factors. California provides an excellent example of a situation in which the physical environment has played a significant role in historical events and cultural development. Geography has shaped people and events at each point in California's evolution. Aboriginal lifestyle was intimately related to the environment; European settlement patterns were conditioned by distance, weather, and nature. The offerings of the land, from fertile soil and climate to gold and other minerals, drew Spaniards, Mexicans, and Americans alike. The size of the state, the abundance of its natural resources, and the general mildness of its weather have continued to play key roles in the economic, social, and cultural life of the state, especially as manifested in agriculture, transportation, oil exploration, and movie-making. The same factors that brought early immigrants continue to attract modern-day newcomers to the state. Thus, the continuity of California's development

43

can, in significant part, be attributed to the realities of climate and geography, with economics playing an increasingly major role.

The Original Californians

Evidence of human presence in California is among the oldest reported in the United States. We know that as early as 29,000–34,000 years ago, primitive people began a Southern California tradition on Santa Rosa Island by barbecuing a dwarf mammoth. There is some additional evidence of human presence in California as early as 50,000 years ago. Evidence of established societal or tribal existence, however, does not appear until much later.

The aboriginal population of California constituted the first wave of immigration to the state. Anthropologists conclude that this immigration came about as a natural result of nomadic wandering through the Bering Strait, across a land bridge, and onward down the North American continent. By the time of the first Spanish forays into California, the Native American population had grown significantly. Although estimates vary widely, it is generally agreed that at least 133,000 and perhaps as many as 300,000 tribal peoples populated California at the time of the first Spanish contact. It is widely accepted as well that California at that time reflected the greatest population density of any *nonagricultural* area in the world and had the greatest population density in North America.

The nature of the California Indians can be described as diverse and varied. In language alone, there were at least 135 dialects in no fewer than 6 different parent language stocks as diverse as Penutian, Hokan, Shoshonean, and Athabascan (Figure 3.1). Similarly, dress, habitation, and physical stature all varied widely. The only common element, in fact, has been the way California Indians have been characterized historically. Perhaps the most widely circulated image is derived from the term gratuitously placed on them by the American forty-niners—"Diggers," which implied root grubbers and insect eaters of low intelligence and primitive demeanor. Applied indiscriminately as a label of

contempt for all California aboriginals, it was a gross oversimplification.

Individual tribal groups in California exhibited vast creativity and frequent sophistication in certain areas of their lives. The Mojave and Yuman tribes were far more than casual agriculturalists, with economies strongly reliant on flood farming. The variety of crops cultivated by these groups included maize, squash, tepary beans, riverine plants, and marsh grasses. The artistic ability of some tribes was especially evident in their intricate basketry, which has subsequently been described as perhaps the best in the world. Coastal tribes became quite adept at harvesting resources from the ocean by using tar-caulked canoes, harpoons, and shell fishhooks. Various groups valued religious ceremony, introspection, and spiritual pursuits. Telling stories, playing with games and toys (tops, dolls, etc.), gambling, and other recreational activities were not uncommon in most California tribes. Indeed, the notion that California Indians were simple and dull witted was highly inaccurate.

Although generalizations about such a diverse group are difficult to make, certain common traits are evident. Certainly, measured against European and American standards of aggression, technology, and materialism, California Indians were less "developed." They were peaceful rather than warlike. They were well adjusted to their environment, living in harmony with it rather than destroying it. By modern standards, they lived an uncomplicated, simple life instead of a hectic and competitive existence. Their religious life was generally highly developed, and their social life apparently was much more complex than previously believed.

The aboriginal inhabitants of California were attuned to their environment and reflected many of the characteristics we now associate with the state. Isolated by mountain and desert from the rest of North America, they developed a culture of appreciation for and reliance on nature. Largely individualistic, living in villages or *rancherias* of approximately 130 people, they built dwellings suited to their surroundings. Planks were used in the northwest, bark in the central regions, earth on the coast, and brush in the deserts. Dress was, for the most part, "early suntan" attire, consisting of topless two-piece apron skirts for the women and *au naturel* for the men. In cooler weather, a blanket or cloak of hides would be added. Nature provided the food in the form of acorns, game, fish, and other natural bounty. Only with the coming of the Spanish friars did organized agricultural production become common with non–Colorado River tribes.

The absence of European-style farming, however, did not mean these tribes were uninvolved with plant resources. Increasingly, experts are recognizing that California Indians engaged in effective indigenous horticulture. Methods of plant management included coppicing (pruning), weeding, annual burning, selective harvesting, soil manipulation, and tillage. These methods required intimate

Figure 3.1 Main aboriginal groups/tribes in California.

knowledge of their consequences, because plant resources were critical to tribal diets, weapons, tools, shelters, medicines, and other vital lifestyle items. One of the reasons why European immigrants (and scholars) erroneously concluded the California Indians were backward and unintelligent was that their methods involved less overt modification of (or violence to) the natural environment.

Singing, dancing, and chanting were familiar pastimes, as were gambling games, athletic contests, and storytelling. Public health and sanitation were accomplished by the use of sweathouses and periodic burning of the living quarters, and herbal medicines were in common use. The dominant form of religion was *shamanism*—the belief that supernatural spirits work for the benefit or to the detriment of humankind through the intervention of the priest, or shaman.

In retrospect, the California Indian appears much more well adjusted, efficient, harmonious, and admirable than generally depicted in the past. If environment does help form social styles, then it may be said that the society of these peoples accurately reflected their surroundings. The mild California environment demanded no excesses from its inhabitants, and they did not seek unnecessary complications for their lives—although, in all fairness, some California tribes (e.g., the Mojave) did not live in gentle environments.

The Spanish, Mexican, and American waves of immigration brought major changes, and their collective and progressive impact on the California Indians was destructive. Disease (various children's diseases, pulmonary ailments, venereal diseases, and other similar maladies), starvation, and violence, such as organized slaughters (with bounties paid for each dead native), caused a drastic reduction in the native population. Similarly, imposed changes in living patterns disrupted the flow of tribal life, whether it was the forced relocations of tribes into mission dormitories by the Spaniards or the forcible ejection of tribes from whole areas by the Americans. Victimized in turn by each incoming group, the California Indian was, as the respected California historian W. H. Hutchinson put it, "vital to Spain, useful to Mexico, and an annoyance to the United States." The discovery and colonization of California by Europeans truly spelled the end to the simple Native American culture, which could not hope to resist "civilization."

Ultimately, and ironically, these once despised native peoples may have had a better sense of social balance and environmental responsibility than the groups that were to follow. As modern society increasingly recognizes the problems associated with environmental degradation, the California Indians and their ecological practices must be seen in a much more positive light. The respect and reverence that these "Native Californians" held for the earth have encouraged practices just now being adopted by a concerned population. Whereas European settlers tended to reduce the populations of plants and animals, the California Indians generally adhered to the concept of sustained yield and multiple uses for both—an approach currently receiving attention in California and much of the rest of the world.

Thus, although the California Indians became victims of the times, many of their ideas and concepts (especially as they relate to stewardship of the earth) have reemerged as sensible and coherent approaches to life in California.

European Exploration

Spanish Dominance

The discovery, exploration, and eventual settlement of California by the Spaniards came about in the aftermath of initial explorations of Mexico by Hernando Cortes and others in the early 1500s. Curious about the nature and extent of this new territory, Cortes, Coronado, Ulloa, and other explorers undertook several expeditions to the north. It was Juan Rodriguez Cabrillo who first entered what would be called Alta California when he sailed into San Diego Bay and on up the coast in 1542. Prophetically enough, he named the coastal area near Los Angeles and Santa Monica "the Bay of Smokes" for the profusion of campfires in the vicinity. Cabrillo died during the voyage, and the reports of the expedition generated little interest in the region among Spanish officials.

Although Spain provided the initial impetus for exploration, California did have other European visitors. Reportedly, Britain's salty sea dog Sir Francis Drake first set foot on California's shoreline in 1579. He spent a month on shore establishing England's sovereignty (so he claimed), engaging in trade and communication with the native peoples, and repairing his ship, the *Golden Hinde*. Having made his ship seaworthy enough to carry the treasure he had stolen from Spanish towns and galleons in the New World, he set sail for England, undoubtedly filled with tales of his adventures and romances in sunny Nova Albion.

Concerned by Drake's activities, the Spaniards sent another expedition up the California coast to seek "harbors in which galleons might take refuge" from sea rovers such as Drake. Thus, between 1584 and 1602, Francisco Gali explored the coast, Sebastian Cermeno sailed along the shoreline noting likely harbors, and Sebastian Vizcaino made an extended excursion, stopping in San Diego, Catalina, Santa Barbara, and Monterey Bay. These expeditions increased Spain's knowledge of the area but not its immediate interest. Thus, Alta California remained largely an ignored outpost until the 1760s.

One other notable foreign incursion involved the Russians at a somewhat later date. Fort Ross in present-day Sonoma County was founded by the Russian-American Company in 1812. Although primarily concerned with the fur trade in Alaska, operating in concert with Americans, the company did extend southward in search of both food and other fur sources (primarily sea otter). With the decline of the sea otter and the fur seal populations, however, the Russians sold out to John Sutter in 1841 and withdrew permanently from California.

Mission Settlement Patterns

As with many colonizations, the settlement of California was largely inspired by international competition. Spurred on by the commercial activities of the Russians and the presence of English and Dutch privateers along the Pacific coast, the Spaniards recognized the need to protect their holdings from foreign rivals. The most effective way to accomplish this goal was to initiate the tripartite policy of mission, presidio, and pueblo.

According to the Spanish concept of empire in the region of Alta California, the *missions* were intended to "Christianize" the native populations while at the same time establishing and reinforcing loyalty to the Spaniards. The *presidios* were frontier forts designed both to control the native populations and to protect Spanish interests from foreigners. Thus, they were located in strategic sites to enable the Spaniards to protect the missions and to permit control of critical ports or trade centers. The *pueblos* were the civil arm of the colonization effort, designed to draw civilian settlers into the region and further reinforce Spanish rule. The pueblo residents interacted with the presidios by selling them food products and by acting as a form of "reserve militia."

Under the energetic leadership of Jose de Galvez, the inspector-general of New Spain, expansion was initiated. With the guidance of Captain Gaspar de Portolà and Father Junipero Serra, the first presidio and the first mission in Alta California were established in San Diego in 1769. During the next 50 years, 20 more missions were founded, some accompanied by presidios, some by pueblos, some by both. Approximately 48.28 kilometers (30 miles) apart (one day's travel by horseback), they formed a chain strung from San Diego to Sonoma, connected only by a dusty path known as El Camino Real (the royal road). These missions were remarkable institutions that assumed the key role in California life. From the viewpoint of the civil authorities, they were inexpensive extensions of the empire, requiring only a few padres, a handful of soldiers, and infrequent restocking of supplies.

Each mission reflected a certain adaptation to California realities. Structurally, the missions evolved an architecture of red tile roofs, massive buttresses, and thick walls as protection against brush fires and earthquakes, the heat and sun of summer, and the cold of winter (Figure 3.2). Agriculturally, each mission operated on a huge tract of around 100,000 acres located in fertile coastal areas supplied with both ample water and native population. The dedicated but paternalistic Franciscan friars intended to bring civilization and progress to the California Indians. Unfortunately,

Figure 3.2 Santa Barbara Mission—a jewel in the chain of California missions.
(Image © RonGreer.Com, 2009. used under license from Shutterstock, Inc.)

more often they brought dependence bordering on slavery, diseases for which the native population had no immunity, and early death for many.

Economically, the missions were probably highly successful. The missions developed a crude form of industrial-manufacturing enterprise, in which weaving, blacksmithing, cattle raising, farming, tanning, masonry, and carpentry became common. Thus, the California Indians learned from the friars how to toil long hours for their food and livelihood, whereas before the padres' arrival they had merely lived successfully off the natural bounty of the land.

Considerable historical debate has focused on the ultimate "morality" of the mission and its impact on the native California population. As a convenient personification of the Spanish church, Father Junipero Serra has been proposed for canonization by some and reviled as an imperialist exploiter by others. Regardless of the characterization, the impact of the missions on California culture and populations was substantial, permanently transforming the face of California.

The presidios and pueblos played a less vital role in the settlement of California, although they were important in extending secular authority to the frontier. The presidios at San Francisco, Monterey, Santa Barbara, and San Diego served to guard against foreign invasion and native uprising. With their limited capabilities, however, it was fortunate they were not put to any severe test. Despite inducements of land, stock, and implements, the establishment of pueblos was not a major success. The pueblos of San Jose and Los Angeles are the only enduring evidence of the civil townships created by Spanish authority (Figure 3.3).

Figure 3.3 Pueblo de Los Angeles at Olvera Street.
(Image © Jolanda, 2009. Used under license from Shutterstock, Inc.)

The Mexican Period

The End of the Mission

The mission period ended suddenly with the disintegration of the Spanish empire in the Americas in the early 1820s. The geography of the mission colonization effort inevitably resulted in its separation from the Spanish empire. All of Spain's American colonies were largely independent of the mother country. Thus, in the early years of the nineteenth century, Spain's American colonies began seceding en masse. After 10 years of sporadic struggle, the Republic of Mexico was formed in 1821. Always a distant outpost at best, California had been little affected by these independence efforts; Mexico, however, asserted its natural trade and political rights as heir of Spain, and California became an official part of the republic.

During the mission period, the establishment of private ranchos had been successfully opposed by church authorities, with fewer than 20 land grants being made to private persons. After the Mexican revolution, however, the number of private land grants increased dramatically, often at the expense of the missions.

The turbulent Mexican period brought several changes to California life. Isolated as it was from Mexico, California did not respond well to governors sent to rule from distant Mexico. The *Californios* (the name given to the non-Indian settlers and rancheros who made their homes in the region of modern-day California) had been accustomed to being left alone, and a series of unpopular and incompetent governors helped formalize a tradition of conspiracy, insurrection, and rebellion, as well as bring about major socioeconomic upheavals.

Although initiated earlier, perhaps the most dramatic event was the secularization of the missions by order of the Mexican government in 1834. Reflecting a concern over the traditional power held by the church and its historical ties to Spain's government, the revolutionary Mexican government decided that the vast holdings and influence of the church should be reduced. Under government edict,

missions were to become local churches, and mission lands were to be divided up among tribal families. The actual mission buildings were to be reduced to the role of parish churches, with the mission friars converted to parish priests. Ostensibly, such changes would have brought treatment of the California Indians in line with the new Mexican government's republican principles. However, corruption, avarice, ineptitude, and fraud deprived the California Indians of their intended legacies. The vast mission holdings were divided up, eventually becoming private ranchos, and for a brief time, the colorful and romantic rancho characterized California life.

The Romance of the Rancho

Of all historical periods, that of the California rancho is perhaps the most romanticized. The image of the cool, tile-roofed rancho, with *ollas* of water dripping peacefully in languorous breezes blowing through enclosed patios, is indeed compelling. The colorfully dressed, devil-may-care *caballero* riding or dancing with reckless abandon also provides a dramatic historical figure. Undoubtedly, these images have been overdone, reflecting as they do only a small percentage of the population during a very limited period. Nevertheless, they represent a colorful evocation of a way of life that held sway for some people. Cattle ranching *was* the predominant economic underpinning of Mexican California. The entire population owed its social and economic existence to this activity, which had replaced the mission almost entirely by the late 1830s. The secularization of the missions brought about an increase in private rancho grants from 20 in 1821 to more than 600 by 1846. This covered most of the appropriated lands up to and including the Sacramento area.

The rancho was a natural outgrowth of several economic and geographic factors in California. Perhaps most important was the fact that the California land and climate were ideally suited to raising cattle and horses. With little or no attention, the California ranchero could let the herds roam the open range, multiplying in wild profusion. *Vaqueros,* or cowboys, engaged in lively cattle roundups twice a year, once for branding and once for slaughter. The best of the wild horses were culled from the herds to provide the predominant means of work, play, and transportation. Because the ranchos ranged in size from 4,500 to more than 100,000 acres, there was plenty of land for this form of enterprise.

By relying simply on beneficial climate, abundant land, and biological reproduction, the ranchero could prosper without artificial or unreasonable efforts. Indeed, with cheap or free land grants and with plentiful Indian and vaquero labor, many of the ranchos were hugely successful. This prosperity has produced the nostalgic image now associated with the period of the rancho. Dashing horses and handsome vaqueros, bright costumes and joyous fiestas, dances and songs, lively rodeos and lovely senoritas—all have come to characterize this period in the minds of many people. If the reality of life in that period was not always exactly as imagined, it nevertheless has provided a fascinating legendary background for the state.

Foreign Incursions and Early Statehood

Mountain Men, Sailors, Pioneers, and Heroes

A long-standing consequence of California's geographic isolation from Spain and Mexico was the constant influx of foreign visitors. Early restrictions on commerce with outsiders were never well enforced in California, especially given that Spain itself could seldom provide the trade items, news, and exchange desired by the *Californios*. Thus, early American incursions into California took the form of Yankee ships that entered its ports for trade and commerce.

American fur traders and explorers also blazed trails into California in the early 1800s, beginning with Jedediah Smith and followed by others such as James Pattie, William Wolfskill, and John Fremont (Figure 3.4). Although these trappers, mountain men, and sailors were not welcomed by the government of California, they began a tide of American immigration that would

soon lead to a significant shift in demographic and political patterns in the region.

Attracted to the relaxed lifestyle, pleasant weather, business opportunities, and pretty senoritas, American sailors and trappers stayed in California to become influential rancheros themselves. Through letters and by word of mouth, California soon became known as an idyllic and profitable place for settlers. A steadily growing number of Americans began to arrive in the area, and American influence increased far in excess of actual numbers settled in the area. By the early 1840s, hundreds of Americans were entering California annually by various overland trails, with the Bidwell and Donner parties being only two of the more famous groups. With growing immigration, the groundwork was laid for the formal political control of California by Americans.

There is strong evidence that even before the Mexican–American War, the administration of President Polk had its eye on eventual annexation of California. With the outbreak of the war in 1846, American forces seized various points in the region. The American desire to annex and the general enthusiasm of the American settlers in California for this move were aided by the general indifference of the California population to the American takeover. Although a few skirmishes occurred in the Los Angeles area, the American occupation was for the most part peaceful, and California formally became a part of the United States in 1848 with the signing of the Treaty of Guadalupe Hidalgo. Spurred on by the influence of the new mining population, leaders adopted a constitution in 1849, and the state of California was admitted to the Union in 1850.

The Gold Rush

If any anomaly of geography or nature helped California become a vital part of the United States, it was the presence of a malleable yellow metal in its streams and mountains. The discovery in 1848 of gold precipitated monumental changes in the area. Population, travel patterns, socioeconomic shifts, racial balances, and myriad other changes took place as a result. The

Figure 3.4 Routes into California and the Mother Lode region.

impact of the gold rush was not lost on America; *Harper's Weekly* observed as early as 1859 that it was "the most significant, if not the most important event of the present century connected with America." This is an observation widely shared by scholars up to the present.

John Sutter did not consciously or directly seek gold. He was interested in the wealth to be gained through another medium: commerce. Convinced that lumber would be a valuable commodity to immigrants, Sutter commissioned James Marshall to survey and construct a sawmill on the American River at modern-day Coloma. In the process of building, Marshall happened to spot gold flecks in the tailrace (a canal in which water flows to and from a mill wheel). Slowly at first and then like lightning, the news spread, bringing would-be miners from all parts of the world. In a year's time, more than 10,000 treasure seekers were scattered over the gold country, with hundreds of thousands yet to come.

The peak of gold production was reached by 1852, but the effects of gold on the state endured. By the 1850s, California had become the most populous region in the western United States. Also, gold drew people away from the coast and into the interior of the state, a demographic pattern not previously encountered.

Although the lode deposits were found in a strip running from San Diego to Siskiyou County, the bulk of the gold was located in the Sierra foothills, in a region that became known as the Mother Lode region (Figure 3.4). The tenor of the times is captured in the names given to the camps and towns: Rough and Ready, Whiskeytown, Angels Camp, Hangtown, Fiddletown, French Gulch, Chinese Camp, Mormon Bar, Drytown, Oroville, Lazy Man's Canyon, Chile Gulch, and Poverty Flat, to name just a few (Figure 3.5). These names reflected ethnic influences as well as cultural humor; many of the towns were later to achieve fame through the works of Bret Harte and Mark Twain.

More significant in the long run, San Francisco, Stockton, Eureka, and Sacramento experienced expansion of agricultural and mercantile activity in response to the needs of the mining regions. The gold rush thus helped awaken California to more than mineral wealth alone. It acted as a magnet for new arrivals. It provided a stimulus to service industries, such as agriculture, cattle raising, shipping, and trade. Finally, it drew to California the kind of Yankee aggressiveness needed to develop the state. As Californio Mariano Vallejo observed, "The Yankees are a wonderful people. . . . If they emigrated to hell itself, they would somehow manage to change the climate." If the physical climate was not changed in California as a result of gold, certainly the political, economic, and social climates were affected in permanent and dramatic fashion.

The Decline of the California Indian Population

The first years of California statehood were turbulent, dramatic, fast moving, tumultuous, and violent. The sleepy legacy of Spain was forgotten as the state forged ahead to face the challenges of a modern world.

The American impact on the tribal peoples of California was devastating and permanent. If the Spanish and Mexican periods represented earlier phases in the brutal treatment of the California Indian population, the gold rush was almost the final blow. With their simple lifestyles, the California Indians could not hope to cope with avaricious gold seekers who tore up their lands and societies looking for the elusive metal.

There is no history of savage warfare here, as with the Plains Indians, but the genocidal effects were just as pronounced. Alcohol, measles, bullets, and culture shock all helped bring about the decline of the California Indian population. The few instances of "warfare" were cruel, one-sided events that included wholesale massacre of tribal women and children by white settlers. Gradually, the natives were killed off or pushed farther into high-mountain or stark desert country. This sad chapter in California history is partially captured in Helen Hunt Jackson's *Century of Dishonor.*

One of the few instances of the Indians' physical resistance was known as the Modoc Wars. Initially, the Modoc tribe had lived in the northeastern part of the state, near Mount Shasta, in a region of timber, lakes, rivers, and lava beds. As more white emigrants passed through the region, the native inhabitants became increasingly

Figure 3.5 Bodie, known as one of the most wicked of the gold rush towns, is now a California State Historic Park maintained in a state of "arrested decay."
(Image © Bob Reynolds, 2009. Used under license from Shutterstock, Inc.)

hostile. In 1852, after tribesmen massacred an emigrant party, local ranchers and miners immediately retaliated. Sporadic battles were waged for the next 12 years; finally, in 1864, the remaining Modoc Indians were moved to a reservation following a tenuous peace agreement. Placed on the reservation with their traditional enemies the Klamaths, a group of Modocs under a chief named Captain Jack made their way back to the Lost River area. Talked into returning to the reservation once, the group finally decided to make a stand on the ancestral homelands. When an army force was sent to capture them, the group took to the lava beds and, relying on their intimate knowledge of the terrain, fought the army to a standstill. Eventually, the shortage of water led to the surrender or capture of Captain Jack and his people in 1873, ending the only notable "Indian war" of California history.

The harsh attitudes of the Americans and the policy of eliminating California Indian opposition resulted in a tragic decline in the native population. Estimated at around 100,000 in 1850, by 1916 this population had dropped below 20,000. Deprived of land, dignity, spirit, and lifestyle, the California Indian became a victim of the state's "progress."

In many respects, the fate of the California Indian is epitomized by the story of Ishi, the "last wild Indian" in America. Discovered in a corral in Oroville in August 1911, Ishi was the last of a small band of Yahi Indians who had fled from contact with civilization and had lived in the Mount Lassen region. As the last survivor of this small band, Ishi eventually was able to reveal much about the last days of the California Indian to his protector and friend, Alfred Kroeber, an anthropologist at the University of California. Ishi lived and worked with Kroeber until he died of a white man's disease, tuberculosis. His true story is a touching and sad indictment of the treatment of the California Indians, encapsulating much of the tragedy of the decline of the native inhabitants of the Golden State. (Kroeber's wife, Theodora, chronicled this story in her book *Ishi in Two Worlds*. Ishi, she wrote, "was the last wild Indian of North America.")

It should be added that recent legal actions have successfully underscored the fact of the mis-treatment of the California Indians. Beginning in the late 1930s, various court decisions have held that they are entitled to both land and financial compensation. Somewhat belatedly, California Indians are receiving part of the bounty of the land that was taken from their ancestors.

As noted earlier in this chapter, the recent appreciation for the contributions of the early native Californians is a final irony. Although the disastrous decline of the native tribes cannot be corrected, misperceptions about their culture can. One example of this can be found in recent scholarship concerning the relative technological sophistication of certain of the tribes. Where once these groups were thought to have developed little technological capabilities, it is now known that some groups, such as the Nicolinos, created an aqueduct system to collect and dispense water. Likewise, anthropologists have pointed out that the Colorado River tribes, such as the Chemehuevi, used irrigation systems as a basic part of their agricultural practices. Thus, the modern water transfer systems that serve cities like San Francisco and Los Angeles (see Chapter 8) may be seen as the technological "descendants" of the "water works" of native Californians, completed long before whites set foot in the state.

Similarly, research by authorities such as John P. Harrington and Thomas C. Blackburn reveals rich, multifaceted tribal cultures. Ranging from tribal stories to pictographs and petroglyphs to gigantic geoglyphs in the Mojave, the spiritual, cultural, and artistic contributions of these peoples are just now beginning to be fully acknowledged and appreciated—further underscoring the tragedy of their decline.

The Rise of the Beef Industry

One benefit of the gold rush was a diversification of the state's economic base. With the initial need to feed growing hordes of miners, the cattle industry was able to expand significantly. Although cattle, raised for their hides, had always been a staple of the rancheros in the southern part of the state, the gold rush provided a source of eager customers for beef, which by the 1860s numbered more than 3 million head. The decline of the gold rush, a serious drought, and high interest rates brought problems to the industry, and other agricultural pursuits assumed equal or greater importance.

Eventually, the economic and geographic realities of ranching in the state caused the evolution of certain forms of cattle production. Large cattle operations became the rule, as the less successful small ranchers were phased out. Fenced-in ranges began to replace free-roaming range cattle operations. Crossbreeding and improvement of the meat and dairy aspects of cattle raising became more critical and widespread. Like the state itself, changing conditions led the cattle industry to diversify, broaden, and improve itself. The beef industry evolved into a key agricultural activity, largely avoiding competition with irrigation farming by grazing cattle on the non-irrigated portions of agricultural lands.

In the context of cattle, agriculture, and landholding, the impact of the Land Act of 1851 was critical. Following California's admission to the Union in 1850, the status of the Spanish and Mexican land grants was placed in doubt. Although the Treaty of Guadalupe Hidalgo had guaranteed protection of the rancheros and land grants, many new American immigrants felt justified in claiming parts of these lands by squatter's rights.

Because of imprecise boundaries, confusion over historical sources of titles, conflicting claims, and financial difficulties, Congress passed the Land Act of 1851. During the next several years, the rancheros and other claimants were required to present to land commissions and courts proof of title before their titles would be recognized. This requirement placed a tremendous burden on many rancheros, whose proofs of title often required extensive and costly searches of records in Mexican and Spanish archives. The prolonged legal expenses often led to the loss of ranchos by debt-ridden Californios.

In the whole story of the land title debacle, legality and morality were largely absent. The long-term impact was to shift much of the California agricultural and ranch lands from the hands of the legitimate Californio owners to the hands of the recently arrived American squatters and others who benefited greatly from the questionable largesse of the state and federal governments.

Transportation and California's Evolution

One of the most important factors in the development of modern-day California has been the transportation revolution. Just as transportation, or lack of it, was instrumental in shaping early Californian customs, culture, and society, it was to have the same impact in the American period. The creation of new means of transportation brought the state into significant contact with the rest of the nation, thus affecting all aspects of life in the Golden State.

The Gold Rush acted as a tremendous spur to the evolving transportation system. It soon became evident that mule trains and stagecoaches, clipper ships and ferryboats, and dispatch riders could not meet the growing needs of the state. Although the names Wells Fargo and Butterfield flash large in the imagination, those stagecoach lines were essentially primitive stopgaps for the real transportation links that were to come. Indeed, the overland stage operations enjoyed only a brief ascendancy, flourishing during the 1850s and 1860s, although some stage operations continued in the Bodie region into the 1880s. Meanwhile, a real transportation revolution loomed on the horizon.

The story of California's development is, to a large degree, a tale of railroad domination. Politically, economically, and geographically, the Central Pacific Railroad Company and its successor, the Southern Pacific, helped mold the state more than any other single entity. The extent of this influence is described in Frank Norris's *The Octopus,* a novel whose title captures the company's pervasive presence.

Although local lines had begun to develop as early as 1855, the critical event was the connection of these local lines to a transcontinental system. Such a system had been discussed in the 1850s, but the Civil War and the immensity of the task delayed its achievement. However, when the federal government sweetened the stakes with legislation granting 20 sections (square miles) of public land for every mile of track laid by the railroad companies, in addition to loans and other encouragements, the project became a reality. The climax of the story is the now-famous meeting of the Central Pacific and the Union Pacific at Promontory Point, Utah, on May 10, 1869, sealed with the driving of the golden spike connecting the two lines.

The California involvement revolved around the efforts the Central Pacific under the leadership of the "Big Four"— Mark Hopkins, Leland Stanford, Collis P. Huntington, and Charles Crocker. These four Sacramento businessmen established the Southern Pacific as a holding company to acquire the Central Pacific's assets and develop rail lines within the state. Through their sometimes unscrupulous practices, the Big Four fixed rates and built a rail transport monopoly that extended to every railroad line coming into the Golden State. The exorbitant rates charged by Southern Pacific led to occasional violence on the part of farmers and businessmen. Because the railroad barons also dominated the political scene, however, little real relief occurred until the rise of the reformist Progressive movement in the early 1900s.

The lasting effects of this railroad revolution on the development of California are immense. The whole complexion of population growth and settlement was colored by railroad activity. Transportation and growth patterns largely mirrored desires, expanding where Southern Pacific wanted, or was paid, to go and bypassing those areas the railroad wished to ignore. Thus, because Southern Pacific had little stake in real estate in the south, development of that part of the state had to wait until competitor lines, such as Santa Fe, were able to break the monopoly and penetrate the state (Figure 3.6).

In terms of land ownership alone, Southern Pacific was (and still is) one of the largest

Figure 3.6 Burlington Northern Santa Fe train in California; in an interesting irony, Southern Pacific Railroad ended its rail operations in 1996 and Santa Fe (currently Burlington Northern Santa Fe) now dominates rail logistics in California.
(Image © Richard Thornton, 2009. Used under license from Shutterstock, Inc.)

landowners, with more than 10 million acres, in the state. Because its business activities, including construction of additional lines and land sales, depended on supplies of people, the railroad also influenced ethnic balances with its recruiting of immigrants from both China and Europe. Its efforts in encouraging immigration resulted in increased labor supplies and more customers for business, farmers for the ample land, and tradespeople for the ever-expanding population. On a positive note, Southern Pacific helped link the state to the rest of the nation and the world through trade and population mobility. On a negative note, it provided a focal point for racial hatred, greed, and questionable business and political dealings. Its excesses led businessmen and farmers alike to organize against it and provided reformer writers like Frank Norris fertile grounds for their works.

The railroad has played a significant role in the evolution of modern California. Without the efforts of Southern Pacific and the Big Four, California would be a different place than it is today. In an era of "robber barons," the Big Four acted as many others around them did and at no time apologized for their methods. They brought progress to the state and placed a lasting imprint on its social, political, economic, and geographic landscape. One need only examine the persistence of the names of these four men to realize how deeply this influence has been felt.

Dry Farming and Irrigation Colonies

Agriculture remains a primary economic activity in California. If any one word describes the agricultural pattern of the state, it is *diverse*. With the wide range of climatic and soil conditions throughout California, diversity is almost mandated in crop production. The hundreds of crops grown in the state have solidified its reputation as the country's leading agricultural producer. (California currently ranks first among all states in the United States, and if it were a country, it would rank between fifth and tenth in the world.)

The nature of California's land cultivation has evolved through two stages—dry farming and extensive irrigation. This evolution has brought with it many changes in the economic and human elements of agriculture in the state.

The early period of California established dry farming as the dominant agricultural mode. Certainly, the native Californian inhabitants of the region engaged in little agricultural innovation, relying primarily on hunting and gathering rather than organized cultivation. With the arrival of the Spaniards, farming assumed greater importance, especially through the efforts of the missions. For the most part, this agricultural development took the form of basic grain, vegetable, and fruit crops that could be grown without extensive irrigation, although the Franciscan friars did introduce some irrigation projects in the southern part of the state. The Mexican period did not further crop growing as a major economic activity because most emphasis was placed on livestock.

With the impetus of the gold rush and the American presence, organized agriculture assumed a key role. Grain became the dominant crop in the state, with California ranking second in the nation in wheat production by the early 1870s. The success of these efforts led farmers to more diversification of crops, expansion of orchard cultivation, and experimentation with vineyards.

Additional stimulus for California agriculture was provided by both the transportation revolution and the federal government. Railroad development permitted widespread marketing of the expanding volume of crops, and government policies provided further encouragement and assistance for farm diversification and growth.

The increasing demand for California food crops brought with it a necessity to free agriculture from the whims of nature. Although dry farming had been a highly profitable endeavor, irrigation proved to be a more dependable and efficient means of production. And although the missions had tried to develop irrigation systems, it was left largely to growers in the American period to capitalize on the potential of large-scale irrigation agriculture.

The first irrigation colony of any significance in California was developed in the mid-1850s. This was a part of a Mormon settlement in San Bernardino, where more than 4,000 acres were brought under direct irrigation. Other irrigation colonies followed, including some in the San Joaquin Valley and in Riverside County. From this beginning, irrigation techniques spread throughout the state, fostering fruit crops, grapes, and other high-value agricultural products.

The legal foundation for irrigation, the Wright Act of 1887, established the concept of "appropriation and beneficial use" of waters and led to the development of recognized irrigation districts. This, in turn, led to massive efforts to cultivate huge tracts of land such as the Imperial and Coachella Valleys. The eventual success of those efforts underscores the impact of irrigation on California's history, economy, and geography.

With the growth of tensions between residential users and growers, and with the reevaluation of water use priorities, *techniques* in irrigation farming are undergoing serious scrutiny. Extended drought conditions in the 1980s and 1990s brought the era of low-cost, government-subsidized, unlimited-use agricultural water to an end. The basic *concept* of irrigation farming, however, appears to be a reality for the foreseeable future. California remains dependent on such irrigation farming to retain its position as the leading agricultural state in the nation. Thus, with the growth in the population and the gradual preemption of agricultural lands for residential purposes, high-yield irrigation farming has become even more vital. A trend toward consolidation and large-scale production, use of efficient machine labor, and high-density growing will undoubtedly continue, as will efforts toward further irrigation of formerly unusable acreage. Although the manufacturing and industrial character of the state has assumed greater importance, agriculture will continue to play a key role in the life of the state.

The "Black Gold" Rush:
The Rise of the Petroleum Industry

When Gaspar de Portolà made his first exploratory march up the coast toward Monterey in 1769, he happened upon a spring with large marshes of pitch and tar. This was the spring of the Alders of San Estevan—now known as La Brea Tar Pits. Although the presence of oil was recognized quite early in California, not until the late 1800s was any organized effort expended to recover and use this resource. Certainly, both the native Californians and the early Spanish settlers had used the crude tar from oil seeps to caulk canoes, baskets, and the roofs of adobes, but it remained for American entrepreneurs to put this California resource to significant use and turn a profit.

In its own unique way, the petroleum boom was every bit as romantic and significant as the gold rush in California. Based on a steady, if undramatic, growth, commercial oil production began around the 1860s. Pico Canyon is frequently cited as the location of the first *commercial* oil well in California. Not until the 1890s, however, did the first boom period begin. Measured against later production, the state's first "boom" in oil was more of a snap-crackle-pop. Nevertheless, this initial period was similar to its earlier gold counterpart, ranking as a "black

gold" rush. Through the late 1800s, independent oil companies engaged in a variety of activities, mostly in the south. By 1895, the combined output had risen to more than one million barrels a year. With the further stimulus of the developing automobile, as well as increased rail and manufacturing usages, oil production became even more important.

The second major oil boom began in the 1920s with the discovery of vast fields in Huntington Beach, Signal Hill, Santa Fe Springs, Torrance, and Dominguez. Certain factors urged this development along, including improved drilling technology, increased industrial and heating uses, and the growing popularity of the automobile. With the easy availability of petroleum, the automobile became a sort of unofficial state emblem, a position it retains today. The boom of the 1920s was sustained by these developments and further nurtured by the military requirements of the nation in World War II.

The latest chapter in the oil saga is the modern period, dating from approximately the early 1950s. Exploration for new wells has continued on land, primarily in the southern portion of the state. The Tidelands Act of 1955, however, ushered in a new era of offshore exploration, drilling, and development (Figure 3.7).

With this new activity has come rising controversy. Environmentalists have pointed to the dangers of oil leakage and spills, such as the highly

Figure 3.7 Offshore oil rigs at Huntington Beach, California.
(Image © Ocean Image Photography, 2009. Used under license from Shutterstock, Inc.)

publicized 1969 Santa Barbara oil spill, and have engaged in an ongoing battle with both the state and the oil companies to prevent development. For their part, the oil companies have increasingly attempted to reduce friction through aesthetic masking of offshore sites, public relations campaigns, and safety precautions against environmental pollution. Meanwhile, oil production in the state has remained a key part of the economic picture and promises to remain so as long as Californians continue their long-standing love affair with the automobile and the oil supply holds out!

The Ascent of the Western Stars: The Making of the Movie Capital

As California began to mature, many businesses and industries in the state were also growing. One of the most widely recognized symbols of the Golden State, the glamorous movie industry, emerged at the turn of the century. In many ways, the geography of the region contributed heavily to this development. Several factors made Southern California extremely attractive to the newly expanding independent moviemakers. The mild climate meant that less money had to be spent filming on location, open space was readily available in such sleepy towns as Hollywood, and the Southern California region was about as far away from New York and other major Eastern cities as these producers could get. This last factor was crucial, because many of these enterprising filmmakers were engaged in running battles with patent holders and the so-called movie trust back East. California offered both distance and the Mexican border if the legal climate grew too warm.

Soon after the filmmakers arrived, the Hollywood area achieved dramatic success, in terms of both economics and publicity. By the early 1900s, California had come to be known as movieland, and by 1915, Hollywood was the self-proclaimed "film capital of the world." With the rest of the country ripe for this type of entertainment, the state had acquired a new major industry. The movies employed a wide variety of persons, including actors, carpenters, designers, researchers, writers, and publicists. Indeed,

drawn by the popularity of early westerns, such famous (or infamous) characters as Emmett Dalton (formerly of the train- and bank-robbing Dalton Gang) and Wyatt Earp (formerly marshall of Tombstone, Arizona, and principal player in the gunfight at the OK Corral) tried their hands with the film industry in Hollywood.

With the development of sophisticated equipment and sound and color technology, the industry brought even more attention to the state. To some people, Hollywood came to represent California; to others, California came to represent America. With a system that rewarded stars such as Mary Pickford and Charlie Chaplin with weekly salaries in excess of $10,000, the "good life" of California became even more appealing.

Through the peak period of the 1930s and 1940s, the movie industry significantly affected the state as well as the rest of the world. Internally, at least three key factors were in evidence. First, the tourist attractions of Hollywood drew millions of people and their money to the state. Second, a major center of industry was created in the state, spinning off secondary economic activities such as cosmetics, electronic technology, broadcast enterprises, and clothing. Third, the magic aura of the Golden State was artificially enhanced, luring still more people to the West Coast.

Ultimately, the television industry reduced the role of Hollywood as prime dispenser of dreams and entertainment for America, and many movie companies disappeared. Although the filmmaking industry has subsided substantially from its peak period of the 1930s and 1940s, it retains a preeminent role in the life of the state. The majority of films made in America still originate in Southern California. West Los Angeles, Century City, Universal City, Culver City, Studio City, and Burbank have joined Hollywood as movie entertainment cities (Figure 3.8). Even though television has surpassed movies as the most important dispenser of popular arts, television itself has become *the* major customer of the movie industry. Furthermore, the lure of Hollywood continues, as evidenced by the throngs of tourists who annually

Figure 3.8 The famous Hollywood sign recognized throughout the world.

(Image © Charlie Hutton, 2009. Used under license from Shutterstock, Inc.)

trek to view Hollywood Boulevard's "walk of the stars," Mann's (Grauman's) Chinese Theater, and Universal Studios Hollywood Theme Park.

From the first commercial film produced in California in 1908 (*The Count of Monte Cristo*), to the first "talkie" (*The Jazz Singer*), to the most recent thriller, comedy, action-adventure, or dramatic spectacular, the California movie industry has molded attitudes, economics, and social styles in an astonishing fashion. Like its host state, the movie industry has built upon a golden legend. In the process, California has added one more romantic attribute to its identity.

World War II: Enter Defense Plants, Exit Japanese Americans

California enjoyed glamor, success, expansion, romance, excitement, and action during the course of its development. The gold rush, oil boom, movie industry, transportation revolution, and agricultural bonanza all brought progress and change to the state.

The Great Depression of the 1930s, however, brought a different tone to California, the rest of the nation, and the world. Unemployment in the state rose drastically, while crops rotted in the fields and orchards for lack of markets. At the same time, lured by its climate and its movie-created reputation, thousands of unemployed workers migrated to California seeking the golden dream. Sadly, the state could not

live up to its legend and the economic situation remained bleak, as did the lives of hundreds of thousands of Californians, new and old.

The agricultural industry was particularly hard hit. Conditions of life for farm workers were deplorable at best. John Steinbeck's anguished novel, *The Grapes of Wrath,* gave accurate voice to the despair of these workers, underscoring the depth of the Depression in both spirit and social reality. Although efforts were made to provide relief, not until America's entry into World War II did California and the nation climb out of economic depression.

In many ways, World War II had a profound and overwhelming influence on the modern development of the state. The war brought with it a significant shift to urbanized, industrialized lifestyles, as well as an increasing racial ambivalence reflected in the roles of Japanese Americans, Mexican Americans, and African Americans in the life of the state.

Well before the attack on Pearl Harbor in 1941, relations with Japan had been seriously strained, a fact reflected in California in both legislation and popular attitudes. This reaction in California had its roots in a variety of motives, both genuine and contrived. Certainly, patriotism and legitimate concern over security played roles in the hostility and suspicion that was generated toward Japanese Americans. Equally important, if less noble, was the historical sense of jealousy, greed, and competition that the successful Japanese businessmen and farmers engendered in the minds of "native" Californians.

With a tradition of hostility toward Asians and other foreigners, the California response to the Japanese bombing of Pearl Harbor was predictable. Foreigners in general, and Asians in particular, had been forced out of the state before. Here was a natural reason to gather, incarcerate, and reject the Japanese again. (It should be noted that although Hawaii was also under martial law, Japanese Americans there were not rounded up and quarantined from the rest of the population.) Under pressure from Californians and military zealots, President Franklin Roosevelt on February 19, 1942, authorized the War Department to take military control of "enemy aliens."

Thereafter citing military necessity, the head of the Western Defense Command, General John L. De Witt, issued relocation orders resulting in the forced "voluntary" internment of approximately 110,000 Japanese Americans from California, Washington, Oregon, and Arizona, two-thirds of whom were *native-born* Americans.

Treated as untrustworthy, ordered to assembly centers, and then transferred to "relocation centers," Japanese Americans were sheared of their basic civil liberties and forced to live in stark barracks encircled by barbed wire and patrolled by armed guards in barren locations such as the Owens Valley (Figure 3.9). Remarkably, the majority of Japanese Americans retained their sense of patriotism to the United States—a patriotism that saw many of the young men join Nisei units of the US Army and serve with unusual valor.

The costs of this shameful policy were immense. Economically, Japanese Americans suffered losses of at least $365 million in the forced sales of homes, businesses, and land. Psychologically, the humiliation and hatred they suffered was a bitter indictment of American attitudes in general, and Californian racial attitudes in particular. Politically, this suspension of constitutional guarantees was a frightening precedent inspired largely by hysteria and racial hatred. Finally, in a period of intense demand, some of the best business owners and agriculturalists in the state were removed from the production lines. In retrospect, this episode in California's history is a chilling reminder of where prejudice and racism can lead. (As a final comment, belated recognition of the injustice of internment occurred in a series of US Supreme Court rulings lasting into the 1980s. Then the Civil Liberties Act of 1988 mandated a redress grant of $20,000 and a formal letter of apology to each of the surviving internees [about 65,000 of a total 120,000 were still alive]. The first of these payments and apologies took place in October 1990.)

At the same time that Japanese Americans were being removed from the life of the state, massive population movements into the state were occurring. World War II demanded a maximum output from California. Defense plants in the south and shipyards in the north required enormous numbers of workers. With the exodus of Japanese Americans, large numbers of migrant Mexican laborers and blacks from the southern states moved into the vacated agricultural and industrial jobs. There was also a massive influx of military personnel. California held strategic value as a military area because of its open land, transportation facilities, and convenience as a debarkation point for the Pacific theater. Vast training bases drew huge numbers of people to the state, implanting in their minds the images of palm trees and balmy climate. Although most of these people were only temporary residents during the war years, many were haunted by their memories and returned after the war to become permanent citizens of the state.

The impact of the war on the state's economy was tremendous. The aircraft industry, already established in Southern California, experienced an unprecedented explosion and became the largest growth industry in the state. By 1941, orders for aircraft made this

Figure 3.9 Stone guardhouses are bleak reminders of the Manzanar Detention Camp in the Owens Valley near Lone Pine.
(Image © Lowe Llaguno, 2009. Used under license from Shutterstock, Inc.)

the key industrial base, with companies such as Northrop, Lockheed, Douglas, North American, and Hughes employing hundreds of thousands of workers. The population impacts on the state were as great as the economic effects. This period saw the beginning of a flow of immigrants from other states that barely slackened after the war was over. Indeed, the industrial profile of Southern California was well fixed during this period. In terms of both population and economics, the war years created massive changes in the state, with a built-in potential for even greater changes to come.

Postwar California:
American Suburbia

The population boom that began during the war continued in the postwar years. In fact, the state population rose by more than one million in the years immediately following the war, and the total growth exceeded 50 percent between 1940 and 1950. Much of the growth could be attributed to returning servicemen who had seen California during the war and had returned with their families. The attractions of the Golden State were also compelling enough to retain large numbers of workers who had migrated to the state during the war.

It is not surprising that the population growth figures exceeded those of most other areas of the country. The desirability of California as a place to live created new settlement patterns for the state, transforming much of the landscape from rural to predominantly urban settings. The necessity to house this growing population helped develop a somewhat unique California style, the "ranch house" in the uniform tract suburb. The checkerboard tracts in those areas came to typify the population sprawl of California (Figure 3.10).

In terms of the economic and industrial growth of the state, the postwar years also saw significant growth. The actual reconversion to a peacetime economy had begun in the state before the end of the war. The state government attempted to plan for this future by creating

Figure 3.10 Urban sprawl and tract homes in San Diego County.
(Image © JustASC, 2009. Used under license from Shutterstock, Inc.)

economic planning commissions and expanding tax bases and other revenues.

Several factors helped ease this transition and make the shift less traumatic. The population growth itself carried with it the germination of new economic activities. New housing stimulated the construction industry, and even if assembly-line production of housing was monotonous, it was nonetheless profitable.

The immediate drop in defense activities was largely offset by a retooling for consumer products such as furniture, clothing, and automobiles. The heavy wartime industries had recognized earlier the need for diversified production capability and were already moving into needed peacetime activities. Former war plants began to engage in civilian-oriented steel production, food processing, and other useful manufacturing activities. Construction and transportation-related industries alone accounted for a large share of this postwar surge.

Except for a brief lull, the defense industry continued to hold a strong position in the economic growth of California, at least into the 1980s. Although World War II had ended, the Cold War provided a continued impetus to military spending and development. California's defense industries soon began exploring uses of nuclear energy, missile systems, aerospace technology, and electronic weaponry. Federal defense contracts supported huge programs in research and development, and "think tanks" such as the Rand Corporation and Pasadena's

Jet Propulsion Laboratory were funded by these grants. By the early 1960s, California's share of federal expenditures for military purposes was around 25 percent of the nation's total. By the late 1960s, at least one-third of the state's industrial production involved defense and space activities.

This military-related economic growth had both positive and negative effects. It placed heavy reliance on federal government largesse for continued favorable employment figures. It converted a large proportion of the California workforce into white-collar workers. But when there was a sharp downturn in government spending, a concomitant slump occurred in the aerospace and defense-related industries. Thus, unemployment figures climbed during the 1960s before other industrial growth helped pick up the slack.

A similar and perhaps more lasting slump began in the late 1980s and continued into the 1990s. With the winding down of the Cold War, companies like McDonnell Douglas, Northrop, Lockheed, and General Dynamics were forced to scale back their activities. Employment in defense-related industries declined by up to two-thirds, contributing to a severe recession that dominated both the public and private sectors. Clearly, reliance on defense-related industries alone could be dangerous; however, such reliance was far lower than many believed, with only about 5 percent of California workers employed in these industries by the early 1990s.

Significantly, other diversified industries balanced out uneven employment in the aerospace business. For example, the petroleum industry remained a key employer in the state, as did transportation-related businesses. Agribusiness also held a strong place in California's economy. Likewise, the large population acted as a constant source of demand for consumer goods and services. The growth of the computer industry, especially in the area of microprocessors, gave rise to the description of the San Jose–Santa Clara Valley area as "Silicon Valley."

The ability of former defense industries to shift to commercial production in the future is a partial index to the state's economic fortunes.

In the face of earthquakes and urban riots, the state's overall economy has remained relatively productive—at least in contrast with much of the rest of the world. Although the unemployment rate in the state tends to rise and fall cyclically, it is probably more accurate to say that economic readjustments, rather than true declines, are occurring. Certain regions have experienced much higher economic dislocations than others, many of the types of available jobs have shifted more toward service and trade industries, and stratification of the workforce has been accentuated by the continued steady influx of unskilled foreign immigrants. Overall, however, California has continued to be successful with people, money, and attitudes. The climate, the lifestyle, and the promise of economic opportunity have all contributed to the appeal of the state. In addition to the goods, produce, and entertainment exported to the rest of the world, California has continued to export its image and style to millions of people. Despite future uncertainties and past mistakes, it is still seen by many as the Paradise of the West.

Patterns for the Present and Future

One might conclude that, more than anything else, accidents of history combined with unique geography have shaped California's development. This conclusion provides a base for viewing the future evolution of the state. Many patterns begun in past years are likely to persist, while positive changes may result from the lessons of the past.

On the positive side, California seems destined to have an even greater influence in the life of the nation. Socially, lifestyles and behavior in the state, from recreation to fashion, seem to fascinate and set the pattern for the nation. As the massive tourism, variety of recreation activities, and significant retirement-oriented industries indicate, the Golden State caters to myriad interests.

Politically, the state has assumed a key role in the nation's power structure. California's voting numbers, political leaders, and political trends are watched throughout the nation. One classic example is the "taxpayers' revolt" that began in California with passage of Proposition 13 and spread to the rest of the nation. Similarly, the governor of California, regardless of party, now is a perennial force to be reckoned with in national politics.

Economically, the state is a key indicator of the health of the country, and any economic slump in California is a problem of national import. Patterns of population growth and migration to California have changed. Now, rather than reflecting the dominance of the American Midwest, trends in migration and population growth indicate dramatically increasing Latin American and Asian immigration. With California poised on the Pacific Rim, this shift is entirely logical and predictable, adding new dynamics to the cultural milieu of the state. The potential for both economic prosperity and sociocultural confusion cannot be ignored.

On the negative side, the state has yet to face up to some of its most pressing problems. Steady population growth has brought ever-increasing urbanization and congestion. Growth in certain areas such as Riverside, Sacramento, San Diego, and Contra Costa Counties has frequently outstripped the ability of those localities to serve the needs of the population.

Although industrial and business enterprises dedicated to providing luxury goods and services have assumed a major position in the state's economy, the *basic* needs of many groups are still not being met. Particularly in the more urbanized areas of the south, nonwhite racial groups are gradually emerging as significant forces. These groups, as well as other Californians, are demanding access to better schools, housing, hospitals, and jobs. With their growth in numbers, these minority groups will be able to exercise more authority—a position unique in California history. Certainly, as the Los Angeles riots of 1992 demonstrated, issues of justice and perceived equality (or lack thereof) will loom large. With census figures revealing the shift to a dynamic and diverse mix of peoples, ethnicities, and backgrounds, and with no single dominant majority culture, empowerment has become the watchword.

California also must come to grips with its long-standing tendency toward extremism of all sorts. Many people in the state have been striving to understand the extremes of devotion afforded to such groups as white supremacists or inner-city gangs or guerrilla environmentalists. Certainly, these groups have lent California a somewhat suspect air in the minds of many. One wonders whether their existence is a symptom or a cause of California's uncertain role in the future.

The certainty for California's move into the future is that problems do exist and must be solved. The effects of growth on the landscape and the habitats of the state must be addressed and controlled. The quality of life in the state must be stabilized in terms of environmental health, economic security, and social equality. Issues like smog, congestion, urban decay, ecological waste, water distribution, and costs of government must be dealt with directly and responsibly. Clearly, this provides a major challenge for the people of the state—a challenge that must be met successfully if California is to remain a "golden" rather than a tarnished state.

The diversity of the state's geography and culture guarantees that many creative solutions will be pursued. Some insight into the possible, and perhaps unique, approaches that will be taken may be found in various aspects of the cultural landscape of California, which we examine in Chapter 5.

Summary

California is rich in its geographic and historical development. Although landscape does not totally control culture, it certainly has a significant influence. As various groups have sequentially occupied California, each has left its mark in different ways. Native peoples had their own distinctive views and approaches to the physical environment. The Spanish and Mexican residents of California framed their lives around the realities of weather, location, and land. The arrival of American settlers, initially drawn by the geography or gold, brought new ideas and ways of doing things. The resulting interaction of cultures and peoples was not always smooth in the face of competition for the same land and resources. The underlying attraction of the state's benign climate, location, and lifestyle, however, remains a powerful magnet for ongoing immigration into California, and with newer demographic trends, the mix of cultures promises to be more dynamic, more varied, and more balanced.

Selected Bibliography

Atherton, G. (1914). *California: An Intimate History*. New York: Harper and Bros.

Bancroft, H. H. (1890). *History of California (1886–1890)*. San Francisco: The History Co.

Blackburn, T. C., and Anderson, K. (Eds.). (1993). *Before the Wilderness: Environmental Management by Native Californians*. Menlo Park, CA: Ballena Press.

California State Government website (www.ca.gov).

Caughey, J. W. (1953). *California* (2nd ed.). Englewood Cliffs, NJ: Prentice Hall.

Cleland, R. G. (1951). *The Cattle on a Thousand Hills: Southern California, 1850–1880*. San Marino, CA: Huntington Library.

—–. (1944). From *Wilderness to Empire: A History of California, 1542–1900*. New York: Alfred A. Knopf.

Hutchinson, W. H. (1969). *California: Two Centuries of Man, Land, and Growth in the Golden State*. Palo Alto, CA: American West Pub. Co.

Jackson, H. H. (1939). *Ramona*. Boston: Little, Brown.

Kroeber, T. (1961). *Ishi in Two Worlds*. Berkeley and Los Angeles: University of California Press.

Lewis, O. (1938). *The Big Four*. New York: Alfred A. Knopf.

Miller, C., and Hyslop, R. (2000). *California: The Geography of Diversity*. Palo Alto, CA: Mayfield Publishing Company.

Norris, F. (1901). *The Octopus: A Story of California*. Available in various editions.

Rawls, J., and Bean, W. (2003). *California: An Interpretive History*. New York: McGraw-Hill.

Rolle, A. F. (1969). *California: A History*. New York: Thomas Y. Crowell.

Sinclair, U. (1972). *Oil*. Cambridge, MA: Robert Bentley.

Steinbeck, E., and Wallsten, R. (Eds.). (1975). *Steinbeck: A Life in Letters*. New York: Viking Press.

Steinbeck, J. (1967). *The Grapes of Wrath*. New York: Viking Press.

Twain, M. (1872). *Roughing It*. Available in various editions.

US Census Bureau website (www.census.gov).

West, N. (1939). *The Day of the Locust*. Cutchogue, NY: Buccaneer Books.

Climates of California

"Climate lasts all the time and weather only a few days."

–Mark Twain

Introduction

California is one of only a few places on the earth that have the most favorable climatic condition for human settlement. However, the notion that the entire state is located in a mild climatic zone with plenty of sunshine is not true. The favorable climatic condition in California exists only along the coastal areas. In inland California, extreme climatic conditions—cold, hot, dry, and stormy—are frequently present. From its humid, cooler climate in the northwest coast to its dry, hotter climate in the southeast region, California provides samples of climates found all around the world.

Climate is one reason people come to California. The mild, comfortable climate in coastal areas provides shelter for more than 25 million people; one out of four Californians lives in the coastal areas. In the Central Valley, the warm, dry conditions and the ever-present sunshine provide a long growing season for one of the most productive agricultural regions in the world. The tourism industry also banks on the state's diverse climate. Ski resorts draw thousands of visitors to the spectacular snow-covered mountains on winter weekends, and people from all over the world visit California beaches to enjoy cool breezes under a tinted summer sun. Other industries—as diverse as the

aerospace manufacturers and the entertainment business—have selected California as their home base partly because of the favorable climate.

Climate is an important part of the state's natural resources. Although California's image is of great sunshine and dryness, water from rain and snowfall fulfills most of the state's water demands. The topic of water as a resource is further presented in Chapter 8. The ample sunshine and the developed water supply systems form the foundation for the agricultural development in the state. This topic is further discussed in Chapter 9. The topic of alternative energy resources, which is closely related to the climatic resources, is explored in Chapter 11. Finally, in everyday life, most Californians enjoy comfortable living conditions in their homes without paying huge bills for heating and air conditioning thanks to the mild climate.

We now start to examine each of the main characteristics of climate—temperature, precipitation, wind, clouds and fog—and then look into the different climate types in California.

Temperature

If asked what the temperature is like in California, many would answer that it is mild and with little seasonal change. This is true, however, only to a certain degree. The isotherm maps, maps that show temperature distribution with lines of equal temperature (Map 6), show that only in the coastal areas are the temperatures mild with less than 8°C (15°F) variation. This is further illustrated by Figure 4.1, that the annual temperature range (temperature difference between July and January) is greater than 10°C (18°F) for most part of the state. The mild temperature region is mostly in the southern coastal areas, where monthly temperatures average from the mid 10°C to the lower 20°C (upper 50°F to upper 60°F) year round (Figure 4.2).

For the vast inland areas in California, the climatic conditions are nowhere near mild. The January isotherm map shows a decrease in temperature from coastal to inland areas, and the July isotherm map shows an increase in temperature from coastal to inland areas. The combined result is a significantly increased seasonal variation in the inland areas compared with the coastal areas. Winter can be very cold in the snow-capped mountains, with temperatures frequently dipping below freezing. In contrast, summer can be extremely hot in the desert valleys, with temperatures easily soaring above 50°C (120°F) (Figure 4.2).

The overall patterns of temperature distribution in California show that the isotherms run somewhat parallel to the coast with a northwest to southeast trend (Map 6). The reasons behind such a pattern are further explored in the following sections.

Controls of the Temperature

If we were looking at isotherm maps like the ones in Map 6 for another state—say, Tennessee—we would see only a few isotherms on each map and the lines running west-east parallel to the latitude. What, then, contribute to the wide temperature variation in California? Let's examine the climatic control elements to find how these elements interact to create California's unique temperature distribution pattern.

Latitude

Latitude is a major control factor of temperature. Latitude determines the amount of solar radiation a place receives because of day length and sun angle variation. Temperature variation corresponds directly to the amount of energy a place receives from the sun. Lower latitudes (those closer to the equator), with a year-round high sun angle and relatively consistent daylight hours, receive a relatively higher amount of solar energy year-round; the result is a higher average temperature with less seasonal variation. Higher latitudes (those closer to the poles), on the other hand, have a relatively low sun angle and significant daylight hour differences between winter and summer; the result is a lower average

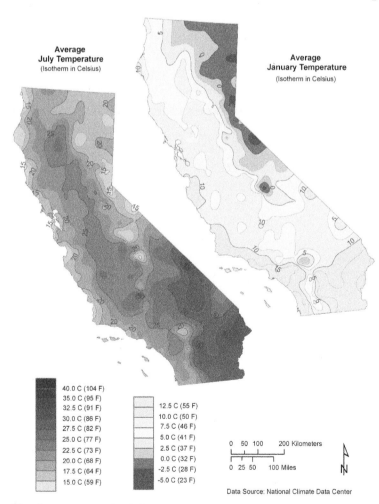

Average
July Temperature
(Isotherm in Celsius)

Average
January Temperature
(Isotherm in Celsius)

| 40.0 C (104 F) |
| 35.0 C (95 F) |
| 32.5 C (91 F) |
| 30.0 C (86 F) |
| 27.5 C (82 F) |
| 25.0 C (77 F) |
| 22.5 C (73 F) |
| 20.0 C (68 F) |
| 17.5 C (64 F) |
| 15.0 C (59 F) |

| 12.5 C (55 F) |
| 10.0 C (50 F) |
| 7.5 C (46 F) |
| 5.0 C (41 F) |
| 2.5 C (37 F) |
| 0.0 C (32 F) |
| -2.5 C (28 F) |
| -5.0 C (23 F) |

0   50 100   200 Kilometers

0 25 50   100 Miles

Data Source: National Climate Data Center

Map 6 January and July temperature distribution.
(Source: Lin Wu)

temperature with significant seasonal variation. If latitude were the only control element, lines on an isotherm map would run parallel to the latitude. This is somewhat true on a generalized temperature map of the world, which often shows that the isotherms follow the latitude with some modification by other control elements. This is not the case in California. Although California is a north–south elongated state that crosses 10 latitude degrees (32°N to 42°N), all these latitudes fall within the midlatitude temperature zone. Temperature within the same zone changes only slightly with the latitude variation. Figure 4.3 shows that January, July, and the annual average temperature decreases only 5°C to 6°C (10°F) along the coastal areas from 32°N near San Diego to 41°N near Klamath (Figure 4.3). The average July temperature stays nearly the same moving north and south along the California coasts.

In contrast to the latitude, two other temperature control elements—topography and land/water variation—have a northwest–southeast trend in California (the coastlines and the mountain ranges run northwest–southeast), and their combined effects outweigh the influence of the latitude. The result is that the isotherms in California run northwest–southeast and are further modified by latitudinal variation (Map 6). Let's take a closer look at these two factors.

Topography

The overall topographic influence on temperature can be described simply: the higher the elevation, the lower the temperature. Air temperature decreases with an increase in elevation within the *troposphere*. The decrease in air temperature with the increase in elevation is caused by the increased distance from the atmosphere's primary heat source—the ground. Although the sun is the atmosphere's ultimate heating source, air has limited ability to absorb *solar radiation* directly. Only about 20 percent of the solar radiation is absorbed directly by the atmosphere, most of which is not at the ground level; another 30 percent of the solar radiation is reflected back to the space. The 50 percent of the solar radiation that reaches the

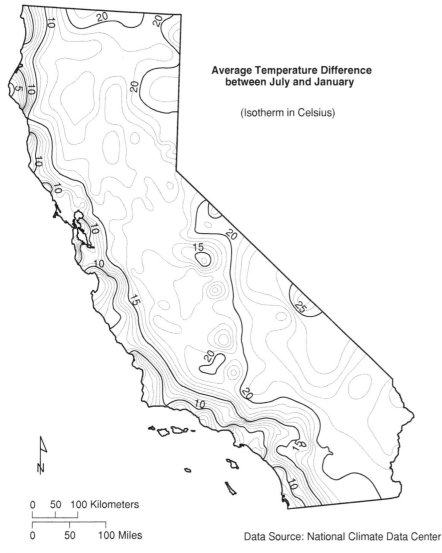

Average Temperature Difference between July and January

(Isotherm in Celsius)

0   50   100 Kilometers

0   50   100 Miles

Data Source: National Climate Data Center

Figure 4.1 Average temperature difference between July and January.
(Source: Lin Wu)

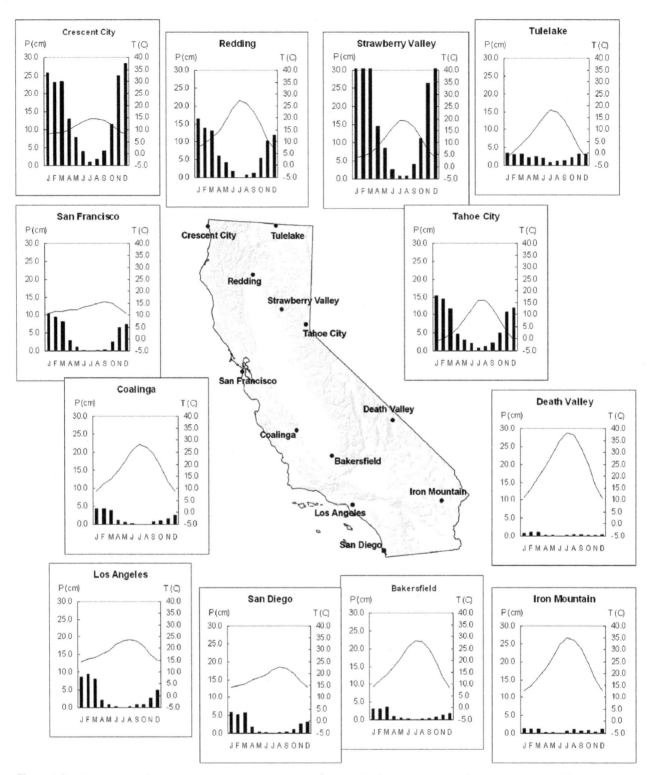

Figure 4.2 Climographs for representative stations in California. The bars on the graph represent monthly average precipitation, and lines represent monthly average temperature. Conversion: °F = 1.8°C + 32, and 1 inch = 2.54 centimeters.

(Source: Lin Wu, Data: National Climate Data Center)

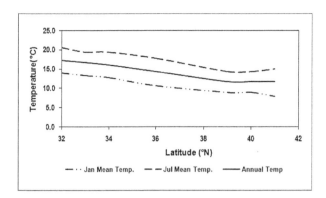

Figure 4.3 Latitudinal temperature variations along the coastal areas.

(Source: Lin Wu, Data: National Climate Data Center)

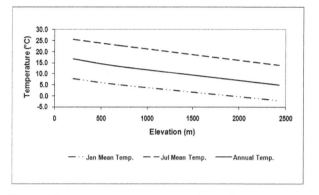

Figure 4.4 Temperature variations along the western slope of the Sierra Nevada.

(Source: Lin Wu, Data: National Climate Data Center)

ground is absorbed by the ground. The ground then emits *terrestrial radiation* into the atmosphere, which is almost completely absorbed by the atmosphere. So the surface of the earth acts like a converter that converts the solar radiation into terrestrial radiation absorbable by the atmosphere. Therefore, the closer the air is to the ground level, the more energy it gains and the higher the temperature it will be. The rate at which the temperature decreases with elevation is called the *environmental lapse* rate; it averages 6°C per 1,000 meters (3.5°F per 1,000 feet). This number can vary considerably under different atmospheric and ground conditions. In some cases, a negative number could occur, meaning that temperature increases with an increase in elevation. This is called *temperature inversion* and is discussed later. Figure 4.4 illustrates the effect of elevation on temperature. The average annual temperature at the foothills of the Sierra Nevada is 17°C (62°F) at Camp Pardee, with an elevation of 200 meters (658 feet). It decreases to 13°C (55°F) at an elevation of 716 meters (2,355 feet) at Tiger Creek and reaches the low point of 5°C (41°F) at an elevation of 2,432 meters (8,000 feet) near Twin Lakes.

Water and Land

Differences in physical properties of the water and the land result in different responses to the same energy input. Under the same amount of exposure to solar radiation, water surfaces tend to remain at a more constant temperature than land surfaces. When solar radiation increases, the temperature of a water surface rises slower

and with fewer degrees than a land surface given the same solar radiation input. This is also true for the cooling process. When solar radiation decreases, the temperature of a water surface cools more slowly and by fewer degrees than a land surface. The differences are caused by variations in the physical properties of land and water. Here are two of these properties: (1) Water has a higher specific heat than land (specific heat is the amount of energy needed to raise the temperature of one unit of a substance one degree); (2) Water moves and is, to a certain degree, transparent. The result of these physical property differences is evident: a place near the ocean or a large lake tends to have less temperature variation during a day, whereas a place surrounded by land surfaces tends to have colder nights and warmer days. Figure 4.2 shows that temperatures along the coastal areas are mild with limited seasonal variation. Average monthly temperature differences between summer and winter in the coastal areas are only a few degrees. For example, the July and January temperature difference is only 7°C (12°F) at San Diego in the southern coast, 4°C (7°F) at San Francisco in the mid coast, and 6°C (11°F) at Crescent City in the northern coast. The seasonal variations increase quickly as distance from the coast increases or as the coastal influences are blocked by mountain systems. Downtown Los Angeles, about 30 kilometers (18 miles) from the coast, has a 9°C (16°F) temperature difference between winter and summer. The difference doubles to 20°C (36°F) in Bakersfield, located on the northeastern side of the Coast and Transverse Ranges where the ocean influence is blocked by the mountain systems. At Death Valley, the difference reaches 27°C (49°F), with an average July temperature at 38°C (100°F) and average January temperature at 11°C (52°F). The highest temperature ever recorded in Death Valley was 57°C (134°F), in July 1913; this is one of the highest recorded temperatures on earth. The lowest winter temperature recorded in California, also in the inland area, was –43°C (–45°F), recorded on January 1937 in Boca, located on the eastern side of the Sierra Nevada. Although the precision of these historical records is often questionable, the temperatures nevertheless illustrate the extreme conditions in the inland areas.

Ocean Currents

In addition to latitude, elevation, and water/land influences, cold and warm ocean currents have a profound effect on the temperature in places with close proximity to the ocean. Figure 4.5 shows a comparison between Charleston, South Carolina, and San Diego, California. Both places are located near an ocean and are along the same latitude zone, but one is on the east coast and the other is on the west coast. It is noticeable that San Diego has a cooler summer than Charleston. The average July temperature in Charleston is 28°C (82°F), and in San Diego it is only 22°C (71°F). This difference is a result of the influence of the cold ocean current near the California coast and the warm ocean current along the

east coast below 40°N latitude. Although the nature of an ocean current does not change seasonally, cold ocean currents tend to have greater influence on summer temperatures, especially in the lower midlatitudes.

Precipitation

Like temperature, isohyets, lines of equal precipitation, in California also have a northwest–southeast trend, but the patterns are often interrupted by influences from regional and local environments (Map 5). The heaviest precipitation in the state occurs in two regions: the northwest coast area and the western slopes of the Sierra Nevada. Precipitation in the northwest coast reaches more than 200 centimeters (80 inches) per year. Most of precipitation in the northwest coastal areas goes back into the Pacific Ocean. Precipitation in the western slopes of the Sierra Nevada also reaches more than 200 centimeters (80 inches) annually, a large portion of it in the form of snow. Water in the forms of rain and melting snow from the Western Sierra feeds into the river systems, underground aquifers, and developed water diversion systems. These are the major water supplies for agricultural and domestic consumption in California. The high yearly precipitation in the northwest and the Sierra Nevada results in forests and a lush environment, which contrast sharply to the harsh, barren environment in the southeast desert areas. The amount of precipitation decreases drastically toward the south and east. Along the southern coastal areas, annual precipitation averages 25 centimeters (10 inches) in San Diego to 35 centimeters (14 inches) in Los Angeles. In the driest regions in the Mojave Desert, annual average precipitation is less than 5 centimeters (2 inches).

Most precipitation in California falls during a few winter months, leaving the majority of the state with little or no rain for more than half a year (Figure 4.2). The rainy season is slightly longer in the northern coastal areas and gets shorter toward the southern coastal areas. In southern California, except for occasional localized and short-lived thunderstorms, summer is essentially free from rain. For those who have

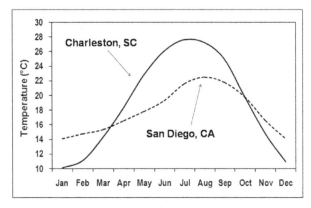

Figure 4.5 Temperature comparisons between Charleston and San Diego.
(Source: Lin Wu, Data: National Climate Data Center)

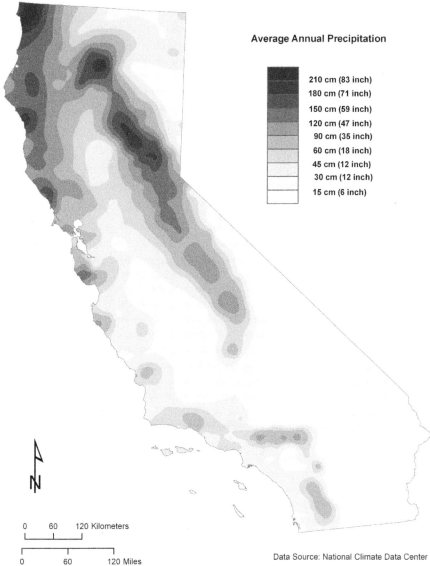

Average Annual Precipitation

210 cm (83 inch)
180 cm (71 inch)
150 cm (59 inch)
120 cm (47 inch)
90 cm (35 inch)
60 cm (18 inch)
45 cm (12 inch)
30 cm (12 inch)
15 cm (6 inch)

0   60   120 Kilometers

0   60   120 Miles

Data Source: National Climate Data Center

Map 5 Precipitation distribution in California.
(Source: Lin Wu)

tickets to the Hollywood Bowl open-air concerts held during the summer season, it is rarely a concern that a concert might be canceled or postponed because of weather conditions.

Precipitation varies not only spatially and seasonally but also from year to year, especially in the southern part of the state. Figure 4.6 shows annual precipitation variation over 100 years in Los Angeles. Although the annual average rainfall in Los Angeles is 35 centimeters (14 inches), in very few years was the rainfall close to that average. The annual precipitation amount varies from 11 centimeters (4 inches) in 2001–02 to 97 centimeters (38 inches) in 1883–84. This is typical in the dry areas of the state: that the "average" rarely occurs. In a wet year, the amount of precipitation could easily be double the average; in a dry year, the amount may not even reach half the average. In the part of the state where

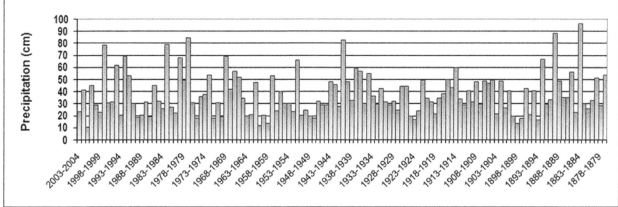

Figure 4.6 Historical annual precipitation variations in Los Angeles.
(Source: Lin Wu, Data: Los Angeles Times)

average annual precipitation is relatively high, the year-to-year change is relatively small and the numbers are closer to the average. Whereas the causes of the wet and dry years are complicated and some are not fully understood, some explanations are offered, including the El Niño effect, a global/regional phenomenon of unusually high precipitation or dry condition (dependent on location) caused by changes in the ocean's surface temperature. These are further discussed in Chapter 10.

Controls of the Precipitation

Pressure Belts and Storm Tracks

Most of the rain and snow in California comes from the Pacific *cyclonic storm* systems. This is especially true of the northwest region. The cyclonic storm system is part of the global pressure and wind system. The earth's surface wind and pressure distribution system is divided into seven pressure belts and six wind belts, with three in northern hemisphere, three in southern hemisphere, and one low pressure belt along the equator (Figure 4.7a). Precipitation distribution follows the pattern of the wind and pressure belts in general, with some deviations associated with regional and local conditions. California, with its latitude running from 32°N to 42°N, falls under the influence of the subtropical high pressure, westerly wind system, and the polar low pressure system. The latitudinal zones of these systems are not fixed; they shift seasonally. In winter, the global wind and pressure belts shift southward and the northern part of California enters the core of the storm track. In Crescent City, more than 80 percent of the yearly precipitation falls in the winter months, from October to March. Although the southern part of the state is not in the core of the winter storm track, it is within the occasional visiting path, resulting in rainy weather from time to time. In summer, the global wind and pressure belts shift northward; the southern part of the state is out of the influence of the cyclonic storm and is constantly under control of the dry subtropical high pressure system, resulting in dry

weather conditions. The northern part of the state, however, is still in the outskirts of the storm path and experiences occasional storms in the summer (Figure 4.7b).

Orographic Process

Mountains, combined with wind effect, create *orographic precipitation*, precipitation caused by topography. When moist air is pushed by constant wind (westerly wind in California) uphill in a mountainous area, its temperature drops because of a physical process called adiabatic process. Lowering the air's temperature increases the air's *relative humidity*, a function of both air temperature and moisture content. When the relative humidity reaches its saturation level (100 percent), clouds form, eventually leading to precipitation. The elevation where the precipitation starts varies depending on the location and atmospheric conditions. Precipitation continues as the air moves up along the windward side of the mountain until it reaches the top or loses most of its moisture. Maximum precipitation tends to occur where air reaches its saturation point but still has abundant moisture content. After the air moves past the windward mountaintop, it moves down the opposite, *leeward* side of the mountain. As the air moves down, its temperature increases and relative humidity decreases because of adiabatic warming. Precipitation stops, resulting in consistent dry conditions on the leeward side of the mountain. Because of this dryness, the leeward side of a mountain is also called the rain shadow (Figure 4.8). Figure 4.2 and Map 5 show the contrast of high precipitation on the western side of the Sierra Nevada and the rain shadow on the eastern side.

Convectional Precipitation

The flood warning signs in the dry deserts of California often seem out of place. Most of the time, there is hardly any trace of water around these signs. People who live in a desert, however, know well that when the occasional storm hits the region, the consequences could be devastating. Although the desert air often appears dry with low relative humidity, the *absolute humidity*, the actual amount of water vapor in the air, can be

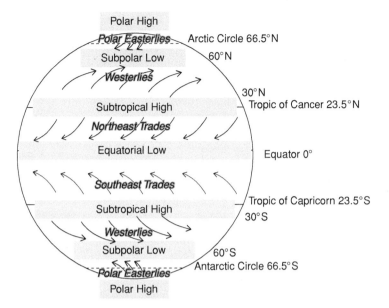

Figure 4.7a Global wind and pressure belts.

(Source: Lin Wu)

A model of the seven pressure and six wind belts on the surface of the earth without considering the land and ocean effect and seasonal variations. These belts shift northward during the summer of the northern hemisphere and southward during the winter of the northern hemisphere.

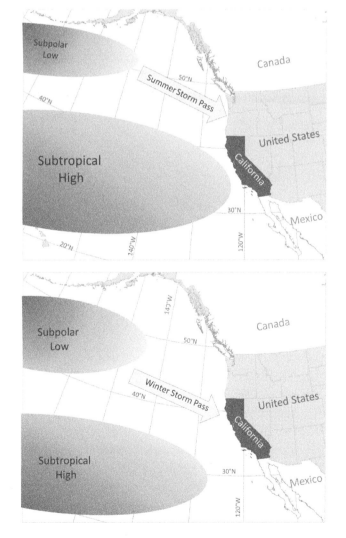

During the summer time, the pressure and wind belts shift northward. The influence of the subtropical high pressure dominates most part of California. The northward shifted summer storm pass brings almost no rain to Southern part of the state and occasional storm to the northern part of the state.

During the winter time, the pressure and wind belts shift southward. The influence of the subtropical high pressure also shifts southward and becomes less dominant. The southward shifted winter storm pass brings frequent rain to the northern part of the state. Some of these storms reach the southern part of the state.

Figure 4.7b Seasonal shifting storm path.

(Source: Lin Wu)

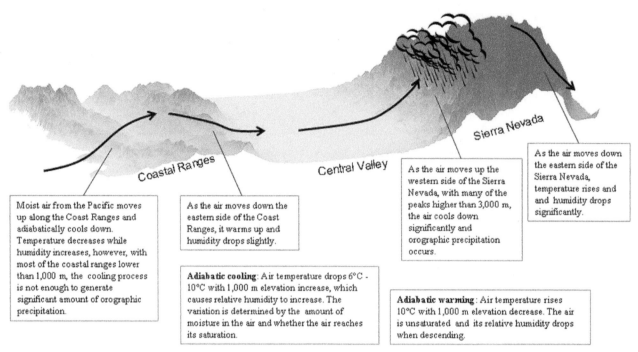

Figure 4.8 An illustration of orographic precipitation.
(Source: Lin Wu)

Coastal Ranges

Moist air from the Pacific moves up along the Coast Ranges and adiabatically cools down. Temperature decreases while humidity increases, however, with most of the coastal ranges lower than 1,000 m, the cooling process is not enough to generate significant amount of orographic precipitation.

As the air moves down the eastern side of the Coast Ranges, it warms up and humidity drops slightly.

Adiabatic cooling: Air temperature drops 6°C - 10°C with 1,000 m elevation increase, which causes relative humidity to increase. The variation is determined by the amount of moisture in the air and whether the air reaches its saturation.

Central Valley

As the air moves up the western side of the Sierra Nevada, with many of the peaks higher than 3,000 m, the air cools down significantly and orographic precipitation occurs.

Sierra Nevada

As the air moves down the eastern side of the Sierra Nevada, temperature rises and and humidity drops significantly.

Adiabatic warming: Air temperature rises 10°C with 1,000 m elevation decrease. The air is unsaturated and its relative humidity drops when descending.

Figure 4.9 A flooded road near Death Valley.
(Source: Image © Massimiliano Lamagna, 2009. Used under license from Shutterstock, Inc.)

high in warm deserts. Most of the time, the moisture stays in the air; however, when a very strong upward movement in the air occurs because of convection (warm air rises, cold air descends), the moisture condenses quickly and produces short-lived torrential rain. The desert surfaces are often exposed and without dense vegetation cover. With few plants intercepting the rain and the dry, compacted ground surface blocking infiltration, the water from the heavy rain quickly congregates into channels and can generate heavy flooding in a matter of minutes (Figure 4.9).

Fog and Low Clouds

In addition to precipitation, fog and clouds are also a liquid form of water in the air and are an important part of the climate. Fog and clouds have many characteristics in common—both are a liquid form of water suspended in the air, both indicate a high concentration of moisture in the air, and both cause low visibility in the atmosphere. However, the two are different in causes and where they occur.

Fog

The world-famous image of the Golden Gate Bridge is often wrapped in a layer of white and gray fog plumes (Figure 4.10). Fog is a frequent weather feature in California that softens the appearance of the coastline. The presence of fog is often an indication of temperature inversion. As discussed earlier, temperature inversion is a phenomenon in which temperature increases with the increase of elevation. Many conditions can cause temperature inversion. In the coastal areas, temperature inversion is often caused by cold marine air moving inland, resulting in a cold air layer near the ground. Fog forms in this cold layer when the relative humidity reaches 100 percent. This type of fog occurs year-round and is more frequent near the coastal areas than inland areas. In some of the coastal areas, the constant presence of the marine fog resulted in a unique environment such as the redwood forest in northern coastal area. The redwoods depend heavily on the water from the fog during the dry season in the summer. In other regions, such as the coastal areas in Ventura County, the fog provides the moisture for the best lemon production sites in the United States.

Another type of temperature inversion that often results in fog is the radiative temperature inversion. Radiation inversion occurs frequently during clear winter nights when the ground surface becomes colder than the air above. During clear winter nights, outgoing terrestrial radiation from the ground surface is stronger than the terrestrial radiation from the sky. The loss of energy causes the temperature to drop at the surface level. Radiation temperature inversion can be very strong in valleys because of the accumulation of cold air sinking to the valley bottom. When that happens, fog often forms in these cold inversion layers. Radiation fog often disappears after sunrise when the ground surface warms up (Figure 4.11). In valleys, the fog can become severe at times, cre-

ating traffic and other problems. In the Central Valley, this type of fog is called tule fog.

Photochemical Smog

Often associated with fog, photochemical smog (*smoke* and *fog*) is a special type of air pollution observed in the Los Angeles basin. It is caused by chemical reactions in the air when certain air pollutants are activated by the sunlight. Most of the chemicals that can cause photochemical smog are emitted from fossil fuel-powered vehicles. Photochemical smog is an especially prominent phenomenon in the Los Angeles area because of the physical and human settlement patterns (Figure 4.12). Los Angeles Basin, with one of the most extensive freeway and road systems in the world, is a main air pollution source in the region. When pollutants released from the vehicles are trapped by the mountains surrounding the area and capped by the frequent temperature inversion layer, they cannot dissipate quickly. Smog then forms when the frequent, if not constant, sunny conditions in Southern California activate the accumulated and trapped pollutants (Figure 4.13).

Figure 4.10 Coastal fog in the San Francisco Bay area.
(Source: Image © CAN BALCIOGLU, 2009. Used under license from Shutterstock, Inc.)

Figure 4.11 Dense fog on a winter morning dissipates after sunrise. Larger water bodies and vegetation areas often enhance the foggy conditions. These two photos were taken from approximately the same vantage point three hours apart on a January morning near Puddingstone Reservoir.
(Photo: Lin Wu)

Figure 4.12 A visible temperature inversion/smog layer against mountains surrounding the Los Angeles Basin as viewed from Dodger Stadium during a baseball game.

(Photo: Samuel Yang)

Los Angeles Basin

1. Westerly wind brings cool air from the ocean into the Los Angeles Basin.

2. The marine air is colder than the air above it, creating a layer of temperature inversion.

3. Pollutants from vehicles running on one of the most extensive freeway systems other sources in the region are trapped and cumulate under the stable temperature inversion layer and the surrounding mountains.

4. The intense sunshine in Southern California interacts with the chemicals in the air creating photochemical smog.

Temperature Inversion

Warmer air layer

Colder air layer

Elevation

Temperature

Figure 4.13 The formation of photochemical smog in Los Angeles Basin.

(Source: Lin Wu)

Chapter 11 further explores the air pollution problems in the state and efforts to mitigate these problems.

June Gloom

People visiting Southern California in late spring or early summer expect brilliant sunshine and warm weather. They are often surprised, however, by cool, overcast conditions in the Los Angeles Basin and coastal areas—a condition Southern Californians call "May gray" or "June gloom." The gloomy skies, covered by low clouds and occasionally accompanied by drizzle, often prevail in the morning hours throughout the region and are gradually replaced by hazy to bright sunshine in the afternoon. Days dominated by these conditions vary year to year from a few days to a few weeks. These overcast conditions in late spring and early summer are often associated with cooler ocean temperatures along the coastal areas and a counterclockwise low pressure circulation pattern off the coast areas that brings the *stratus clouds* and fog-laden marine air inland.

Regional and Local Winds

Westerly Winds

On a global scale, California is located in the westerly wind belt and on the eastern side of the Pacific high pressure cell (Figure 4.7). With this background, westerly wind, which comes from the west and blows toward the east, is the most frequent wind in California. Under different regional and local weather conditions, wind directions may change drastically; some even exhibit daily cycles.

Santa Ana Winds

One of the most influential regional wind systems in Southern California is the Santa Ana winds, named after the Santa Ana Canyon right at the pass of the winds. Nicknamed "devil's breath," the Santa Ana winds are dry, hot, strong winds that occur mostly during late summer and early fall, although they could occur any time during the year. The temperature of the Santa Ana winds can reach higher than 40°C

(100°F), and relative humidity may drop below 10 percent with gusty winds exceeding 30 meters per second (70 miles per hour). Although Santa Ana winds cause unseasonably hot conditions, they originate from the cold conditions developing in the desert area in late summer and early fall. As Figure 4.14 illustrates, a high pressure system in the high desert develops as a result of cooling in late summer and early fall. Relative to the high pressure over the land, a low pressure develops over the ocean near the coast. Between the high pressure in the inland and the low pressure over the ocean, air is forced from the high pressure to the low pressure, resulting in an easterly wind pattern (air moving from inland out to the ocean). When the easterly winds squeeze through the mountain passes, their velocity increases significantly, creating strong, gusty winds. The high temperature and low humidity of the Santa Ana winds are caused by the air descending from an elevation higher than 1,000 meters (3,000 feet) to the Los Angeles Basin and Orange County coastal areas with elevations close to sea level. The over-1,000-meter (over-3,000-foot) drop in elevation causes a roughly 10°C (18°F) degree temperature rise and a decrease in the air's humidity because of the adiabatic processes explained earlier.

The biggest problem associated with the Santa Ana winds is that wildfires often burn out of control under the conditions of the Santa Ana winds. The extreme low humidity, strong wind, and high temperature associated with the Santa Ana winds enable wildfires to spread quickly. In addition, Santa Ana winds often occur during the fire season, when vegetation is dried up after a long, dry summer and turns into fuel for the fire. These conditions, combined with a spark of fire, can trigger deadly consequences. In October 2003, during Santa Ana winds, 15 wildfires were burning around the same time in Southern California. The largest of the 15 was the Cedar fire in southern San Diego County. The fire, one of the worst wildfires in US history, started in the Cleveland National Forest on October 25, 2003. Fanned by the Santa Ana winds, it quickly spread out of the control and burned for more than two weeks, consuming everything in its path. In the

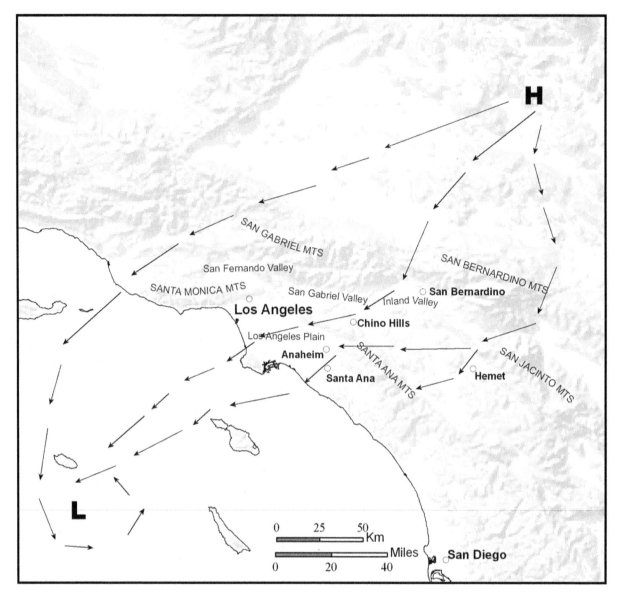

Figure 4.14 Pressure and wind patterns during Santa Ana winds condition.
(Source: Lin Wu)

end, the fire burned 1,134 square kilometers (280,278 acres) of land, destroyed 2,820 homes and other structures, and caused 14 fatalities.

Even if no fire is burning, the strong wind may still cause problems. At times, trucks close to mountain passes can be overturned by the strong winds. The winds can also damage utility lines and homes. The extreme hot and dry conditions may trigger medical problems for people with allergies, asthma, or other medical conditions.

One of the few benefits associated with the Santa Ana winds is that it sometimes brings unusually clean air to the region. The strong winds blow air pollutants toward the ocean and improve the local air quality. Air quality may be worsened, however, when a fire is burning in the area or when sand and dust are kicked up. In such areas, the solid pollutants in the air increase under Santa Ana winds conditions.

Sundown Wind of Santa Barbara

A *sundown* wind, also called a sundowner, is a wind system that occurs in the Santa Barbara coastal area (Figure 4.15). Sundowners share many characteristics with Santa Ana winds: they are hot, dry, offshore winds that can blow more than 30 kilometers per hour (70 miles per hour). The two wind systems also share

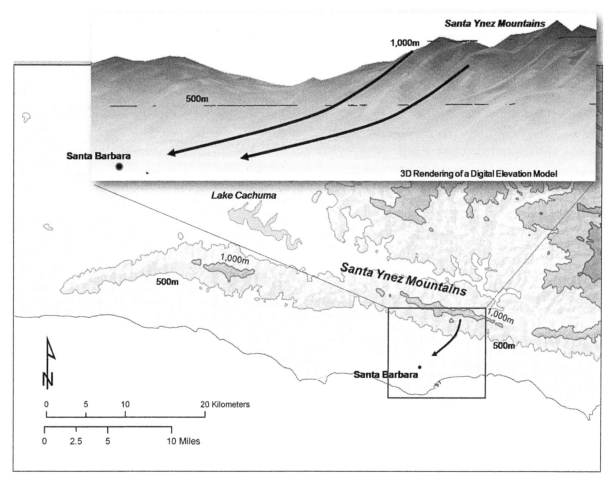

Figure 4.15 Sundown winds in the Santa Barbara area.
(Source: Lin Wu, DEM data: US Geological Survey)

some similar causes: both wind systems come from mountain areas, and the fast elevation drop triggers the adiabatic warming process that results in hot and dry conditions. Differences between the two systems are also significant. As the name indicates, the sundown wind occurs in late afternoon, at sundown, caused by the cooling of the ground surface at the mountain area. It is an extreme version of a mountain wind as discussed below. Unlike Santa Ana winds, which are often in the weather forecast, the onset of the sundown winds can be very fast, without warning. Temperature and humidity can drop drastically in a matter of hours if not minutes. For example, one strong sundown wind event occurred on July 20, 1992. Records show that, during midday, the air was calm with temperatures in the mid-20s°C (70s°F) and relative humidity in the 80 to 90 percent range. By evening, temperatures had reached over 40°C

(105°F) and relative humidity had dropped to 14 percent!

Because of the fast temperature and humidity changes, sundown winds can cause serious medical problems. The winds can also spread wildfires quickly if a fire is burning when the wind starts. During the May 2009 devastating wildfire in the Santa Barbara area, the sundown wind played a significant role in spreading the fire; numerous homes were burned down and thousands of people had to be evacuated.

Sea and Land Breezes

People who have visited California beaches in the summer may have noticed cool breezes blowing in from the ocean on a sunny afternoon, providing a release from the heat for beachgoers. People who have walked along the beaches during a summer night may have noticed cool breezes blowing from inland toward the ocean.

The former is called a sea breeze (from the sea toward the land); the latter is called a land breeze (from the land toward the sea). Sea and land breezes form because, as discussed earlier, the land and the ocean have differing thermal properties. During the daytime, the relatively cold ocean surface has a slightly higher pressure than the land, which causes the air to move from ocean to land. During the night, the land surface becomes relatively colder and the process reverses. Sea breezes can reach as far as 30 kilometers (18 miles) from the coast if not restricted by coastal mountains or other barriers. Land breezes are less consistent, compared to sea breezes, and are usually not as strong as sea breezes. The processes of land and sea breezes are illustrated in Figure 4.16. Land and sea breezes are present only when the regional wind systems are relatively calm.

Winds in the Mountains and Valleys

Mountain and valley winds are other types of local wind systems. Their formations are similar to those of sea and land breezes; however, their characteristics are different. During the day, the valley surfaces become warm. The warmer air, with its lower density, rises and forms an upward valley breeze on the slopes. Unlike sea breezes, which bring cool air, valley breezes are composed of warm air. During the night, the surfaces become cool. The cooler air, with its higher density, slides down slope forming downward mountain breezes (Figure 4.17).

Missing Hurricanes and Rare Visits of Tornadoes

Devastating hurricanes and tornadoes often make news headlines nationwide. These violent weather systems are usually absent in California, however. When gale winds, torrential rain, and pounding

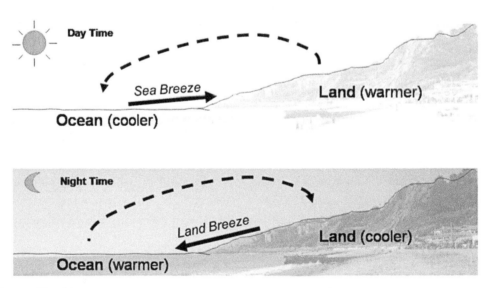

Figure 4.16 Sea and land breezes.
(Source: Lin Wu)

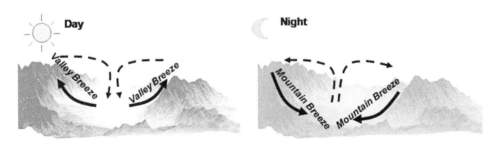

Figure 4.17 Mountain and valley breezes.
(Source: Lin Wu)

ocean waves invade the East Coast in late summer and early fall each year, California's beachgoers are often sitting in the shimmering sunshine, watching ocean waves rolling back and forth with ease. Why are places on the East Coast often battered by hurricanes while places with similar latitudes on the West Coast are usually free of these destructive systems? The answer lies within the paths that hurricanes or tropical storms follow.

Most hurricanes or tropical storms form over warm ocean surfaces near 10°N or 10°S latitude. Storms that originate over the Atlantic Ocean move toward the East Coast, following the easterly wind on the southern side of the Atlantic high pressure system (Figure 4.7). They sometimes encroach on the land, causing disasters in the south and southeast coastal regions of the United States. Storms that originate over the Pacific Ocean also move westword, to East Asia and Southeast Asia, leaving California coasts behind them.

Occasionally, Southern California may experience cloudy and windy conditions from the effect of a tropical storm when it forms close to the West Coast off Mexico. These experiences, however, never grow into a hurricane landing experience.

A *hurricane* is a regional scale system with a diameter of hundreds of kilometers. Tornadoes, on the other hand, are small systems with a diameter often less than a kilometer but could be equally destructive. California, along with most of the western states, is relatively free of tornadoes. The average number of tornadoes in Texas is more than 124 per year, whereas in California, with a land size about two-thirds that of Texas, the average number of tornadoes is only 5 per year, none of which are as strong and devastating as the ones in Texas and its neighboring states. Tornado paths are often marked on large flat land surfaces, so the complicated mountain systems in California and other western states are obstacles for tornado development. However, smaller scale, localized tornadoes, or microbursts, tornado-like phenomena that are not strong enough to be categorized as tornadoes, do occur in relatively flat areas of the state. The destruction caused by these systems is usually minor compared with that caused by the devastating tornadoes in "tornado alley"—the region with strong and frequent tornadoes in the relatively flat regions of the Central United States.

Climate Regional Variations

In the previous sections, we explore different climate variables, their unique characteristics in California and the processes behind them. In the last section of this chapter, we explore the *climate*—the collective presence of these variables over a long period of time. But before we explore the climates in different regions of the state, let's take a look at another example of the application of a scientific research method in geography—classification of the climate.

Climate Classification

To facilitate learning and application, climates are often classified into groups and subgroups. In fact, classification is one of the most frequently used research methods in almost all scientific fields. Classification methods are often based on the characteristic similarities and distinctions of the objects under consideration. Because climate is a result of many combined factors and processes, drawing dividing lines between these multifaceted elements is often difficult. Therefore, none of the existing climate classification methods satisfies all the applications. The classification used in this book is a modified version of the Köppen classification, a widely used method that classifies climate based on temperature and precipitation. Modifications are made based on topography and other environmental conditions that best reflect the nature of the climate in the state. The resultant classification map (Map 4) divides California into three major climatic regions, and each is further divided into two subtypes.

Mediterranean Climate

Mediterranean climate (Cs) is found along the coastal areas of California. The climate is characterized by mild seasonal changes with a distinctive dry summer and moderate winter rain. Mediterranean climate is considered one of the most desirable climates for human settlement. It is located in mid-latitude around 35 degrees north and south latitude on the west coast areas. Only a few places in the world have this ideal location: places surrounding the Mediterranean Sea, where the name of the climate originated; coastal areas in central Chile; the southern tip in South Africa; and two Southwestern tips of Australia. In California, typical Mediterranean climate conditions exist in northern and central coastal areas. Moving south and inland, the mild conditions gradually give way to more extreme temperatures that hardly qualify for this climate type, although the winter rain and summer dry conditions still prevail. Subdivisions of Mediterranean climate are based on summer temperatures. Areas close to the ocean have the warm summer Mediterranean (Csb) subclimate type characterized by a mild, warm summer with monthly average temperatures below 22°C (72°F). A long stretch of the California coast from the Los Angeles Basin in the south to Crescent City in the north has the warm summer Mediterranean climate (Map 4). The hot summer Mediterranean climate, characterized by above 22°C (72°F) average summer monthly temperature, occurs toward inland areas and in the valleys of the coastal areas.

The mild seasonal change in temperature of the Mediterranean climate can be explained by the mid-latitude and coastal location. The characteristics of summer dry and winter rain patterns are caused by the seasonal shifting of the wind and pressure belts. During winter, the global wind and pressure belts shift southward bringing frequent storms from the North Pacific Ocean to the northern California coastal areas. From time to time, these storms reach the southern coastal areas. During summer, the wind and pressure belts shift to the north and the state is under the dominant influence of the subtropical high pressure system, resulting in consistent dry weather conditions.

The mild Mediterranean climate provides comfortable living conditions indoors and outdoors throughout the year. As we discuss in Chapter 11, the favorable climatic conditions reduce the need for active cooling (air conditioning) and heating (heaters) during summer and winter and make California one of the states

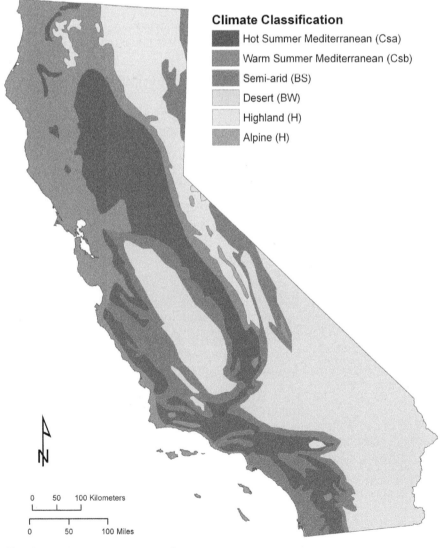

Climate Classification

- Hot Summer Mediterranean (Csa)
- Warm Summer Mediterranean (Csb)
- Semi-arid (BS)
- Desert (BW)
- Highland (H)
- Alpine (H)

0 50 100 Kilometers

0 50 100 Miles

Map 4 California climate classification.
(Source: Lin Wu)

that consume the least energy on a per person basis. As the population continues to increase in the region, the challenges of living in the Mediterranean climate region also intensify. One problem is the continuous urban sprawl into natural vegetation areas under the Mediterranean climate: the chaparral, a type of vegetation that is full of fuel for fire. Fire may be a healthy process in the natural environment under the Mediterranean climate, but it is hazardous for people living in the area. Another problem is the lack of water supply for the region. As a result of the dry summers, annual precipitation of the Mediterranean climate is limited. With an increasing population, the natural water supply in the region is not enough to support the large population and so must be brought in from other regions.

Desert Climate

Desert climate is distributed over the eastern and southeastern part of the state. It is a dry condition characterized by the amount of potential evapotranspiration (primarily determined by the temperature) exceeding the amount of precipitation. It is typical that the precipitation of the desert area averages less than 6 centimeters (2 inches) a year; however, this average is usually highly variable. Figure 4.18 is a scene in the Anza Borrego Desert State Park. It receives an average annual precipitation over 16 centimeters (6 inches). With an average annual temperature over 22°C (72°F), its potential evapotranspiration exceeds its precipitation; making it a typical dry desert climate.

Temperatures in desert areas vary, depending on elevation and location; most of the areas have significant daily and seasonal variation because of their distance from the ocean. Temperatures in desert valleys can be extremely hot in the summer and cold in the winter (Figure 4.2). Subclimate types in the desert region are defined by the degree of dryness and the temperature. Areas in the transitional zones between the humid Mediterranean climate and the dry desert climate are classified as the semiarid climate region, with an annual precipitation slightly higher than in the desert areas.

Figure 4.18 A scene of desert climate in the Anza Borrego Desert State Park.
(Photo: Lin Wu)

The dryness of the desert climate in California is caused by the subtropical high pressure, rain shadow, and long distance from the ocean. Deserts in the southern part of the state are under the influence of subtropical high pressure year-round. The descending air in the core area of the subtropical high pressure creates constant dry conditions and gives little room for convective precipitation to occur or to allow storms to move in. Deserts in the eastern side of the mountain are created by the rain shadow effect, discussed earlier. In addition, these regions are all relatively far from the moisture sources of the ocean.

Highland Climate

Highland climate is located over the Sierra Nevada, the Klamath Mountains, and the Southern Cascade region, with a few small areas in the Southern California mountains. It is a humid climate with a relatively cold temperature year round. Winter can be very cold. Most of the region's precipitation falls in the winter in the form of snow. The amount and seasonal precipitation distribution vary significantly in highland climate because of differing latitude, elevation, and other locational factors. The low temperature of the highland climate is caused by the environment lapse rate: temperature decreases as elevation increases. One subclimate type within the highland climate is alpine climate, which is defined by average monthly temperatures below 10°C (50°F) in every month of the year.

Climate Change

As we learned by now that climate is the average weather conditions at a given location over a period of time. The average weather conditions are determined by temperature, precipitation, humidity, wind, etc. At a given location, these measurements change from day to night and from season to season. The daily and seasonal changes are part of the characteristics of climates. *Climate change* refers to changes in these measurements over millions of years, thousands of years, hundreds of years, or tens of years. It is important to examine climate change in these different time scales. Long-term climate change, changes over

millions or thousands of years, could be caused by various natural factors, including changes of the solar condition, changes of the earth's orbit, changes of the tiltation of the earth's axis, massive volcanic eruptions, etc. However, many of the atmospheric scientists believe that human activities, such as the release of carbon dioxide from fossil fuel burning, have caused climate change in alarming rates over the twentieth century and into the twenty-first century. These changes are evident in California as well.

Based on studies released in 2013 by the California Environmental Protection Agency (CEPA), annual average air temperature in California has increased by about 0.8°C (1.5°F) since 1895, with the minimum temperature increasing more than 1°C (2°F) and the maximum temperature increasing 0.5°C (2°F). In most regions of the state, the increase in average temperature has accelerated over the last three decades (1980–2010). The increase in minimum temperature has a great impact on winter snowfall and snow accumulation, which in turn affects the water supply of the state.

It was also observed that extreme heat waves, measured as the intensity, frequency, duration and regional extent of heat patterns, have increased since 1950. Nighttime heat waves have increased all over California, and the Mojave Desert and north coastal areas also observed increased daytime heat waves. Wind chill time, which is critical for fruit growing in the state, has decreased. The increase in temperature has caused the freezing level elevation, the altitude where air temperature drops to the freezing point, to increase. It has been observed that the freezing level elevation at Lake Tahoe has risen 150 meters (500 feet) over twenty years (1990–2010).

According to CEPA, a study of seven glaciers in the Sierra Nevada found that the areal extent of the glaciers in 2004 was only 22–69 percent of the areal extent in 1900. In addition, the average sea level at San Francisco and La Jolla has been observed to have risen 15–20 centimeters (6–8 inches), and the ocean water temperature has increased 1°C (1.8°F).

One of the most evident consequences of climate change in the state is the increase in wildfires both in frequency and in size. Of the top five largest wildfires by size in the state, four of them occurred from 2003–2013. One of the largest fires is the Rim Fire near Yosemite National Park in 2013. The fire started in mid-August and did not reach full containment until the end of the October. The fire burned more than 250,000 acres and caused serious damage to the ecological system in the area.

These are not the only examples of climate change observed; some of the other evidence associated with climate change is discussed in other chapters, such as Chapter 11, which also explores possibilities of and practices in reducing the environmental impact.

Summary

The diverse landforms in California, its ten degree latitudinal extension, and its coastal location create three distinctive climate regions in the state: the Mediterranean climate, one of the most comfortable climates in the world; the desert climate, one of the driest climates in the world; and the highland climate, one of the coldest climates in the world. This chapter explores the different characteristics of these climates starting from the basic climate variables—temperature, precipitation, moisture, and wind—and the processes behind them, such as solar and terrestrial radiation, seasonal shifting of wind and pressure system, orographic process, etc. Also discussed in this chapter are the concept and the evidence of climate change. The topics explored in this chapter recur throughout the book in different contexts because climate is such an important element of geography that its influence goes way beyond natural process.

Selected Bibliography

Blier, W. (1998). The sundowner winds of Santa Barbara, California. *Weather and Forecasting* 13:702–716.

Bowden, L. W., Huning, J. R., Hutchinson, C. F., and Johnson, C. W. (1974). Satellite photograph presents first comprehensive view of local wind: The Santa Ana. *Science* 184(4141):1077–1078.

California Climate Data Archive. *CalClim.* Retrieved August 15, 2013, from http://www.calclim.dri.edu/ccda/about.html.

California Environmental Protection Agency. (2013). *Indicators of climate change in California.* Retrieved August 15, 2013, from http://oehha.ca.gov/multimedia/epic/pdf/ClimateChangeIndicatorsReport2013.pdf

Hess, D., and Tasa, D. (2011). *McKnight's Physical Geography: A Landscape Appreciation.* Upper Saddle River, NJ: Prentice Hall.

Jet Propulsion Laboratory (JPL). (2001, June 11). When gloom blooms in June, is Catalina Eddy the reason for the season? Retrieved August 15, 2013 from http://www.jpl.nasa.gov/releases/2001/catalina_eddy.html.

Lantis, D. W., Steiner, R., and Karinen, A. E. (1989). *California, the Pacific Connection.* Chico, CA: Creekside Press.

McKnight, T. (2003). *Regional Geography of the United States and Canada.* Upper Saddle River, NJ: Prentice Hall.

Oliver, J. E., and Hidore, J. J. (2009). *Climatology: An Atmospheric Science.* Upper Saddle River, NJ: Prentice Hall.

State of California. *Climate Change Portal.* Retrieved August 15, 2013, from http://www.climatechange.ca.gov/.

University Corporation for Atmospheric Research (UCAR) and National Center for Atmospheric Research (NCAR). *Spark Science Education Learning Zone.* Retrieved August 15, 2013, from http://spark.ucar.edu/resources.

Contemporary Folkways, Cultural Landscapes

"In Los Angeles, all the loose objects in the country were collected as if America had been tilted and everything that wasn't tightly screwed down had slid into Southern California."

–Saul Bellow

"Secretly, I think everyone who makes fun of California really does want to be in California."

–Zooey Deschanel

Introduction

California has acquired a reputation not only for diversity but also for unmitigated strangeness. One of the oft-quoted gibes is that if the whole country were tilted up on end, all the loose nuts would end up in California. This theme, echoed with some consistency throughout the state's history, continues to be a favorite of East Coast writers. But it is a judgment that both exaggerates and oversimplifies the astonishing variety of geographical and cultural personalities found in the Golden State. Within the confines of California can be found much of the best and worst of all 49 other states combined. Large-scale immigration has contributed to the phenomenon, as has the wide diversity of climate and landforms. Thus, it may safely be said that California does have something for everyone.

This diversity of culture takes many forms. Clearly, the various regions of the state differ, one from the other. The redwood country of the north is a different world from the concrete-covered south. The quiet calm of the Mother Lode towns is in distinct contrast to the fast-paced life of the southern beach communities, as San Francisco is to Los Angeles. Diversity is more than regional, however.

The range of entertainment forms in the state is phenomenal. Californians can choose from many forms of theme parks, ersatz historical locations, zoos, wildlife parks, and nightspots. The choice of restaurant styles and formats adds yet another dimension to the leisure-time activities of the state's residents. Apartments for active singles vie with retirement communities for space and attention, while artsy cemeteries offer marriage and burial at the same site.

Ethnic populations vary from locale to locale, district to district, and county to county. In all, the state provides an anthropologist's heaven for the study of varied cultures, peoples, lifestyles, customs, and behaviors.

Cultural Geographic Oddities

Isolation or Uniqueness

What makes California different from other states or other places in the world? Isolation is frequently mentioned by geographers to explain cultural differences and variations. This is certainly true of California. At least as far back as the time of Ordóñez de Montalvo's mythical island, the region has attracted continued references to its relative isolation from the rest of the country (and at times the world).

In 1579, the explorations and subsequent efforts at cartography by the famous English adventurer Sir Francis Drake, as well as those of other European geographers, resulted in the widespread conviction that California was indeed an island—and so it was depicted on the maps of the early period. Not until the explorations of Father Eusebio Francisco Kino over a hundred years later was California recognized to be a part of the North American mainland.

Although the Spaniards discovered that they could reach the area by a land route, this did not substantially affect the concept of California's isolation. Removed from the settled areas of Mexico, the area was difficult to get to by both sea and land owing, respectively, to adverse coastal headwinds and currents and forbidding barren desert wastes. Thus, once settlers arrived in California, they were not much inclined to engage in active travel back and forth to Mexico. This very real *insulation* contributed to their sense of *isolation*.

The Mexican period saw little change in the patterns of settlement in the area first established by the Spaniards. Not until the discovery of gold did outsiders make a concerted effort to reach the region. Yet, even with the lure of gold to draw them on, travelers from the States had to contend with the arid expanse of the Great Basin between the Rocky Mountains and the Sierra Nevada, the imposing barrier of the Sierra itself, the barren desert wastes of the Mojave, and the semiarid regions of the Central Valley. The alternative was a long, costly, and time-consuming ocean voyage around Cape Horn or an ocean-land-ocean route ultimately ending in San Francisco. In either case, the trip to California constituted a significant separation in terms of time and distance. Without the impetus of gold, the journey took on even greater overtones of sacrifice. Thus, until the completion of the transcontinental railroad, the isolation of the California settlers was very real and very obvious.

The initiation of overland mail, the development of the telegraph and the railroad and eventually the invention of radio, television, airplanes, and autos did reduce the physical isolation of the West Coast, but often not the psychological or perceptual isolation. The fact remains that California still is somewhat distant from the population centers of the East and Midwest. Even today, time and distance may be factors of importance to many visitors to the state. Likewise, most Californians must give serious thought to a vacation back East, especially when everything they could desire, from beach to mountain to desert to river recreation to resorts to amusement parks, is readily available within the state itself. As air travel becomes more accessible to all socioeconomic levels, the time and distance factor may subside. However, as economic factors push the cost of air travel upward, fewer travelers are able to opt for this rapid connection, and surface journeys eastward will continue to involve extended commitments of time, energy, and money.

Thus, as some authorities have noted, California in many ways continues to be "an island upon the land," maintaining its uniqueness and character as a distinctive region within the country. What may be said about the uniqueness of California as a whole from the rest of the country may also be said about its individual regions.

The Perceptual Regions of California

Although the concept of regional differences has long fascinated scholars in many fields, geographers in particular have traditionally been committed to the idea that certain cultural regions or cultural landscapes can be identified and recognized as distinctive. The unique interplay between the land and the people who inhabit it was what French geographer Paul Vidal de la Blache described as the "spirit of place"—a regional "personality" that helps us recognize (without written clues) that we are not in Los Angeles when we drive down the streets of Bakersfield.

As might be expected, many different schemes have been developed to explain the concept of "region." In one very simplified and straightforward approach, we can arrange California's regions into two categories: (1) nodal and (2) uniform. A *nodal* region takes its character primarily from some central place or "node." The entire region thus becomes identified with the interactions that take place within the region *with* the focal point or node. Trade or commercial activities, media, employment focus, and similar phenomena tend to tie all parts of the region to the node. Thus, when someone asks you where you are from and you answer, "the Bay Area" or "LA," you are implicitly recognizing the appropriate nodal region (even if you actually live in Concord or Glendora). By contrast, a *uniform* region has no focal point, node, or central place that gives the region its character. Rather, the regional identity is taken from a widely shared set of values, ideals, lifestyles, and social and physical characteristics common to the entire area. Thus, for example, wine country may not have a nodal point to define it, but the commonality of style and interests throughout the area certainly helps to provide a "spirit of place" that is quickly recognizable.

Some observers are bothered by the lack of precision in delimiting the boundaries of these regions. However, as prominent American geographer Wilbur Zelinsky has noted, such regions are often perceptual in nature. People living within (and outside) the regions create a self-perception of their "sense of place" and act accordingly. That the boundaries may not be precise does not diminish the importance of such regional awareness for demographers, marketers, social planners, and others. Certainly, the long-standing debate over the possible division of the state reflects an awareness that regional differences do exist. Indeed, in 1993, voters in 27 central and northern counties approved the concept of dividing the state in two, and the California Assembly Rules Committee approved a statewide advisory vote for 1994 to

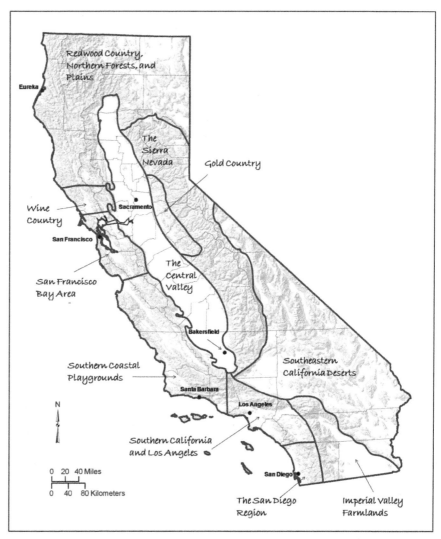

Figure 5.1 Perceptual (cultural) regions of California (Wu and Hyslop).

ascertain the attitude of the people toward dividing the state into Northern California, Central California, and Southern California. Although that effort was not successful, this was not the first time such an effort took place, nor is it likely to be the last.

Putting aside the political, economic, and social implications of dividing the state, regional differences clearly do exist in the minds of most Californians; the fascinating challenge is to determine what complex interplay of geography and culture creates these different cultural landscapes. The correlation between geography and culture cannot be ignored or dismissed. Although geography by itself does not fix the cultural personality of a region or people, its impact on lifestyles, economic patterns, and social characteristics is nonetheless readily apparent. California provides an interesting study of varying cultural identities that can be traced in significant part to the shifts of geographic, climatic, and demographic conditions from one part of the state to another. Although the generalizations that follow necessarily pass over individual divergences, it may still be agreed that some common identity does exist. This collective personality or culture of a region does provide some fascinating insights into the motivations of people and the ways they interact with the environment in which they live.

Logging Paul Bunyan Style: Redwood Country, Northern Forests, and Plains

One of the first associations that comes to mind in connection with Northern California is giant redwoods. Certainly, the redwood forests of the north are dramatic, unique, and picturesque. The role of lumber in the state, however, does extend beyond that one imposing species. California's timber production once ranked among the highest in the country, with fir, pine, and redwood constituting the major forest regions of the state. However, in recent years, timber harvesting has drastically fallen, in part due to cheaper competition from Canada and Oregon, and in part due to a changed cultural attitude toward cutting down forests—especially

specialty forests such as redwood. Compared with most of the rest of California, the northwestern portion of the state is a green and wooded world of its own (Figure 5.2). Increasingly, the value of this forested region is found in its environmental and tourist appeal.

Another rural element of the north is the presence of livestock, with ranching playing a key role in the lives of many citizens. The open ranges of the northeast, in particular, foster a way of life that is close to the land and reliant on its bounty. Clearly, the prime element and predominant theme in the northern portion of the state is nature.

Given the closeness of the people to nature and the land, it is not surprising that controversy has arisen over the issue of proper use of the land and resources. One of the bitterest controversies has centered on the expansion of national and state parks. Environmentalists argue for the need to preserve and protect the priceless forestland, while logging and local economic interests advocate for jobs, economic growth, and the protection of historic livelihoods.

There is merit to the argument of each side. Few visitors to the extensive federal or state-owned parks, forests, campgrounds, or recreation areas would question the value, beauty, or utility of these areas. Yet with lumber-related employment on the decline in regions such as those around Eureka, and with the demand for forest products constantly rising, the timber industry's position can at least be understood. An uneasy balance will, undoubtedly, continue

Figure 5.2 The famous Redwood forest in northwest California.

(Image © WellyWelly, 2009. Used under license from Shutterstock, Inc.)

to exist for some time. Ultimately, the people of the north will be the critical decision makers, with their opinions and behavior determining the balance among competing uses.

Because this region has a higher percentage of native-born residents, a strong sense of local identity prevails. The population base is relatively low, and rugged individualism is still a cherished ideal. With an economy based primarily upon agriculture, tourism, recreation, dairy farming, ranching, lumber, and commercial fishing, the spirit of the people embraces nature and reflects individual responses to it. There are few cities of any size, Redding being the largest north of Sacramento. This demographic factor has contributed to the conservative, self-reliant attitude of the people, which, in turn, has occasionally taken the form of opposition to and hostility toward development in many parts of the north. There has been genuine reluctance to see expansion or improvement of the state highway system, as well as a related desire to protect and promote local control over undeveloped regions.

This is not to say that the north is bereft of any desire to improve, advance, or mature. The presence of Humboldt State University in Arcata has assured the region of a concentrated and respectable effort to promote growth in areas such as forestry, fisheries, and oceanography. Likewise, seashore development has not been totally lacking, as Sea Ranch in the Mendocino area demonstrates. Similarly, Chico, with its state university campus, and Redding, with shopping, entertainment, medical facilities, and other amenities, have both experienced growth. However, they are more the exceptions than the pattern for the region, which continues to be characterized by slow growth and evolution. But the decline in the lumber and fishing industries has forced the north to examine alternatives and consider some changes.

Nevertheless, from the foggy northwestern seacoast to the inland volcanic moonscapes, the predominant spirit is deliberate and calm. If expansion and change are to occur, the people are determined to control such development themselves. Accustomed as they are to a quiet, natural lifestyle, they are unlikely to see

a need for precipitous change in the near future (Figure 5.3).

Argonauts and Ghost Towns: Gold Country

It is amazing that the most populous state in the Union also contains a backcountry of astonishing proportions. Ghost towns and Old West communities abound, carrying the colorful frontier past of the state into the present. For the most part, this region is concentrated in the highland block of California known as the Sierra Nevada, in the communities strung along routes US Route 395 and California State Route 49, and in parts of the Mojave Desert (see Figure 5.1).

Part of the historical romance of California stems from its past role as the home of boisterous gold camps and rugged mining sites. The names of present-day sleepy communities give eloquent testimony to the roistering expansion of the 1850s, and crumbling ruins attest to the transitory nature of these boomtowns. For the historian or romantic, however, the Sierra breezes sigh like ghosts of the past, reminding the visitor of an exciting period now vanished.

The greatest modern-day treasure of the region may in fact be just this ambience. Certainly, the Mother Lode area cannot be characterized as a bustling manufacturing or population center. Its charm lies in the very fact that it maintains a relatively slow pace of life that evokes the past. Ironically, it is this aspect of the region that

Figure 5.3 Scenic Mount Lassen in northeastern California.

(Image © Bruce Grubbs, 2009. Used under license from Shutterstock, Inc.)

has brought about a modest but steady growth in population since the early 1980s, fueled by retirees and by urban professionals seeking escape from the pressures and frustrations of city life. Because many of these new residents are employed by "footloose" industries that do not require any particular urban location, they have the luxury of seeking a place to live that provides them with a sense of country serenity and historical continuity. The subsequent growth in small towns such as Sutter Creek and Nevada City (in the foothills) and Hemet and Lancaster (in the desert) has been called "the growth of Penturbia" by some experts, referring to a fifth cycle of urban migration to smaller towns and communities.

Although there are some small pockets of high-tech industry in the foothill region, the primary income of the area comes from modest agricultural activities: unromantic rock, gravel, and sand mining; some ponderosa pine and white fir lumbering; and the rapidly growing recreation- and tourist-oriented trade. The latter capitalizes on historical sites, winter sports, and wilderness recreation. The nature of this economic activity does not conflict with the yesteryear sense of simplicity and naturalness. This undoubtedly has contributed to the modest boom in foothill towns in recent years, with retirees and urban refugees finding the atmosphere they seek in the area.

The advantages of living in this region include distance from the turmoil of big cities, availability of open space and nature, and the presence of historically fascinating sites all around. The area encompasses forested foothill and barren desert, each with its own unique characteristics and appeal. The sense of peacefulness, however, pervades both foothill ranches and lonely desert and carries over to the little towns that still exist, such as Mariposa, Downieville, Garlock, Bodie, and Darwin. A visitor with a good imagination can populate the lonely stretches of US Route 395 and California State Route 49 with prospectors and cowboys.

The sense of history is also evident in colorful places such as Fiddletown, Angels Camp, Mormon Bar, and Mokelumne Hill. Columbia and Calico (Figure 5.4) have been turned into

Figure 5.4 Calico—a restored mining town.
(Image © Jim Feliciano, 2009. Used under license from Shutterstock, Inc.)

restored western towns for tourist consumption exclusively, and their success is evidence of the popularity of commercialized history.

Many more of the Mother Lode cities now serve as gateways to the popular wilderness areas of Sequoia, Kings, and Yosemite National Parks and other state and local wilderness settings. During the warm season, providing services for hiking, fishing, hunting, and camping constitutes the mainstay of these towns. During the cold season, skiing becomes the prime attraction. Tourists and temporary visitors move on, but residents continue to enjoy their peaceful environment and colorful past. Architectural treasures of the past, magnificent wilderness, open landscapes, and low population density are attributes that hold the loyalty of the people. They take pride in their frontier heritage, and few would trade their lifestyle for one that required them to move to Los Angeles or San Francisco.

A State without Wine Is Like a Day without Sunshine: Wine Country

Winemaking has become a sophisticated, respected, and palatable business in California. From its early and simple beginnings with vines brought by Franciscan padres, the California grape has matured into a well-traveled and widely consumed product. In fact, grape growing and wine making have become multimillion-dollar industries.

The term "wine country" popularly refers to the Napa–Sonoma area north of San Francisco.

This is a somewhat misleading designation because respectable vineyards and wineries exist in many other locations throughout the state. In fact, the Central Valley and the southern counties now produce most of the state's wine. The Napa–Sonoma region, however, is still "wine country" to most Californians, and the region takes great pride in this identity even though it now produces only about 20 percent of California's total wine output.

The California wine industry is a remarkable success story. Currently holding a share of the total American wine market of about 60 percent and enjoying an extensive foreign export market, the business is solid and growing. Directly and indirectly, viticulture in the state owes much to the contributions of European winemakers. The immigration of experienced vintners brought first of all a new expertise to the state. During the 1850s, the German vineyards at Anaheim and the Italian Swiss Colony at Asti began operations. Contributions by Charles Krug, Etienne Thee, Charles Lefranc, and Agoston Haraszthy took the infant industry further toward fulfillment and demonstrated the significant impact of immigrants. The hard work of these new Californians resulted in expanded vineyards throughout the state, especially in the Napa–Sonoma region.

Indirectly, Europe helped by providing vine cuttings of quality and endurance. The infusion of this established stock enabled California wines to grow in popularity and quality. Continued expansion of the industry was assured, and new wineries sprang up in such diverse locations as Fresno, Madera, Modesto, Cucamonga, Rutherford, and Saratoga. Although Prohibition (1919–1933) caused major financial damage to the California wine industry, the industry returned stronger than ever with the repeal of Prohibition.

Traditionally, wine country reflected mature respectability. Nowhere was this more evident than in the Napa–Sonoma region. In architecture, cultivated hillsides, and temperament, the area exuded a flavor of the Old World (Figure 5.5). Long-time residents prefer a region of cultivated serenity and traditional values. The

Figure 5.5 Napa Valley vineyard at Christian Brothers winery.
(Image © Andy Z., 2009. Used under license from Shutterstock, Inc.)

plethora of family vineyards adds a superficial sense of a stable social order to the area. The feel of the traditional culture is found in the orderly sense by which the long rows of cultivated vines march across much of the available land.

The battle lines increasingly are being drawn between those who advocate growth and those who wish to maintain the Old World ambiance. Sonoma County has seen the greatest development pressures, with Bay Area commuters swelling the population base; wine-related tourist attractions expanding the demands on the infrastructure; and high-tech, white-collar, and service industries increasing their presence in the region. In response, controlled-growth advocates have fought a prolonged and bitter battle to roll back this powerful tide. Indeed, the city of Petaluma acquired some fame in the late 1970s with a test case challenging its anti-growth ordinance (which was ultimately upheld by the courts). Although population growth and pressures for change have not been as intense in Napa, the tourist trade has created major problems for many residents, who resent what they see as exploitive commercialization of viticulture.

What is clear is that the romanticized, genteel, slow-paced wine country has been inexorably mutated. The once tranquil, scenic nature of the area has been significantly altered by the presence of visitors touring the many tasting rooms. Although the courtesy and charm characteristic of wine country is evident in the care with which the old-family wineries treat their guests, the underlying uncertainty over the future character of the region remains. The irony is that much of its supposed historical foundations have already eroded away. For example, if there is an old aristocracy in California, one is tempted to look for it here; however, many of these rustic, homey-looking wineries are now owned by international corporations.

Recently, as noted previously, the controlled nature of the agricultural landscape of the Napa–Sonoma area has been copied somewhat in efforts to prevent uncontrolled residential sprawl. Concern over spread of the Bay Area population has led to zoning efforts to retain the traditional mood, lifestyle, and flavor of the region. With its history of controlled environment, the region is likely to fight hard to retain its romantic and tranquil existence, or at least its facade.

Sophistication: The San Francisco Bay Area

San Francisco has acquired a rare and enviable reputation as an exotic, breathtaking, world-class city. To a certain degree, the entire Bay Area shares this romantic image; it is an environment of ethnic mixes and cultural plenty, refurbished Victorians and cluster housing, self-conscious snobbery and experimental thinking, and scenic landscapes and dramatic skylines (Figure 5.6). It is at once historical and modern, hurried and calm, artsy and staid. It is unique, its self-image one of quintessential sophistication.

Figure 5.6 Golden Gate Bridge.
(Image © Andy Z., 2009. Used under license from Shutterstock, Inc.)

The San Francisco Bay Area is, of course, more than the simplified realm promoted by songwriters or chambers of commerce. Here is a greater metropolitan region that encompasses all or part of nine counties in an interconnected web of bedroom communities, residential tracts, shopping centers, diversified industrial–retail developments, and highways and railways. Although the Oakland–East Bay area and the San Jose–South Bay area form overlapping centers of residential and commercial activity, the main hub of this urban web is San Francisco, which provides the overall identification for the entire region.

The diversity of the Bay Area is phenomenal. It can be seen in the variety of architectural styles ranging from Asian influences to refurbished Victorian splendor to leapfrogging tracts to palatial estates to tidy, middle-class, ranch-style homes to the increasingly trendy high-density urban village developments. It can be seen in the obvious extremes of economic affluence, from wealthy Hillsborough to troubled central Oakland, with every shade in between.

The ethnic makeup of the Bay Area is likewise notable. San Francisco's Chinatown is one of the most celebrated in the country. Oakland and Richmond contain high concentrations of African Americans. Inner San Francisco has enclaves of eastern European groups, Filipinos, African Americans, Japanese and other Asians, and Mexicans and other Spanish-speaking populations.

The climate reflects the theme of diversity in an astonishing fashion. Owing to its watery surroundings, San Francisco itself enjoys climatic conditions entirely different from those experienced by most of the rest of the region. Meanwhile, across the Bay, temperatures and conditions represent a dramatic contrast—sun versus fog, heat versus cool, dry versus damp. Landscapes, too, vary immensely within a very limited geographic region. Urban settings alternate with agriculture; flat marshlands contrast with precipitous hillsides.

San Francisco itself has undergone some significant changes in recent years. Once a major center of commerce, trade, and finance, the city has gradually seen much of this activity shift to other Bay Area cities or to Los Angeles. Although it remains a highly popular tourist destination, the balance of ethnic, economic, and social forces has been replaced by a bipolar structure. San Francisco is now occupied primarily by wealthy, professional property owners (mostly white and Asian), with members of the middle and lower classes forced to commute in and out of the city to service this wealthy population. The effects of this demographic shift are significant, as the dynamics and diversity of new growth now stimulate other locales. This is not to say that San Francisco will not retain importance in the region. Its widespread reputation for charm and sophistication remains well deserved. Its international character endures in the culinary, architectural, and cultural choices available. Its political and social tolerance still fosters the open acceptance of alternate sexual preferences and lifestyles. And its climate, cable cars, gingerbread houses, theaters, restaurants, and fog continue to draw visitors to the city.

But where gentrification in San Francisco is creating a somewhat homogenized "ghetto" of wealthy white and Asian residents, outlying suburbs are recharging their ethnic and economic identities. At least partly because of this diversity, most experts see the East Bay and the South Bay gaining financial and trade dominance.

A brief tour of the region radiating outward from San Francisco quickly reveals this diversity. To the east, across the Bay and connected by the Bay Bridge, is the Contra Costa–Alameda complex. This populous, urbanized region provides much of the industrial production of the Bay Area. Historically, the Oakland–Richmond complex was a key terminus for rail, ocean, and truck transport. However, Oakland has struggled in recent decades to survive the flight of residents and businesses to "ContraCostapolis" and its more attractive environs. Both economics and race play a role in this shift. Although metropolitan Oakland has a large minority population, there is a less obvious ethnic mix as one moves south and west from the Oakland core. Similarly, where Oakland appears alternately crowded and congested or boarded over and abandoned,

shading out and away from the city are typically suburban cities like Walnut Creek, Hayward, Fremont, Concord, and Livermore. The Contra Costa County communities in particular have experienced growth in technical jobs, office complexes, and corporate expansions, as well as a resulting whiter, more middle-class population. The growth has not been without problems: traffic jams, overpriced housing, unmanaged development, and shortages of local services are among the more obvious challenges.

Moving southward from San Francisco along the peninsula, the entire area stretching from South San Francisco to Palo Alto has been a steadily growing corridor. Neighboring peninsular cities such as Daly City, San Bruno, San Mateo, Redwood City, Menlo Park, and Palo Alto have struggled with the problem of expansion. They have had mixed success in controlling the development of the peninsula, which now reflects a blend of residential, retail, and diversified light industry.

To the extreme south of the peninsula begins the cluttered sprawl of San Jose and the Santa Clara Valley. Here is one of the most typically disarranged of California urban areas. A region of mixed development, decaying inner city, spreading tract housing, and bustling industry, the South Bay region is still experiencing problems coming to terms with itself. The successful industrial capacity of this locale was typified by the major aerospace establishment of Sunnyvale and by the explosive growth of the Silicon Valley computer industry. However, with the winding down of the Cold War, the reduction in military spending, and the effects of a prolonged recession, the fortunes of the region have been somewhat moderated. Still, within the greater Bay Area, the South Bay has enjoyed population growth, urban development, and increases in high-tech industrial and overall economic prosperity. Meanwhile, the supporters of the rural, agricultural lifestyle battle to maintain a hold against the ever-expanding residential tract developments. Moving farther south and east from this point, the Bay Area identity begins to weaken until finally the agricultural Central Valley takes over.

To the north of San Francisco, across the famed Golden Gate Bridge, are the North Bay and the Marin County peninsula. The confused and unattractive urban sprawl is less evident in this area, which is a mix of mostly higher-income residential development and open land. The wealthier socioeconomic structure here has helped to preserve much of the open marshland, recreation areas, and beautiful green terrain. A quieter, more peaceful section of the Bay Area, it gradually merges into the wine country, not far from the waters of the Bay itself.

Today, the greater San Francisco Bay region is undergoing some traumatic transitions. Long known for its commercial potential, the area continues to support rail, truck, and port facilities. The challenge of the growing southern portion of the state, however, has created serious competition for this economic base.

Like the rest of the state, the Bay Area continues to face serious questions about its future. Changes in the business climate, international politics, tax bases, demographic profiles, and economic structure present new realities. Also, events such as the Loma Prieta earthquake demonstrate that planning frequently falls short of necessity. With the growing demands of a burgeoning population, the Bay Area also faces many challenges to the old ways and assumptions. Faced with many of the same problems as the south, the region can no longer look with carelessly veiled contempt on Los Angeles. The growing pains of Southern California are now also a reality in the Bay Area. The challenge will be to resolve these problems without mimicking and magnifying the mistakes of the south.

Nashville West: Bakersfield, the Central Valley, and the Farm Belt

Agriculture is an important economic activity in California: a substantial portion of the state's land is dedicated to food production. In few places is this more evident than in the Central Valley. Constituting most of the heart of the state, the Central Valley stretches for about 470 miles from Bakersfield north to Redding and lies between the Coast Ranges and the western Sierra Nevada. Fertile productivity is the hallmark of the region, and long, flat, open expanses of combed and cultivated land are the common landscape. It is a region populated by farmers, truckers, and field hands and characterized by wide expanses of cultivated fields, diverse crops, and agribusiness operations. It is the source of California's leading position in American agricultural production and income.

This is also a region of far more complexity and variation than imagined by the typical urban dweller of the megalopolis of greater Los Angeles or the *galactic metropolis* of the Bay Area (a galactic metropolis is defined as an influential core city surrounded by self-sufficient outlying communities loosely tied to the core city). The astonishing variety of sociocultural realities present within the Farm Belt in general, and the Central Valley in particular, lend a richness and depth to the face of California.

To begin to appreciate the magnitude of this richness, one might drive the length of California State Route 99 or Interstate 5 between Los Angeles and Sacramento. These routes pass through extensive agricultural land and are unbroken for miles and miles by any significant urban setting. The variety of crops along these routes is great enough to astonish the most sophisticated agriculturalist. A world apart from the populous bustle of the major urban centers, much of the Central Valley reflects yet another side of the state in both ecological and cultural terms. There are a few large cities located in the region. Most of them unmistakably reflect the priorities of the farm, although this is gradually beginning to change.

The city of Bakersfield is a perfect case in point. Driving into Bakersfield, one encounters Montgomery Ward, truck stops, farm implement dealers, gas stations, cafés, and honky-tonk bars. The uncomplicated strains of country-western music float through the air from bars and truck speakers. Values in the songs are simple and direct, mirroring the attitudes of the residents of this region. Dubbed "Nashville West," Bakersfield has earned a reputation as the western home of country music (Figure 5.7).

Historically, the city also has served as a focal point for oil production and related industries, which has added to its rough-hewn aura.

Figure 5.7 Agricultural fields near Bakersfield.
(Image © Jennie Book, 2009. Used under license from Shutterstock, Inc.)

Interestingly, too, Bakersfield represents a cultural core of a substantial Basque population drawn to the region to raise sheep in the open spaces of Kern County. All of these activities provided the down-to-earth reputation prevalent throughout much of Bakersfield's history. This has been a city that was "close to the ground," without pretensions. It existed to serve the agricultural needs of the area with its shipping facilities and stores. Its merchants catered to the farm population, providing services, products, and entertainment for evenings and weekends off the farm. However, Bakersfield has begun to add a patina of urbanization and economic diversification to its cultural landscape. Although its country roots remain strong, the city no longer can be accurately characterized as agricultural only. The impact of suburbanization has become very evident, with some industries from Los Angeles moving here, some newer suburbs expanding, and a growing number of retirees choosing to locate in the region.

What is true of Bakersfield is also true of other Central Valley cities, such as Sacramento, Visalia, Fresno, and Stockton. Sacramento through most of the 1970s, although it was the state capital and contained immense government operations, made no pretense of being a cosmopolitan center. In fact, it was an agriculturally oriented city located in the midst of a rich delta farming area. Sacramento was really a farm town; it just happened to be the state capital.

But this began to change in the late 1970s. In the 1980s and 1990s, the region became one of the nation's fastest-growing metropolitan areas. This rapid urbanization has had both positive and negative consequences. Clearly, the Sacramento area has diversified greatly in terms of economic influences, cultural opportunities, and lifestyles; it is no longer dominated by an agricultural mentality. However, with the explosive growth, the trappings of urbanization have arrived as well, including uncontrolled sprawl development, traffic snarls, street gangs, and a bitterly divided local political structure. Indeed, the nastiest "street fights" seem to be between the advocates of suburban development and those who wish to return to the days of agricultural dominance. Thus, the greater Sacramento area fairly well typifies the problems shared by the rest of the Central Valley cities.

Visalia came into being and continues to prosper in part as a result of its role as a farm service community. Likewise, Fresno, although it is a big city that offers many of the benefits and ills of a large urban area, also exists because it serves the needs of the Central Valley farms that surround it. Stockton is equally a product of agriculture; it came into being due to the gold rush but stays alive due to its role as a major agricultural service center.

These cities owe their existence to the agrarian wealth of the state, and most of their residents remain proud of the role of farming in California and the nation. If their values seem a little more basic and old-fashioned than those of the megalopolis, and if their music is the music of the country, and if their lifestyle is more sedate, it is because that is what they cherish.

An interesting irony is that both old and new residents of the Central Valley tend to embrace the concepts associated with farm or country living. However, the rapid spread of suburbia into the Central Valley presents a challenge to the historically agricultural character of the region. Much of the development and growth is fueled by commuter residents who are willing to spend 4–6 hours a day traveling to and from work. The consequence of this phenomenon is that by the 1990s, Central Valley farmland was being converted to suburban uses at a rate of approximately 20,000 acres per year. Furthermore, cities like Fresno, Visalia, and Stockton have experienced tremendous nonagricultural growth in recent decades.

Although not physically part of the Central Valley, the Imperial Valley in the far south shares the same values, attitudes, and priorities. Like the Central Valley, the Imperial Valley is a highly productive agricultural region because of the long growing season, diverse soil conditions, extensive irrigation, and advanced technological applications. Many of the same problems face the area, including urban encroachment and agribusiness consolidation. Economically, socially, and culturally, it is a continuation of the Farm Belt.

Collectively, the farming portion of the state remains quite significant. It is responsible for mass production of cattle and dairy products, poultry, grapes, vegetables, hay, nuts, cotton, and fruit. The hard work and efficiency of the California farmer in producing these and other crops has made the state a world leader. Certain problems do exist, however.

Even with the increased demand for food, urban residential land uses are cutting into productive agricultural land. The trend toward consolidation and incorporation in agribusiness has brought new complexity to the lives of the farm population. Unionization of farm workers and advancing agricultural technology are bringing about a shift in the number of persons attracted to this way of life. Furthermore, serious competition from abroad has cut into the profit margin of California agriculture. Add to this an increasing awareness of environmental issues pertaining to chemical fertilizers and insecticides, and the changing government policies relating to water pricing and the future of agricultural California is challenging indeed. The California farming community, conservative and fundamental in culture, may have to reexamine some

of its goals. But even if some changes do occur, the Farm Belt will continue to exert significant influence on the economy, environment, and culture of the state.

Neon Glitter: Southern California and Los Angeles

On occasion, Southern California has had a rather dubious public image. Characterized as the last stronghold for assorted faddists, nuts, kooks, and misfits, it cannot be accused of being dull! Congestion, freeways, smog, cars, trucks, recreational vehicles, and rows of identical tract houses typify much of the south. The urban tide of Los Angeles has spread in an almost continuous flow down through Orange County, up through part of Ventura County, and across through large portions of Riverside and San Bernardino counties.

The greater Los Angeles metropolitan area takes in a huge portion of the land of Southern California. The city alone covers more than 450 square miles, but the Los Angeles influence extends far beyond even these extensive bounds. In terms of its sphere of influence, the Los Angeles metropolitan region extends to the ocean in the west, to the San Fernando Valley and Ventura County in the north, to Orange County in the south, and to the "Inland Empire" of Pomona, Riverside, and San Bernardino in the east. The problems, concerns, and orientations of these areas are a part of the greater entity of Southern California. Los Angeles itself provides the unifying link and shared bond.

Over half of the state's total population is found in Southern California, with the largest percent in Los Angeles County alone. Although there has been migration outward from Los Angeles proper into the newer suburbs (once called "white flight," which is no longer an accurate title), there is also continued growth within the city and county of Los Angeles (mostly new immigrants). In each location, however, the LA sphere of cultural influence is dominant.

As a culture of diversity, Los Angeles and Southern California represent an astounding range of urban–suburban environments. Within one metropolitan area can be found miles of beautiful beaches and blocks of decaying urban jungles. Cool mountain resorts look out over bleak desert landscapes. The plastic glitter of Hollywood contrasts sharply with the air-conditioned luxury of corporate Century City skyscrapers (Figure 5.8).

The mix of racial and ethnic groups is equally diverse. The south now has large and significant populations of African Americans, Hispanic Americans, Asian Americans, and others. Most of these people are tossed together in an interesting cultural salad bowl called the Los Angeles Basin. Indeed, surveys by the Los Angeles Unified School District in the early 1990s revealed students speaking over 80 different languages. Ten years earlier, the *Los Angeles Times* had reported that 1980 marked the point of ethnic shift in Los Angeles, wherein the Anglo population now numbered less than 50 percent. By 2013, the Anglo population in LA County had adjusted down to slightly over 27 percent.

The southern portion of the state is clearly a land of plenty. There is an abundance of people, money, stores, freeways, motor vehicles, billboards, noise, shopping centers, gas stations, parking lots, restaurants, lights, schools, fast-food chains, theaters, houses, condominiums, apartment buildings, ghettos, barrios, slums, street gangs, deteriorated buildings, office vacancies, paranoia, pollution, random violence, lawyers, and realtors.

The nature of this abundance can be traced to many of the factors that originally drew the population to the region. The defense and

aerospace industries offered jobs; the movie industry promoted glamour; the mediterranean climate promised health. The romance of the state was touted by the tourist and service industries, which relied on masses of people for their success. And once people arrived, they stayed.

In the 1980s and 1990s, many of these "older" immigrants decided to move out from the core—some to the surrounding suburbs, some to other states. However, to massive numbers of both legal and illegal immigrants from Latin America and Asia, the relative wealth of and opportunities in Los Angeles continued to act as a powerful magnet. This was also true of foreign businesses. In recent years, foreign ownership of major downtown LA properties has contributed to the international character of the area. Such prominent landmarks as the Bonaventure Hotel, Crocker Plaza, The Park, Pacific Financial Center, The Biltmore, One Wilshire Boulevard, California First Bank and 800 Wilshire are now or have been owned by foreign corporations.

The Los Angeles area in particular and the Southern California region in general do have problems. People's desire to pursue unique notions and lifestyles has not fostered a commitment to planned or controlled development, and monotonous and drab suburbs attest to this fact. An exaggerated sense of individualism among area inhabitants has encouraged a political environment in which efforts to build mass-transit systems and establish clean-air standards for industry can be thwarted. Thus, air quality has become a crucial issue at the same time that the population clings to its imagined independence by commuting one-to-a-car. The inflated sense of self has made local government decision making a morass of confused, contrary, and involved arguments. The interaction of ethnic and racial groups is uncomfortable at best. The explosion of violence following the initial verdict in the Rodney King beating trial in April 1992 demonstrated the tenuous nature of race relations in Southern California, which have not necessarily improved in the intervening years.

All the clichés concerning the cultural wasteland of Southern California bear careful examination, however. In spite of the crush

Figure 5.8 Century City—an upscale commercial center.
(Image © egd, 2009. Used under license from Shutterstock, Inc.)

and confusion of a booming population, Los Angeles has achieved enviable progress as well. The abundance of quality private and public universities and colleges has made Southern California a recognized leader in the field of education. Here can be found some of the world's most impressive programs of collegiate extension courses. With a multitude of locations, subjects, and time slots, these programs have made higher education truly a public pastime. Southern Californians can study virtually any subject, take educational tours to most parts of the world, and participate in the learning experience for as long as they wish. Also, with the impetus of military and government contracts, the Los Angeles metropolitan area has become a center for scientific research and development, as evidenced by the Rand Corporation and the Jet Propulsion Laboratory in Pasadena.

Unfortunately, the end of the Cold War and the recession of the 1980s and 1990s caused a major reassessment of priorities within both corporate and educational establishments. Downsizing and streamlining became the operational guidelines, and, as a result of tightened tax bases and reduced income, economic growth slowed. This is not to say that Southern California is moribund—quite the contrary, in fact. However, the region has been forced to evolve, with low-tech manufacturing and service-based industries becoming more evident.

There has been a serious commitment to the arts, with the Music Center, County Art Museum, and various dramatic companies that are the envy of much of the rest of the nation. Contrary to popular opinion, the Los Angeles area offers a wide variety of cultural activities. Numerous private museums supplement the public offerings, with the Norton Simon and the J. Paul Getty museums being only two obvious examples. On almost any sunny weekend, one can find outdoor art shows featuring the work of thousands of painters, sculptors, and other craftspeople and artists. Likewise, community support for the arts ranges from local theater groups to shows sponsored by community businesses to local library presentations. A slightly different form of art can be seen in the profuse displays of architectural experimentation found in both private and commercial edifices, from shopping centers to hideaway retreats.

Los Angeles has come to rival New York City as the news capital of the country, both in television and print media. In 1980 *Newsweek* identified the *Los Angeles Times* as one of the three best metropolitan newspapers in the country. Economic and readership changes in the print news industry, however, have brought new challenges—and new ownership—to the *Times*. In the electronic media, Los Angeles has become a top assignment.

The contributions of ethnic groups to the culture of the region are significant. Like San Francisco, the Los Angeles metropolitan area has an extensive mix of ethnic groups and, thus, highly diverse lifestyles. Although the various ethnic cultures are mixed well into many areas of the Los Angeles Basin, there are some notable concentrations that have become largely synonymous with the various ethnic populations.

Traditionally, Watts was one of the sections clearly identified with the African American population. Receiving nationwide attention in the late 1960s for the riots that occurred there, Watts has subsequently attempted to achieve a more cohesive sense of African American identity, self-direction, and pride. Certainly, the annual Watts Festival and other events have focused attention on the contributions of African Americans to the culture of the state and the nation. And although the 1992 Rodney King riots flared strongly in Watts, community leaders exerted a concerted effort to try to prevent further damage to the residents and the reputation of the area. Although African Americans still constitute around a third of the Watts population, Hispanics now account for twice that number in the area. Like many other groups before them, the African American population has dispersed to various other communities in the Los Angeles region.

East Los Angeles is dominated by the growing and important Mexican American community. The influence of this ethnic group is, of course, evident throughout California and the Southwest. East Los Angeles, however, has long

Figure 5.9 Hsi lai Buddhist Temple in Hacienda Heights, reportedly the largest such temple in North America.
(Image © Zach Frank, 2009. Used under license from Shutterstock, Inc.)

Figure 5.10 Skyline of Los Angeles, a major metropolitan center of the United States.
(Image © Konstantin Sutyagin, 2009. Used under license from Shutterstock, Inc.)

been identified as a Spanish-speaking *barrio,* or community, and the area has provided a cultural haven for this population. The growth of the Latino population in the area has accelerated with the influx of undocumented aliens from other Latin American countries who are drawn to this neighborhood by the familiarity of a Hispanic-based culture.

At the other end of the city is West Los Angeles, an area that reflects a strong Jewish influence. Delicatessens and synagogues are more evident here, and the names on stores and buildings reinforce this cultural identity.

Many other sections of the greater Los Angeles region reflect diverse ethnic–cultural influences. A rapidly growing Southeast Asian population has swelled the size of the city's Chinatown and Little Tokyo and has also established itself in outlying regions, such as Orange County and the San Gabriel Valley. Indeed, these refugees were drawn in large numbers to cities such as Westminster, Garden Grove, and Santa Ana. By the 1990s, more Cambodians and Vietnamese had migrated there than to any other urban region in the United States and had established clearly identifiable ethnic enclaves for themselves (Figure 5.9).

Finally, Native Americans, immigrants from South Pacific islands, relocated English and Canadian citizens, and many others help provide the dynamic cultural diversity to be found in Los Angeles.

In all, Southern California has largely outgrown its provincial image and has moved to the forefront of the nation as a true center of culture, politics, economics, ethnic awareness, and power—a true cosmopolitan environment (Figure 5.10).

Beachboys, Boating, and the Body Beautiful: Southern Coastal Playgrounds

Perhaps the most common image of California, as seen in posters, movies, and other media, is the beach-and-seashore scene of the southern coastal playgrounds. A common scene is that of surfers (or lifeguards) dramatically swooping toward shore on foaming waves while bronzed and beautiful bodies lounge in the sand. The glamour of sunny California is also portrayed in pictures of trim sailboats lolling off the coast or bobbing lazily in tidy marinas.

This imagery is not created purely for export consumption; living within this particular culture has its effects on the residents as well. Frequently, the stereotype creates a sense of obligation to fulfill the expectations associated with the image. In spite of warnings about exposure to the sun's ultraviolet rays, what Southern Californian would not feel embarrassed to be seen pale and flabby in shorts or bathing suit? In fact, the region exudes a self-conscious pride in staying beautiful, bronzed, healthy, and trim. It is an attitude that permeates the culture and does not stop with youth, as casual observation at Venice Beach will quickly reveal.

The Southern Californian beach scene is a fascinating subculture with great outward display and little depth. There is a peculiar arrogance that encourages conspicuous displays of wealth (and skin) and that rewards a tan acquired by wasting countless hours lying in the sun and by supporting tanning clinics that preserve the deep-baked tan. It is a wealthy, smug, sensual, gaudy, indulgent, and social subculture.

The area included in the southern coastal playground begins around Santa Barbara and extends southward along the coast through Los Angeles and Orange counties all the way to the edge of San Diego County at Camp Pendleton. This long stretch of coast includes a remarkable variety of beach and coastal headlands, as well as an equally broad range of affluence and styles. Santa Barbara is a wealthy, exclusive, and self-contained community that retains a flavor of old Spain in its "Queen of the Missions" and its Moorish courthouse. Tidiness and order are the style here, and expensive ranchos and urban mansions are typical. The frantic pace of the big city is missing in both Santa Barbara itself and the surrounding countryside. The quaintness increases as one moves inland to the lush, coastal valley farmlands. The Scandinavian-flavored tourist town of Solvang (Figure 5.11) is just one additional interesting landmark of the Santa Barbara region, where stability, homely virtues, and good living are evident.

Other selected communities along the southern coastal shoreline share the exclusive, wealthy security of Santa Barbara. Corona del Mar, Lido Isle, Palos Verdes, and Emerald Bay take quiet pride in their expensive lifestyles. These are not transient, brash beach communities, but rather strongholds of the affluent.

In contrast to these orderly, quiet, wealthy beach towns are the popular beach areas where the fast life prevails. Malibu, Newport, Santa Monica, Venice, and Marina del Rey share a preoccupation with youth, parties, the body beautiful, and novelty. Sports cars, surfboards, sailboats, and sexual tension are evident in all quarters; transience and frantic motion are the norm.

Figure 5.11 Solvang in the Santa Ynez Valley is a quaint Scandinavian-themed village.
(Image © Mariusz S. Jurgielewicz, 2009. Used under license from Shutterstock, Inc.)

Between these two extremes are various other styles of oceanfront atmosphere. Surfing, pier fishing, and oil wells share space with middle-class neighborhoods in Huntington Beach. Tidepools, tourists, art, and scuba diving typify such areas as Laguna Beach and Catalina. All along the coast, high-rise condominiums and apartment complexes compete for space with exclusive restaurants and growing financial centers. The older beach areas of Seal Beach, Manhattan Beach, Dana Point, and San Clemente are feeling the impact of new money on old lifestyles, and shabby dwellings are being replaced by expensive residential and business edifices (Figure 5.12).

The continued growth and increased affluence have inevitably changed the face of the southern coastal area. The frantic pace at which residents party and dance, shop and dine, skate and boat, and try to make a buck has become the norm. Only the persistent pier fishermen and the rugged bluffs of Palos Verdes attest to the peaceful past.

Resort Mecca and Commercial Hub: The San Diego Region

Vacation, recreation, and retirement are the crucial draws of modern San Diego. They have made the city one of California's leading metro areas while at the same time giving it a unique flavor. Building upon its climate, natural variety, and geographic location, San Diego made of them an art, a lifestyle, and a justification for existence. With one of the best ports on the coast, San Diego has a strong maritime orientation, with the extensive naval facilities exerting a strong social, economic, and political influence. Likewise, the port facilities continue to support a significant shipping industry and a commercial fishing fleet. The extensive naval and private marine activities, in turn, have helped foster related light manufacturing industries, as well as a respectable aerospace program.

Typically, this type of economic base is associated with a conservative outlook in the general population—and San Diego is no exception. A traditional no-nonsense attitude is evident in most of the cultural character of the area. Politically, the region has long been known as a conservative stronghold in the state. Economically, the laissez-faire doctrine is still popular. Socially, the values of small-town America are highly cherished.

On the other hand, San Diego has a lighter side to its collective personality—a single-minded dedication to play! The same factors that encourage and support a maritime and naval presence have also created a strong recreational spirit. The seaside environment has nurtured numerous ocean-oriented pastimes and entertainments. Mission Bay has been developed into a posh and scenic locale. Here, Sea World and its tourist shows draw visitors by the thousands. In addition, the extensive marina, hotel, and restaurant facilities have made this area a major recreation center. Coronado Island caters to those who prefer Victorian-style buildings and fancy dining and lodging. Deep-sea sport fishing is a recognized tourist attraction. Recreational boating, water skiing, surfing, swimming, and skin diving round out the pursuits available in the area.

Location and climate are also part of the playground requisites of the San Diego region. Proximity to the Mexican border makes bullfights and foreign trade available to tourists. Mild weather enhances the appeal of scenic Del Mar racetrack, various golf courses, numerous state parks, famous Balboa Park with its Shakespearean Festival and museums, Torrey Pines State Park, world-famous San Diego Zoo, Wild Animal Park, Old Town, and many other attractions. The city and its surroundings cater to the tourist trade. The abundance of motels, hotels, restaurants, picnic grounds, shops, and amusement parks is eloquent testimony to its success (Figure 5.13).

Figure 5.12 Newport Beach is known for its pricey homes, commercial real estate, and fast-paced lifestyle. (Image © Bobby Deal/RealDealPhoto, 2009. Used under license from Shutterstock, Inc.)

Figure 5.13 San Diego has an abundance of attractive sites for tourists and residents alike, such as Balboa Park. (Image © Andy Z., 2009. Used under license from Shutterstock, Inc.)

Historically, San Diego has undergone some interesting changes. From its beginning as a sleepy mission and presidio, it has evolved into a more aggressive and sophisticated locale. In recent years, the expansion of the region has been dramatic, with much of this expansion occurring in outlying areas, such as Escondido, Oceanside, and La Mesa. As a new rapid-growth area of residential settlement, the area has seen a shift in the average age and ethnic pattern. Although retirement-age persons are still attracted to the region, the increase in young families is noticeable. This has created a rising demand for schools, playgrounds, and housing.

Indeed, one of the single greatest fears of many San Diegans is that the rapid growth will turn the region into a southern version of oft-reviled Los Angeles. By the 1980s, both San Diego County and San Diego City reached the number two position in total population for counties and cities in the state (close behind Los Angeles County and Los Angeles City). The problems of growth have led to a strident antigrowth movement, which successfully pushed much of the new development outward along "transportation corridors," particularly in north county areas. The fact that much of the population growth comes from ethnic minorities (mostly Asian and Hispanic) has resulted in instances of racist rhetoric and behavior in the region, including highly publicized actions by white supremacist groups along the border. It is abundantly clear, especially after the passage of the North American Free Trade Agreement (NAFTA) that this infusion of new ethnic groups will not stop. The critical issue for San Diego is how to blend social, economic, political, and cultural diversity into a more smoothly functioning whole—not unlike the state at large.

A final minor footnote can be found in some of the earlier pastoral nature of the region, which can still be found in the backcountry areas. Peaceful farm communities such as Julian, Banner, Aguanga, and Santa Ysabel alternate with picturesque agricultural valleys and hillsides. However, even these areas are feeling constant pressure from growth and development. With the population already exceeding 3 million

people and continued growth projected in the next decade, the area will continue to be dominated by constant expansion. An additional ironic twist is the aggressive expansion of Indian casinos on the backcountry reservations. This has provided a growing stream of people and money into this region. Its role as residential and resort mecca will be augmented by its critical function as a natural hub for culture and commerce.

Mad Dogs and Californians

To state that California has its unique and peculiar cultural phenomena is to express the obvious. Although this aspect of the state has been overplayed by eastern commentators, residents of the state do seem to take pride in forging new fashions and fads, social arrangements, and entertainment. Experimentation has never been a process feared in this state. When these experiments are successful, they are admired by outsiders. When they fail, they are cited as evidence of the basic looniness of Californians. What begins in the Golden State, however, frequently spreads to the rest of the nation and becomes part of American culture.

Creating Dreams: Disney, Knott, and Others

Technology has brought numerous benefits to American society. Advancing technology creates greater leisure time. Likewise, technology usually brings a higher degree of affluence. With leisure and affluence, public literacy and sophistication generally rise. The end in this chain of development is a demand for amusements to fill up the new leisure time. California has certainly met this challenge.

The forms of mass amusement and entertainment in this state are wide-ranging. The traditional forms of television and movies remain popular, of course. In the major urban areas, cable and satellite television have become multimillion-dollar operations, providing first-run movies, sports and entertainment events, and soft-core pornography mixed with various shopping networks. Many Californians also spend untold hours on computers, tablets, and cell phones. But California also has become expert in providing less prosaic forms of amusement.

Theme parks are hugely popular attractions. Visitors have proved willing to spend millions of dollars to enjoy created experiences in nostalgia, adventure, the Old West, futurism, romanticism, or combinations thereof. Although the original Disneyland (and now the added "California Adventure") in Anaheim may be the best known of these parks, it is certainly not the only one. Cedar Fair's Knott's Berry Farm and Ghost Town (Buena Park), advertising itself as "America's oldest amusement park," offers a variety of adventures and experiences. Cedar Fair's California's Great America in Santa Clara offers thrill rides and water parks; Six Flags Magic Mountain in Valencia provides white-knuckle rides and experiences, as well as music and entertainment. Universal City Tours brings the movie world to the customer, or vice versa. Raging Waters in San Dimas and San Jose provides human-managed "natural" thrills. Other parks provide similar adventures in which technology and imagination combine to entertain customers (Figure 5.14).

A particular offshoot of theme parks is the wildlife–nature amusement park. In these semi-natural settings, wildlife and sea life are

Figure 5.14 Santa Cruz Boardwalk Beach Amusement Park.

(Image © Karin Lau, 2009. Used under license from Shutterstock, Inc.)

displayed and often provide shows for the guests. San Diego has its SeaWorld and San Diego Zoo Safari Park. Long Beach has its Aquarium of the Pacific. Such attractions typically attempt to create an illusion of a natural setting, which distinguishes them from the more traditional and still popular zoos (present in most major California cities).

Still another type of popular amusement features historical themes. In many cities, major tourist attractions are "Old Towns" from the Spanish days (Sacramento, San Diego, Pasadena) or "roots" locations (Olvera Street, Cannery Row). Similarly, the California missions build upon this historical sense to draw tourists. Where true history may not be available, enterprising California businesspeople have created a subgenre of commercial kitsch-popularized, if not accurate or aesthetically appropriate, history. Here, stylized plastic history is merchandised in restaurant–shopping–entertainment complexes. San Francisco's Pier 39, Cannery, and Ghirardelli Square are joined by such consciously quaint tourist towns as Sausalito and Carmel-by-the-Sea and by Knott's Berry Farm's replica of Independence Hall and similar cultural expressions.

Finally, there are a variety of other forms of entertainment available in the state. These range from convention center presentations to concerts held in stadiums to programs put on by local theater groups and colleges. Elite culture can be found in opera houses, music centers, symphony halls, museums, and art shows. Various bars, country-western clubs, jazz halls, and rock concerts provide yet another dimension of the multilayered entertainment available in California.

Eating Your Way to Nirvana—By Railroad, Bistro, Pub, and Drive-In

Food is a special form of amusement in California. As with other forms of entertainment, this does not make the state different from the rest of the nation—except in degree. Based on the amount of time and space dedicated to eating, food apparently is one of the primal entertainment urges in American society, especially in its relation to all forms of recreation.

Whether it is drinks at the concert, snacks at the game, or junk food at the mall, food seems to be a necessity for most recreation.

This becomes even more interesting when we consider that the act of eating itself has become major entertainment for huge numbers of people. By the late 1970s, as much as one out of every three food dollars was spent on dining out. According to reports from the US Census Bureau, by the late 1990s over $355 billion was spent annually by Americans on dining out. By the early 2000s around 47 percent of all US food expenditures were from dining out, and California accounted for a gigantic portion of this total. Various factors have contributed to this growth, including increases in leisure time, growth in disposable income, higher percentages of single and childless adults, and the rise in the number of working women. California has some additional unique factors that have made dining out even more feasible and popular.

First, the weather usually does not pose any significant obstacle to an evening out. Second, mobility, an assumed and accepted part of California culture, facilitates the dining-out phenomenon. Third, the casual and experiential lifestyle encourages culinary adventures; because Californians frequently seek or create novelty, the varieties of dining experiences are extensive.

Logically, in a state wed to the automobile, speed, and mobility, fast-food restaurants head the list in popularity. The fast-food establishments have been able to achieve such a prominent position because they fulfill certain needs: the menu is standardized and predictable, the food is relatively inexpensive, and the service is fast and, in many cases, does not even require leaving one's car. Because there are so many fast food spots, a wide variety of food choices are available.

The burger, fries, and soft drink establishments are king, with McDonald's leading the pack. California-style Mexican food is a major competitor, with chains selling what has been called "pasteurized" Mexican food for Anglo tastes. A variety of other fast-food outlets offer fish, chicken, pizza, roast beef, hot dogs, a variety of Asian dishes, and other less familiar dishes. All share the vital characteristic of serving the lifestyle of a people always on the move who desire cheap, fast, and palatable food.

Holding a prominent position as entertainment venues is the vast array of theme restaurants. Here, the entertainment function or gimmick is often as vital as the food. Depending on whim, mood, and funds, diners can choose from railroad cars, sailors' havens, family boardinghouses, English pubs, ranchers' tables, mining grottos, gypsy dens, or plantation mansions. If you can imagine a theme, there is likely to be a restaurant in California catering to the concept. The key is the "experience" more than the food; if the food is tasty, so much the better.

California also reflects some unusual culinary styles that are made possible by its cultural and agricultural environment. Natural or health-food establishments build on the idea of selling a California lifestyle. Thus, Golden State gastronomics support avocados, bean sprouts, alfalfa shoots, fruit, nuts, and grains as part of the healthy California image.

The plethora of ethnic groups in the state has made possible an ethnic restaurant business unsurpassed in the country. In Los Angeles alone, many hundreds of noteworthy ethnic restaurants compete favorably for the dollars of hungry diners.

The state's long coastline has made seafood a vital part of the diet. The choice of seafood restaurants up and down the coast is extensive, ranging from fancy, expensive bistros to stand-at-the-counter fish stalls.

Another category of serious eater is served by the dinner house approach, which usually features limited menu items and large quantities. Buffets also cater to this clientele by letting diners fill their plates with any number of salads or starches for a fixed price.

Plastic and neon are the identifying marks of another familiar category of restaurant. This is the coffee shop or café, which has the enduring advantage of always being open for business. For mobile Californians, this kind of 24-hour establishment provides an oasis in the night.

Another category is the cocktail and candlelight spot, which sells snob appeal and romance.

These places can be recognized instantly by their French names or locations atop tall buildings. A toned-down version of this category is the fine-food-by-formula restaurant, where smiling, college-aged waiters and waitresses serve the specialty (turtle, steak, lobster, or omelet) in a standardized "gourmet" decor.

Finally, it would be remiss to omit mention of the now ubiquitous specialty coffee restaurants. Starbucks is only one of many such establishments that offer gourmet coffee, pastries, and room to socialize for all ages—not to mention wireless computer connections.

Perhaps the most salient characteristic of the restaurant experience in California is the choice available. The state has made dining out a recreational activity equal to that found in any amusement, theme, or wild animal park. Speed, mobility, romance, ethnicity, variety, entertainment, convenience, and image may be purchased at a choice of restaurants usually within a half-hour's drive. All this variety, in addition to food, makes for quite a bargain. If, as the old saying goes, "We are what we eat," some interesting speculations could be made about California culture (Figure 5.15).

Isolating Age: Leisure World and Age Ghettos

Given the specialization of restaurants and amusement parks in the state, we should not be surprised to find specialized arrangements for living as well. One of the most noticeable of these is the segregation by age in communities around the state. In its compulsion to celebrate youth and beauty, California culture seems embarrassed by wrinkles, age, and gray hair. Thus has evolved a form of age ghetto and segregated living. It also seems that, once started on this process of segregating by age, people do not know where to stop.

Most identifiable throughout the state are the various restricted communities catering to the elderly. Formally structured closed societies of the aged, labeled euphemistically as "Leisure World," are found in several parts of the state. Here, locked safely behind security gates and fences, the elderly can pursue their interests free from outside distraction or bother.

Less structured, but equally real, are the informal age ghettos of mobile home parks. These parks are scattered in various locations throughout the state, including stark desert, cool seashore, and dusty urban jungle. Specializing in housing the elderly on fixed incomes, they provide a central gathering spot for the lower income individuals. In these age ghettos, families can visit their grandparents on three or four special occasions a year but be spared the embarrassment and inconvenience of seeing them on a regular basis. Out of sight, the aged can safely be ignored, and California can perpetuate its myth of eternal youth (Figure 5.16).

Another interesting form of age classification is the over-40 adults-only townhouse and condominium development. Growing in popularity, these communities attempt to stratify

Figure 5.15 Encounter restaurant at LAX.
(Image © Carlos A Torres, 2009. Used under license from Shutterstock, Inc.)

Figure 5.16 Mobile home in senior park, San Jose.
(Image © Leifr, 2009. Used under license from Shutterstock, Inc.)

and structure communal life in such a way as to exclude both the old and the young. The ideal residents are the mature couples that have achieved some measure of success, have moved past child-rearing age, and now seek to enjoy some of the pleasures affluence can bring. Together with other people of similar background and interests, these couples can effectively shut out the irritations of youth and childhood on one hand and senility and infirmity on the other. Of course, some of the segregated living patterns are not exclusively driven by age. Rather, the financial resources controlled by members of the over-40 group allow them to exercise selectivity in how and where they live.

A classic form of age segregation is the young adult or singles complex. Often recognizable by the fact that they are named after a tree (The Apple, The Aspen, The Pines), these dwellings are California's answer to the search for life in the fast lane. Characterized by relentless organized socializing, these communities provide a form of family security to otherwise independent youth. It is a snob appeal approach that promotes the healthy, free-and-easy good life of California—to those who live there, at least. Marriage usually spells the end of this residential period, because such a permanent relationship implies that one has become too sedate and staid for this community.

Finally, there is a growing form of age segregation that has not been sought by its residents: the community for families with small children. For those who can afford the rising costs of suburban living, new housing tracts frequently contain a high proportion of small children. For less affluent families, apartment or condominium complexes that will accept small children are becoming less common. In practice, those that do accept children eventually become age ghettos themselves, overcharging families for the privilege of living with small children. Although such segregated patterns have been challenged in the local courts, no definitive answer has been developed.

What the entire process of separating people by age implies about California is intriguing. The concept of the traditional nuclear family appears to be seriously undermined in the state, except among and within certain traditional ethnic groups. Although similar changes in living patterns have been observed throughout America, nowhere is it more apparent than in California. All the factors that might explain this shift are evident in the state, including mobility, technology, expanded recreation, and economic affluence. If California sets styles for the rest of the nation, here is one style in which it has moved far ahead of the rest of the country.

The Art of Artsy California Burial

As might be expected, Californians have found a way to make even the process of death unique, unusual, and stylish. The traditional community cemetery is too tame for many in this state. Rather, California residents can experience the hereafter surrounded by rose gardens, art galleries, statuary gardens, and serene country meadows created especially for the benefit of the deceased. Price is no object!

The most familiar symbol of the California burial must be Forest Lawn. Planned in the early 1900s, this institution set the pattern that others would follow. Rather than conceding death, Forest Lawn and its descendants created a whole cycle of services ranging from marriage chapels to art galleries to final places of rest for the weary in "slumber rooms" and grassy knolls. In California, one is not necessarily buried in a cemetery; instead, one may be enshrined in a memorial garden or a heather-filled moor or a lawn of oaks or a devotion garden. How can anyone view such a prospect with terror, especially with the ever-present sun beaming down?

In a strange way, a California funeral can become a semi-gala event involving an automobile parade and a final chance to use the products for which one was so strongly encouraged to prepay. By selecting ahead of time, one can choose exactly how to spend eternity. This includes piped-in music for those who expect to be bored in their coffins. Where else but in California would such an attitude exist?

There may be another sad side to the funeral business, however. It should be noted that memorial gardens are not limited to people.

A thriving business in pet cemeteries marks the last word in devotion. Pet owners can buy permanent resting places for their dogs, cats, birds, horses, mice, hamsters, turtles, goats, and other exotic animal friends. With names such as Good Shepherd and Pet Haven, these institutions cater to Californians' need to preserve the memories of faithful companions. The graves are frequently decorated with flowers, toys, and trinkets and are visited regularly by the lonely owners. In a culture dedicated to the pursuit of happiness, it is a fascinating commentary that many can find true friendship only with a pet (Figure 5.17).

California's Salad Bowl Culture

Concerning the ethnic makeup of the state, much has changed in the past 50 years. As recently as the 1960 census, the Anglo population constituted 92 percent of the total in California. By the 1980 census, that figure had dropped to 67 percent. Census figures for the 1990s revealed further shifts, with Anglos and African Americans declining as a percentage of total population and Asians and Hispanics rapidly increasing their proportion. The 2000 census continued this trend and by 2012, Anglo population only slightly exceeded 39 percent. With the related rise in numbers of different ethnic populations, Californians have become much more aware of their many contributions to the unique culture of the state. (For more detailed demographic data, see Chapter 9.)

Figure 5.17 Pet cemetery.
(Image © Damian P. Gadal, 2009. Used under license from Shutterstock, Inc.)

In a state as diverse as California, the number of ethnic groups is large. In terms of total numbers alone, however, certain specific groups have become more influential and obvious in the affairs of the state. The largest single ethnic group is of Hispanic origins and includes Mexican Americans or Chicanos, Cubans, and other Latin Americans. Over the past three decades, this group has grown even more due to a significant influx of undocumented aliens who have crossed the border between the United States and Mexico. The 1990 census showed that this group had grown by a rate of over 60 percent since 1970 (when the total California population grew at a rate of 18.5 percent and the Anglo population by less than 5 percent). As of 2000, they accounted for over 32 percent of the state's population, and according to US Census estimates, this number had risen to slightly over 38 percent by 2012.

The Hispanic population tends to be clustered more in certain regions of the state. The highest concentrations of Hispanics can be found in the Los Angeles Basin, the San Diego region, and the San Francisco Bay Area. In part, this reflects the greater appeal of urban areas for Hispanics. The agricultural regions of Imperial County, San Benito County, and the San Joaquin Valley, however, have substantial Hispanic populations that are quite visible and influential in these areas. With its rapid growth rate, this distinctive and proud ethnic group is projected to constitute at least 40 percent of the total California population by 2020.

The second largest ethnic minority group in California is of Asian derivation and includes residents of Chinese, Korean, Japanese, Vietnamese, and other Southeast Asian origins. The 1990 census revealed that this group was experiencing a significant growth rate and accounted for 10 percent of the state's population. By 2000, census figures showed that the percentage had increased to almost 12 percent, and in 2012 that number was about 14 percent, with projections for 2020 rising to over 15 percent. Although the San Francisco Bay Area, Los Angeles Basin, Orange County, and San Diego County account for significant concentrations, people of Asian background can be found throughout the state, primarily in urban settings. This consolidated category of Asian ethnic peoples now makes up a growing percent of the state's total population, a cultural force that will have increasing impact.

The 1990 census revealed that African Americans, with around 7 percent of the total, made up the third largest minority population in the state. This group saw an increase of 33 percent over the 1970–1980 period but has declined in growth since then. The 2000 census placed their percentage of the state's population at 6.4 percent, in 2012 at 6.6 percent, and projections for 2020 remaining fairly stable. As with most other ethnic groups, the African American population tends to be found in greater numbers in the metropolitan regions such as Los Angeles or Alameda County. Although the traditional source of African American immigration to California has been states in the Deep South, this pattern has broadened out somewhat in recent years. With its history in the state, the African American population promises to be a force in the future.

The last ethnic population of statistically important size in California is that of the Native Americans, a group that composes between .53 and 1.9 percent of the state's population total (depending on how their ancestry is calculated). In California, native peoples are spread between discrete Indian Trust Lands (reservations) and urban locales. Holding approximately 450,000 acres of land in the state, the native California Indian population has begun to assert its identity and rights in various legal cases, public forums, political arenas, and economic actions, funded in large part by the proceeds of lucrative gaming (casino) revenues.

The categories noted here are, of course, overly broad clusters that reflect the categorizations of the Census Bureau. Subsumed within these categories are many other distinctive ethnic groups. These broad categories do, however, provide a convenient and simplified focus for discussion of the cultural contributions of non-Anglo Californians.

In spite of the fact that until relatively recently the Anglo population exceeded 90 percent of the California population, the cultural identity of the state reflects many non-Anglo influences. One need only look around to see the impact of various ethnic groups on the life of California.

On the most obvious level, the physical landscape of California reflects many ethnic influences. Much of the architecture of the state is a tribute to these sources. The tile roofs, mission-style buildings, hacienda-themed houses and tracts, neo-adobe constructions, open-patio formats, and similar recreations of the mood of early Mexican California abound. Likewise, Asian influences can be found in modified pagoda-style buildings and houses, in ornamental gardens, and in interior decor. In most major cities in the state, one can find echoes of many cultural styles in the buildings and dwellings.

Certainly, entertainment patterns and styles in California are shaped by various ethnic contributions. Music, restaurants, and dramatic productions can be found catering to every ethnic taste or desire. Thus, Mexican restaurants featuring mariachi music adjoin Jewish delicatessens, and Chinese theaters. Ballet folkl6rico vies for patrons with Japanese Kabuki and African music festivals. Supermarkets and other stores stock a wide variety of ethnic foods and products, and certain neighborhoods in large cities specialize in uniquely exotic markets. Such specialized neighborhoods are particularly evident in the large metropolitan areas of the state.

Many of the businesses in the state reflect an ethnic character in their development and history. The Japanese and Hispanics traditionally have held key roles in the agriculture of California. The Italian influence is evident in the fishing, banking, and wine industries. Southeast Asians are increasingly assuming a larger role in the fishing industry. Native Americans control some important acreage in the state, including parts of Palm Springs, and play a significant role in California casino-style gambling. Entertainment and sports reflect the presence of a variety of ethnic cultures, with the growth in the popularity of soccer the most obvious example of this influence.

The growing impact on the state of California's ethnic populations is also being felt in the areas of law, politics, literature, religion, and language and in lifestyles and general culture. If current demographic trends continue, California's culture will become even more multiethnic.

Finally, we should note the impact of general immigration into the state by foreign-born citizens. Historically, the countries that have contributed the greatest share of foreign born are, in descending order, Mexico, Canada, the United Kingdom, Ireland, Italy, Germany, and the former Soviet Union. Indeed, by 2012 well over 27 percent of California's total current population was foreign born, and the pattern is holding firm. Such a continual influx has only served to enhance the dynamic and varied culture that has made California such a fascinating and ever-evolving society.

Recreation or Else

The Resort Mentality: A Place in the Sun

California as a destination for permanent residence is one of the realities that have influenced the culture of the state. There is another facet to the lure of the state and its mild climate, however.

With its ideal and varied weather, myriad tourist attractions, and wealth of geographic wonders, California has also been a prime setting for those who wish to escape the harsher climate of their native areas for a short time only.

Many parts of the state cater to a transitory population by providing the atmosphere, amusements, and diversions sought by this clientele. Palm Springs is widely known for its luxury hotels and golf courses and for its mild winters. San Diego has created a tourist haven of hotels, aquatic and wildlife parks, beaches, and golf course resorts; San Francisco is a mecca for visitors with its well-known Chinatown, Fisherman's Wharf, and Golden Gate Park; Lake Tahoe is a tremendous draw; and the growth of time-share condominiums at the seashore and on the ski slopes has brought a new meaning to the term *resort community* in California.

The various forms of tourism and recreation are not merely for the outside visitor to the state. The people of California are equally engaged in the pursuit of play. Although some of the more common forms of entertainment in the state were discussed previously, the more active forms have not yet been addressed. Sedate, family-style amusements are certainly vital, but Californians tend to shine most when they are actively pursuing physical excitement. Recreation is a huge business in the state, accounting for millions of hours of leisure and millions of dollars of profit.

Spectator Spectacles: There's More Than One Coliseum

Sports are big attractions statewide. In fact, California has been called the sports capital of the world. At the professional, college, high school, community, and individual level, athletics are widely popular throughout the state. Several things contribute to this advanced position in the world of sports. Climate permits almost year-round competition and activity. Affluence and mobility make most spectator and participant sports easily accessible to the majority of citizens. Some unique geographical characteristics expand the number of sports and recreation activities available. California can support any activity that takes place in a desert, mountain, or ocean locale.

On the spectator level, there is much to choose from. The number of stadiums, arenas, speedways, parks, courses, and tracks in California is stupendous. Big cities and small towns alike offer events loyally patronized by fans. Professional sports draw wide followings in football, baseball, hockey, basketball, soccer, auto and boat racing, boxing, bowling, golf, horse racing, tennis, rodeo, and some lesser sports (Figure 5.18). Alumni, student, and popular support for amateur athletics make possible extensive athletic programs at thousands of schools, colleges, and universities across the state. Competitive "club" sports also account for many hours and many dollars, with competitive gymnastics, diving, dance, and swimming as just a few examples. If spectator sports are one's particular love, there is no shortage of choices to meet that need. In person, or by television, radio, or online transmission, interested Californians can satiate their appetites.

Figure 5.18 Petco Park.

(Image © Christopher Penler, 2009. Used under license from Shutterstock, Inc.)

Active Play: Everybody Is a Star

If spectator sports and activities are a passion, participant recreation is a compulsion. In a culture in which awareness of the body is a fetish and concern for health has become a religion, active play is a necessity. Happily, California can provide almost any recreational experience imaginable.

One broad category of active recreation involves the "cult of the body." This involves those activities that call for displays of the body and awareness of the strenuous exercises needed to keep in shape, including aerobics, weight training, and body sculpting. The widespread appeal of this is evident in the multimillion-dollar industry that has sprung up to satisfy these needs. Health clubs, spas, racquetball courts, and athletic clubs can be found almost any place where busy executives, workers, housewives, and students can slip away to the club for an hour's strenuous workout.

Another form of popular activity utilizes the wealth of natural settings available in the state. Numerous national and state parks offer opportunities to rough it in the open air. Hiking, swimming, camping, fishing, boating, and hunting can be pursued in forests, wilderness areas, deserts, and mountains all over the state—although in some instances reservations several months in advance are required. One can climb Mount Whitney or brave Death Valley, explore Yosemite National Park or wander through Lava Beds National Monument, ski at Mammoth or Tahoe or houseboat on the Sacramento River Delta. Between the federal and state public lands and waters, residents and tourists have a wide choice of natural settings to explore and enjoy.

A more common and accessible form of recreation familiar to California is the localized sports activity. Many Californians regularly schedule weekly golf games or tennis matches. Even more Californians can be seen during early morning or late evening in their jogging togs or on their thousand dollar touring bikes. Neither rain nor sun, smog, Thanksgiving, or Christmas can stop these devotees from doing their daily miles. The relative importance of these sports can be assessed easily by comparing the price of golf, tennis, or jogging shoes with normal street shoes. The price certainly reinforces the importance attached to recreational activities pursued by the majority of Californians.

Of course, other forms of active recreation are available. With the many tastes, styles, personalities, and preferences that exist within the state, no one needs to feel ignored. In the area of recreation, too, California is an eclectic mix.

Summary

The cultural landscape of California presents a vast array of options, perhaps unparalleled anywhere else in the country. Such diversity is facilitated by a fortunate geographic setting.

First, the geography (topography, climate, soil, etc.) varies widely throughout the state. This, in turn, leads to diverse economic and social responses that take advantage of the unique features of space and place. The respective cultural regions of California give eloquent testimony to the variety of settings in the state. Second, the fortuitous location of California on the Pacific Coast and on the Pacific Rim has facilitated a rich influx of new residents from both south of the border and across the Pacific Ocean. Third, the benign climate and relatively healthy economic status of California have encouraged a steady flow of ambitious new residents from the United States and abroad.

The same factors have helped create a culture of experimentation and enthusiastic openness to new things. A year-round climate has also made possible many recreational and social activities, as well as some oddities of lifestyle (from the perspective of outsiders). In all, California presents a rich mosaic of people, place, and time.

Selected Bibliography

California Department of Finance website (www.dof.ca.gov).

California State Government website (www.ca.gov).

California Travel and Tourism Commission website (www.visitcalifornia.com).

Caughey, J., and Caughey, L. (1977). *Los Angeles: Biography of a City.* Berkeley and Los Angeles: University of California Press.

Clarke, T. (1996). *California Fault: Searching for the Spirit of a State Along the San Andreas.* New York: Ballantine Books.

Haslam, G. W. (1990). *The Other California: The Great Central Valley in Life and Letters.* Santa Barbara, CA: Joshua Odell Editions, Capra Press.

McKnight, T. (2003). *Regional Geography of the United States and Canada.* Englewood Cliffs, NJ: Prentice Hall.

McWilliams, C. (1973). *Southern California: An Island on the Land.* Santa Barbara, CA: Peregrine Smith.

Miller, C., and Hyslop, R. (2000). *California: The Geography of Diversity.* Palo Alto, CA: Mayfield Publishing Company.

Rieff, D. (1991). *Los Angeles: Capital of the Third World.* New York: Simon and Schuster.

US Census Bureau website (www.census.gov).

Vidal de la Blache, P. (1918). *Principles of Human Geography.* (E. de Martonne, Ed., M. T. Bingham, Trans.). London: Constable Publishers.

Zelinsky, W. (1992). *The Cultural Geography of the United States.* Englewood Cliffs, NJ: Prentice Hall.

Vegetation of California

"… the venerable aboriginal Sequoia, ancient of other days, keeps you at a distance, taking no notice of you, speaking only to the winds, thinking only of the sky."

–John Muir

Introduction

California is truly a land of superlatives, with vegetation that is wonderfully diverse and complex. From the redwoods (*Sequoia sempervirens*) growing in the northern Coast Ranges to the Joshua trees (*Yucca brevifolia*) of the Mojave Desert, this incredible range of plant life is the result of a variety of interwoven factors—namely, the complex interplay of topography, geology, and climate. California is a big place, the third largest of the United States, covering 411,800 square kilometers (159,000 square miles) and spanning approximately nine degrees of latitude and longitude. In addition, its topography is highly varied ranging from 84 meters (276 feet) below sea level in Death Valley to its highest point of 4,418 meters (14,495 feet) at Mount Whitney. The variation in climate observed across the state is largely the result of this incredible range of elevation and latitude, as well as proximity to the ocean. These factors, along with soil variation, fire, and human impact, give rise to the most diverse vegetation found in temperate North America.

Because of its mild Mediterranean climate, California has an extremely high number of species compared with other areas of North America. The Great Plains states are almost four times the size of California, yet the number of native species is less than half (2,496) that of California's (5,862; after Holland & Keil, 1995; Table 6.1). Only tropical areas, for example Guatemala, have greater species diversity than California (Table 6.1). In addition, California's vegetation has a high degree of endemism, the presence of species not found elsewhere. One of the most famous examples of California's endemic species is the mighty sequoia (*Sequoiadendron giganteum*), found only on the western slopes of the Sierra Nevada (Figure 6.1). This endemic tree is the largest biological species in the world, known to reach heights of more than 75 meters (247.5 feet) and diameters of 30 meters (99 feet); it is believed to be the largest species that has ever lived. Many species of nonnative vegetation have also been introduced to California and, over time, they have become a part of the natural landscape. To begin to make sense of the tapestry that is California vegetation, it is necessary to sort it into meaningful categories. This can be done both biologically, using taxonomic rules, and geographically. In this chapter, we examine California vegetation by using a blend of the two methods, with an emphasis on the geographic divisions.

Geographic Region	Area in square km (square mi.)	Number of Native Species	Percentage Difference from Number of Native California Species
California	411,800 (159,000)	5,862	
Great Plains	1,605,792 (620,000)	2,496	43
Alaska	1,478,883 (571,000)	1,366	23
Guatemala	109,421 (42,248)	7,817	133

Table 6.1 Comparison of Species Diversity.
(Source: Holland and Keil, 1995.)

Figure 6.1 The famous giant sequoia (*Sequoiadendron giganteum*), found only on the western slopes of the Sierra Nevada.
(Source: K. Zecher)

Taxonomic Divisions of California Flora

California is the home to 6,880 species of vascular plants. Of these, 5,862 are considered native (85%), and 1,020 (15%) were introduced by humans during the eighteenth through twentieth centuries. *Vascular plants* are defined as those having a well-developed vascular system to transport water, dissolved minerals, and other substances throughout the plant body. The vascular plants include angiosperms, gymnosperms, ferns, and allies. Because this chapter is an overview of California vegetation, its focus is on vascular plants, as nonvascular plants (fungi, algae, mosses, and liverworts) are a very small portion of the vegetated landscape (Table 6.2).

All the major taxonomic groups of vascular plants are represented in California, with the vast majority being angiosperms (*Angiospermae or Magnoliophyta*). Angiosperms are flowering plants that produce a covered seed. They are one of the major groups of extant (not extinct) seed plants and the most diverse major plant group on earth with roughly 260,000 living species classified into 450 families. Angiosperms vary tremendously in size, shape, and longevity, and occupy every conceivable habitat—except the most extreme environments in the highest elevations and latitudes. In California, they account for 97.3 percent of the total plant species.

Six plant families contribute 40 percent of the total plant species found in California; all six are angiosperms (Table 6.3). (1) The Asteraceae, or sunflower family, has more species than any other family of flowering plants; examples are the golden yarrow (*Eriophyllum confertiflorum*), which has a variety of medicinal and herbal uses; the aromatic California sagebrush (*Artemisia californica*); and the perennial herb California aster (*Aster chilensis*). (2) In the grass family (*Poaceae*), the flowers are small, greenish, and inconspicuous. Examples are *Festuca californica*, known by the common name California fescue, and purple needle grass (*Nassella pulchra*), the state grass of California. (3) The Fabaceae, or pea family, is also known as the

Taxonomic Group	Number of Families	Total Number of Species	Percent of Flora
Angiosperms	154	5,700	97.3%
Gymnosperms	5	60	1.0%
Ferns and allies	14	100	1.7%
Total	173	5,860	100%

Table 6.2 California's Native Vascular Plants.
(Source: Holland and Keil, 1995.)

Family	Genera	Species
Asteraceae (sunflower)	185	907
Poaceae (grass)	106	438
Fabaceae (legume or pea)	44	400
Scrophulariaceae (snapdragon or figwort)	30	313
Brassicaceae (mustard)	56	279
Cyperaceae (sedge)	14	210
	435	**2,547**

Table 6.3 Six Most Common Angiosperm Families Found in California.
(Source: Holland and Keil, 1995.)

Leguminosae family. California examples of the pea family are palo verde (*Cercidium spp.*), which grows in the desert; western red bud (*Cercis occidentalis*), used by Native Americans in weaving baskets; and the drought-tolerant deciduous mesquites (*Prosopis spp.*). (4) Scrophulariaceae, or figwort, is often referred to as the snapdragon family. This family contains three of the most well-known California native plants: Indian paintbrush (*Castilleja pruinosa*), which grows on serpentine soils; Santa Cruz monkey flower (*Mimulus rattanii* ssp. *decurtatus*), which is endemic to Santa Cruz County, and beeplant (*Scrophularia californica*), which is found in wetland areas. (5) Common members of the mustard family (*Brassicaceae*) include numerous invasive old-world mustards (*Brassica* spp.); the desert candle (*Starleya* spp.), found in the Mojave desert; and the sea rocket (*Cakile* spp.), found on beach dunes. (6) The sedge family (Cyperaceae) contains grasslike herbs that grow in wetland areas; members include tules (*Scirpus* spp.) and sedges (*Carex* spp.).

Although only one percent of the total plant species in California are gymnosperms (Table 6.2), they are the dominant species in much of the mountainous areas of California. Unlike angiosperms, which produce their seeds in flowers, gymnosperms produce their seeds on cones. Conifers are the most common type of gymnosperm; examples include pines, firs, and redwoods. Most conifers are needle-leaved evergreens: the structure reduces snow load and transpiration during the winter, and by retaining their needles year-round conifers are ready to begin photosynthesizing immediately in the spring.

California boasts 60 species of gymnosperms, including the very large sequoia (*Sequoiadendron giganteum*; Figure 6.1), the very tall coast redwood (*Sequoia sempervirens*; Figure 6.2), and the very old Great Basin bristlecone pine (*Pinus longaeva*; Figure 6.3). California also has numerous species of cypress, including the Monterey cypress (*Cupressus macrocarpa*; Figure 6.4), found on the Monterey Peninsula; firs, such as the rare bristlecone fir (*Abies bracteata*), found only on the slopes of the Santa Lucia Mountains; pines, such as the endangered Torrey pine (*Pinus torreyana*); and spruce, such as the weeping spruce (*Picea breweriana*) of northwest California.

Endemic and Relict Species

One amazing aspect of California vegetation is the large number of species endemic to the state. An *endemic species* is one that is limited to a certain region or habitat; for our purposes, we consider an endemic species to be one that grows only within the boundaries of California. Of the 5,860 species native to California, 1,415 are endemic (24 percent), quite high in comparison

Figure 6.2 Coast redwoods (*Sequoia sempervirens*) in the Lady Bird Johnson Grove, Redwood National Park.
(Source: Lin Wu)

Figure 6.3 Ancient bristlecone pine (*Pinus longaeva*), White Mountains.
(Image © Andrzej Gibasiewicz, 2009. Used under license from Shutterstock, Inc.)

Figure 6.4 Endemic Monterey cypress (*Cupressus macrocarpa*), Monterey Peninsula.
(Image © Lynn Watson. Used under license from Shutterstock, Inc.)

to other temperate regions of North America. Some endemics have a narrow range, like the Monterey cypress (*Cupressus macrocarpa*), which is found only along the coast near Monterey. Others are more broadly distributed, such as the blue oak (*Quercus douglasii*), which rings the Central Valley. The presence of so many endemic species is the result of the great variety of specialized microclimates, a wide range of edaphic (soil) conditions, and growing cycles adapted to a Mediterranean climate. *Relict species* are endemic species that were once more widespread, but long-term climate changes have caused their distribution to shrink. The coast redwood (*Sequoia sempervirens*) and the Monterey pine (*Pinus radiata*) are examples. Some of these relict species, like the big cone spruce (*Pseudotsuga macrocarpa*) are even considered to be on the way to extinction.

Geographic Divisions of California Flora

To discuss how and why the many species of plants are distributed across California, it is necessary to impose geographic divisions on the landscape, making it easier to make sense of the complex mosaic that is California vegetation. The first geographic division is due to the presence of the Sierra Nevada Mountains and the Southern Cascade Range (Map 3). These large mountains create a natural north–south barrier and give rise to what is known as cismontane California, the 75 percent of the state to the west of the mountains, and transmontane California, the remaining 25 percent to the east. The presence of this large barrier cuts off cismontane California from the rest of North America, forming an ecological island. That isolation coupled with a complex topography and a mild Mediterranean climate results in a region with a very high level of both species endemism and diversity. By comparison, transmontane California has much lower levels of both species endemism and diversity.

Floristic Provinces

The vegetation found in cismontane and transmontane California can be further categorized into floristic provinces. A *floristic province* is a geographic area that has plant communities composed of species that are characteristic of, and best developed in, a particular area or region (Map 7). Botanists have divided all the continental landforms into floristic provinces. In North American, there are 12 floristic provinces, four of which occur in California: the *Californian*, the *Vancouverian*, the *Sonoran,* and the *Great Basin*, although all four extend beyond the political boundaries of the state. Cismontane California comprises the Californian and Vancouverian floristic provinces; these two areas contain 85 percent of the vascular plant species and have far greater species diversity and a higher percentage of endemic species (50 percent) than is found in transmontane California. The *Sonoran* and the *Great Basin* floristic provinces combined cover approximately 25 percent of the state (the transmontane region), contain only 15 percent of the vascular plant species, and have only 2 percent endemism. Within each of these floristic provinces, the types of vegetation can be further divided into specific plant biomes and plant communities.

Plant Biomes and Plant Communities

A *biome* is a region of distinctive plant and animal communities adapted to the area's climate and terrain. In California, the diversity of vegetation is a response to the interaction of plants in the many different environments. For example,

Map 7 Floristic provinces of California.
(Source: Bryan Wilfley)

large conifers (e.g., redwoods) grow where orographic precipitation is greatest, whereas other species (e.g., Joshua trees) survive in harsh desert conditions with rainfall at or below 25 centimeters (10 inches) per year. California has five major biomes that we will examine. Four are contained in the Californian and Vancouverian floristic provinces, and the fifth is found in both the Great Basin and Sonoran floristic provinces (Table 6.4; Map 8). Each biome contains numerous plant communities, where plant species are organized into assemblages that interact with themselves and their environment. A *plant community* is defined by the physical appearance of the dominant species (e.g., trees, shrubs, or grasses), the occurrence of a few dominant species, and common physical and environmental characteristics.

Californian Floristic Province

The California floristic province is the smallest in North America (Map 7), yet it has the highest species diversity of comparable areas in North America, half of which are endemic. It covers much of the cismontane region of California including the Central Valley, the Peninsular Ranges, the Transverse Ranges, the southern Coast Ranges, some of the northern Coast Ranges, parts of the Klamath Mountains, parts of the southern Cascade Mountains, and some of the foothills and valleys of the Sierra Nevada (it also extends into northern Baja California). It is characterized by a Mediterranean climate, and the plants found here exhibit classic adaptations to the hot, dry summers and cool, wet winters. This floristic province contains three of the five major plant biomes found in California: the oak woodland biome, the grassland and marshland biome, and the chaparral and coastal sage scrub biome (Maps 7 and 8). Currently, only about 25 percent of the original vegetation of this floristic province remains in pristine condition; in particular, the grassland and marshland biome of the Central Valley has been largely replaced by agriculture.

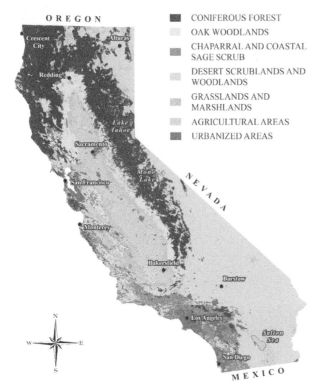

Map 8 Distribution of the five primary biomes of California vegetation.
(Source: Ari and Bryan Wilfley)

Vancouverian Floristic Province

The Vancouverian floristic province, found in the higher elevations of cismontane California, includes the Sierra Nevada, southern Cascades, Modoc Plateau, much of the Klamath Mountains, and parts of the northern Coast Ranges. This area contains the coniferous forest biome, the well-watered forests of mixed

Floristic Province	Climate(s)	Geomorphic Province(s)	Biome
Californian	Semiarid, warm/hot Mediterranean	Foothills/valleys of Sierra Nevada, Central Valley, Coast Ranges, Transverse Ranges, Peninsular Ranges	Oak woodlands
	Semiarid, warm/hot Mediterranean	Transverse Ranges, Peninsular Ranges, Coast Ranges, foothills around Central Valley, S. Cascade Mts., Klamath Mts., eastern Sierra Nevada	Chaparral and coastal sage scrub
	Arid/semiarid, warm/hot Mediterranean	Central Valley	Grasslands/marshes
Vancouverian	Highland, alpine, warm Mediterranean	N. Coast Ranges, Klamath Mts., S. Cascade Mts., Sierra Nevada, Transverse Ranges	Coniferous forests
Great Basin	Semiarid, arid	Modoc Plateau, Great Basin	Desert scrublands and woodlands
Sonoran	Arid	Mojave Desert, Colorado Desert	

Table 6.4 Comparison of California's Floristic Provinces with Climates, Geomorphic Provinces, and Biomes.

	Sequoia (*Sequoiadendron giganteum*)	Coast Redwood (*Sequoia sempervirens*)
Range	Sierra Nevada from Placer to Tulare County	Narrow band along the Pacific Coast from southern Oregon to central California
Elevation	1,524–2,438 meters (5,000–8,000 feet)	Sea level–1,097 meters (sea level–3,600 feet)
Height	84 meters (275 feet)	110 meters (360 feet)
Age	3,200 years	2,200 years
Bark	79 centimeters thick (31 inches)	30.5 centimeters thick (12 inches)
Trunk	6–9 meters diameter (20–30)	2.4–5 meters diameter (8–15 feet)
Reproduction	Seed only	Seed or root sprout
Cones	Shaped like an egg	Shaped like a large olive
Foliage	Small, overlapping, awl-shaped needles	Single needles that fall off in sprays

Table 6.5 Comparison of a Sequoia (*Sequoiadendron giganteum*) and Coast Redwood (*Sequoia sempervirens*).

evergreen and conifers that include some of California's better-known vegetation, the aptly named *Sequoiadendron giganteum* (sequoia), the world's largest living organism, and its taller cousin *Sequoia sempervirens*, (coast redwood; Table 6.5). The primary climate types are highland, alpine, and warm summer Mediterranean (Maps 7 and 8).

Great Basin Floristic Province

The Great Basin floristic province covers the western United States from the eastern side of the Sierra Nevada and the Cascades throughout Nevada, western Utah and portions of southern Idaho, southeastern Oregon, and northwestern Arizona. In California it is limited to the deserts of the Modoc Plateau, north of Owens Lake and Death Valley, and covers only six percent of the state. This area contains the desert scrublands and woodlands biome; it is often referred to as a "sagebrush ocean" as it is dominated by Big Basin sagebrush (*Artemisia tridentata*). This area is a high-elevation semi-desert that lies in the rain shadow of the Cascade Mountains and the Sierra Nevada; it contains both arid and semiarid climates (Maps 7 and 8).

Sonoran Floristic Province

The Sonoran floristic province is located south of the Great Basin floristic province and also contains the desert scrublands and woodlands biome. It covers a large part of the southwestern United States, extending from the eastern flanks of the southern Sierra Nevada and both the Transverse and Peninsular Ranges into southern Nevada, Arizona, and northwestern Mexico. It covers approximately 20 percent of the state; both the Mojave and Colorado Deserts are contained in this province. The arid climate is unpredictable from year to year, and rainfall is scarce (Maps 7 and 8). Plant species characteristic of this province include Joshua trees (*Yucca brevifolia*), fan palms (*Washingtonia filifera*), and creosote bush (*Larrea tridentata*).

Coniferous Forest Biome

The *coniferous forest* is the primary biome in the Vancouverian floristic province and is found in the high elevation mountainous areas where orographic precipitation is heaviest. This includes the Sierra Nevada, the southern Cascades, the Klamath Mountains, and the northern Coast Ranges (Maps 7 and 8; Table 6.4). Conifers (the most common type of gymnosperm) are the dominant species in the higher elevations. Common types of evergreen conifers include cedar, cypress, fir, juniper, pine, spruce, and redwood. A coniferous forest is structurally composed of two layers: the trees themselves and an understory comprising grasses, herbaceous perennials, and shrub species (Figure 6.5). The climate of this biome changes rapidly with elevation from the coldest alpine climate at the highest elevations to highland to warm summer Mediterranean; importantly, adequate precipitation is necessary to sustain these large trees. While there are many individual plant communities contained within the coniferous forest biome, to simplify, this biome is divided further into (1) the north coastal forest, (2) the montane coniferous forest, and (3) the subalpine forest plant communities.

The *north coastal forest* community is found in the northern Coast Ranges and the Klamath Mountains; outside California, it continues north along the Pacific coast into southeastern Alaska. As the name implies, such a forest is limited to coastal regions. The maritime climate of these areas, with its abundant rainfall, fog and moderate temperatures, produces very large trees. One or more conifers are the dominant species throughout the north coastal forest including Douglas fir (*Pseudotsuga menziesii*), western hemlock (*Tsuga heterophylla*), Sitka spruce (*Picea sitchensis*) and the most well-known tree of California (indeed, the state tree!), the coast redwood (*Sequoia sempervirens*;

Figure 6.5 Example of a coniferous forest biome.
(Image © javarman, 2009. Used under license from Shutterstock, Inc.)

Figure 6.2). Hardwoods also grow in the north coastal forest and include tan oak (*Lithocarpus densiflorus*) and big leaf maple (*Acer macrophyllum*). Because the montane forest occupies such a large area of the mountains and such a wide range of elevations, roughly 1,200 herbaceous plants make up the understory.

The coast redwood (*Sequoia sempervirens*) is an example of a relict species. Its range was once much larger, covering areas around the Pacific Rim in Asia and North America, but it is now limited to the Pacific Coast from southern Oregon to central California and extends not more than 80 kilometers (50 miles) inland. Only three genera of the redwood family remain; the rest are extinct as a result of long-term climate change. The two other remaining species are the dawn redwood (*Metasequoia glyptostroboides*), found in central China, and the giant sequoia (*Sequoiadendron giganteum*), found on the western slopes of the southern Sierra Nevada.

The distribution of the coast redwood is correlated with that of the thickest part of the fog belt where, during the summer, cool fog moves off the ocean and onto the land, giving the redwoods a source of moisture during the dry summer months. Its current distribution is also determined by the trees' limited tolerance for climate variations. Coast redwoods require not only abundant moisture, both winter precipitation and summer fog, but also moderate temperatures (cool summers and mostly frost-free winters). Thus, their distribution is limited to wet, coastal valleys, particularly in the southern portion of their range. Few other species grow with the coast redwood; this is due to the fact that the redwoods cast considerable shade, their root systems invade the ground under other trees, and their needles create a deep organic layer, all of which impede the growth of other species. In addition to being a relict species, the coast redwood is nearly endemic to California; only a small area exists outside the state in the southwestern corner of Oregon.

The *montane forest* community is found in the Sierra Nevada, as well as in inland portions of the Klamath Mountains and the northern Coast Ranges. It can begin at elevations as low as 600 meters (2,000 feet) and extend up to

2,400 meters (8,000 feet). The most common conifer species are the Douglas fir (*Pseudotsuga menziesii* var. *menziesii*), ponderosa pine (Pinus ponderosa), incense cedar (*Calocedrus decurrens*), white fir (*Abies concolor*), and sugar pine (*Pinus coulteri*). Along the western side of the Sierra Nevada, the giant sequoia (*Sequoiadendron giganteum*; Figure 6.1) is found. Like the coast redwood, it is a relict species and is completely endemic to the state. While its cousin, the coast redwood, is taller and more slender, the giant sequoia is one of the oldest living trees on earth. It is also one of the fastest growing trees, reaching 75 meters (250 feet) or more in height, and thrives in California's Mediterranean climate. Sequoias have very thick bark, which is its primary defense against fires and insect invasions. In fact, the main way a sequoia dies is to be struck by lightning and fall over! As John Muir once said, "Of all living things the sequoia is perhaps the only one able to wait long enough to make sure of being struck by lightning" (Table 6.5).

At the highest reaches of the montane forest is a gradual transition to the *subalpine forest*. Subalpine forests are found in the Sierra Nevada and in small patches in the Southern Cascades near Lassen Peak and Mount Shasta. This plant community can begin as low as 1,500 meters (5,000 feet) and extend up to 3,300 meters (11,000 feet). Some of the same species that grow in the upper elevations of the montane forest are found in the lower elevations of the subalpine forest. Commonly found conifers are the lodgepole pine, the mountain hemlock (*Tsuga mertensiana*), and the western white pine (*P. monticola*). Other species include the limber pine (*P. flexilis*), foxtail pine (*P. balfouriana*), and bristlecone pine (*P. longaeva*). Bristlecone pines grow in the White Mountains under very dry conditions, receiving an average of 12 inches of precipitation per year (Figure 6.3). Incredibly, the bristlecone pine is one of the longest-lived species on earth, despite growing in such a harsh environment. Dendrochronology dating of these trees has shown ages of up to 4,600 years, eclipsing the life span

of the mighty sequoia. The climate of the subalpine forests is colder and has heavier precipitation than in the lower elevations of the montane forest. The trees are widely spaced, with herbaceous perennials and shrubs between.

Vertical Cross Section of Elevation

The composition of plant species changes as a function of precipitation, which in turn is dictated by elevation (Figure 6.6) and proximity to the ocean. This is especially true in the *coniferous forest biome*. There is insufficient precipitation at lower elevations to support coniferous forests, so they are not found until elevations of roughly 365 meters (1,200 feet) in the northern part of the state and up to 1,830 meters (6,000 feet) in the south. Upper elevations exhibit what is known as a tree line. Areas above this line are devoid of trees; which cannot grow due to cold temperatures and decreased precipitation. The elevation of the tree line varies from 2,990 meters (9,800 feet) in the northern mountains to 3,300 meters (11,000 feet) in the southern mountains.

Using Figure 6.6 as an example, let's start near the gold rush town of Sonora and hike up the western slope of the Sierra Nevada. In doing so, we encounter successive bands, or zones, of vegetation as we traverse from the foothills up the mountain slope and finally to the tree line. At the lower elevations (below 600 meters

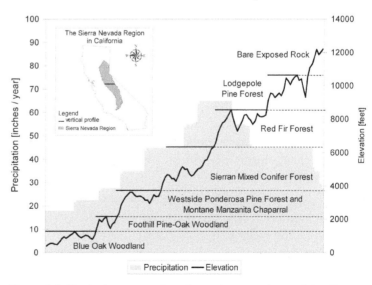

Figure 6.6 Vertical cross section along the west slope of the Sierra Nevada.

[2,000 feet]), we hike through the *oak woodlands* biome (which is discussed later in this chapter). Specifically, in this area we see the native blue oak (*Q. douglasii*), which grows in the lower western foothills of the Sierra Nevada. At higher elevations in the oak woodlands biome, we would see blue oaks growing with foothill pines (*Pinus sabineana*) and perhaps other oaks species, such as the valley oak (*Quercus lobata*) and black oak (*Quercus kelloggii*). These trees are adapted to drought and dry climates and live in the foothills where they receive average annual precipitation of 50 centimeters (20 inches).

At about 600 meters (2,000 feet), we enter the *coniferous forest* biome, and our hike takes us into the lowest-elevation *montane forest* called westside ponderosa pine forest. Here, the forest has an open, park-like feel with ponderosa pine (*P. ponderosa*) growing up to 70 meters (230 feet) in height and a sparse understory of scattered chaparral shrubs and young trees. Next, we reach the sierran mixed conifer forest at around 1,220 meters (4,000 feet). This forest is similar to the westside ponderosa pine forest, but the feel of the forest becomes more dense, with the trees closer together and the crowns often touching. This characteristic is also an indicator that this elevation zone receives higher annual precipitation, a fact we confirm when consulting Figure 6.6.

Finally, at approximately 1,830 meters (6,000 feet), our journey takes us into the lower elevations of the *subalpine forest*, where we enter dense stands of red fir forest (*Abies magnifica*) casting a deep shade. The understory is nearly absent, with abundant litter from needles and branches. This is an area of heavy precipitation, primarily in the form of snow, reaching 3 to 6 meters (10 to 20 feet) of snowpack, some of which remains until June. Finally, our hike nears the approximately 3,300-meter (11,000-foot) tree line and we pass through the upper reaches of the *subalpine forest*, the lodgepole pine forest. Here, the dominant species is the lodgepole pine (*Pinus contorta*), the forest becomes more open, and the trees are shorter in stature. Where precipitation peaked in the red fir forest belt, it now

begins to decrease at higher elevations. There is less snow cover than at lower elevations, and temperatures are lower. Other trees we might see include aspen and mountain hemlock, and the understory varies from quite light to absent, consisting only of some scattered shrubs and herbs.

Oak Woodlands Biome

The *oak woodlands* biome is found throughout the foothills and valleys of cismontane California, ringing the Central Valley and intermixed throughout the Coast Ranges, Transverse Ranges, and Peninsular Ranges (Map 8). This biome comprises 10 percent of the vegetated landscape and is almost completely restricted to the state of California. Many of the dominant species are endemic, such as the valley oak (*Quercus lobata*) and the blue oak (*Quercus douglasii*). The oak woodlands biome is found at elevations of between 90 and 1,525 meters (300 and 5,000 feet), and is transitional between the lower elevation grasslands and chaparral biomes and the well-watered coniferous forest biome of higher elevation. The climate types where oak woodlands are found range from semiarid to both hot summer and warm summer Mediterranean. In general, the climate types are somewhat cooler and moister than the grassland and chaparral biomes below it but are warmer than the coniferous forest above, with an average precipitation of 50 centimeters (20 inches). Because the oak woodlands are so common to and characteristic of California, some people believe it should be declared the state vegetation (Figure 6.7; Figure 6.8).

Oak woodlands are dominated by trees, primarily oaks, and vary in appearance from open savannas where water is limited to a dense, closed canopy appearance where water is in greater abundance. The oak woodlands biome can be further divided geographically and by dominant species into the northern oak, southern oak, coastal live oak, and foothill and valley woodland communities. The *northern oak woodlands* are found in the foothills of the northern Coast Ranges and are dominated

Figure 6.7 Valley oak, member of the oak woodlands biome.

(Image © David Brimm, 2009. Used under license from Shutterstock, Inc.)

Figure 6.8 Hills of Tehachapi, example of oak woodlands biome.

(Image © Richard Thornton, 2009. Used under license from Shutterstock, Inc.)

by the Oregon oak (*Quercus garryana*). The Oregon oak is the only oak species found in California that extends a significant distance beyond the state borders—in this case, all the way to British Columbia. The *foothill* and *valley woodland* community is the best-known and most-widespread of all the oak communities found in California. Notably, this community is found at about 120 to 1,525 meters (400 to 5,000 feet), and rings the Central Valley foothills along the Coast Ranges and the Sierra Nevada. It is also found in the Transverse Ranges, South Cascades, and Klamath Mountains. The dominant species are the foothill pine (*Pinus sabiniana*) and the blue oak (*Q. douglasii*); white buckeye (*Aesculus californica*) is also widespread and

appears as a tree or large shrub. Valley oaks (*Q. lobata*; Figure 6.7) can be found growing on alluvial terraces at low elevations. The valley oak was once commonly found growing along the rivers and streams of the Central Valley, but most have been cut for firewood and to make way for agriculture.

The blue oak (*Q. douglasii*) was an important plant to a number of Native American tribes living in California. The acorns were a highly desirable food staple; harvested acorns were shelled, dried, pounded into flour, sifted for fineness, leached of bitter tannins, and made into mush, soup, or bread. Acorns were also used as bait in traps and in snares to catch birds. The tree's inner bark was boiled and the brew used for relief of arthritis. After a fire, the blue oak regenerates quickly, and the young shoots, which are very straight, flexible, and strong were ideal for basketry and cradleboards. The wood was also used for such utensils as paddles, spoons, ladles, and digging sticks, and oak posts were used to construct houses.

The *coastal live oak woodland*, as it names implies, is found along the coast and is the most mesic (wet, moist) plant community in the oak woodlands biome. The coast live oak (*Quercus agrifolia*) is the dominant species, and often the only tree species found in this community. Other species that can be present include the Madrone (*Arbutus menziesii*), California bay laurel (*Umbellularia californica*), and big-leaf maple (*Acer macrophllum*). The coast live oak is an evergreen species with a dense canopy; thus, the ground is well shaded so understory vegetation is sparse or absent. The dominant species found in the *southern oak woodlands* differs considerably from the oak woodland communities of the central and northern part of the state. The valley oak, blue oak, foothill pine, buckeye and Oregon oak do not extend into this part of the state. Here, the dominant species are the California black walnut (*Juglans californica*) and Engelmann oak (*Quercus engelmannii*). The coast live oak (*Q. agrifolia*) can be found here as well, extending further inland in southern California than in the rest of its range, and the interior live oak (*Quercus wislizenii*) can be found growing on rocky outcrops.

Anthropogenic Impacts on Oak Woodlands

The oak woodlands biome includes some of the largest remaining old-growth forest found in the United States today. Unfortunately, with 80 percent in private ownership, only a relatively small amount is protected. Since the start of California's mission period (1769) and continuing today, grazing by cattle and other domestic animals has taken a serious toll on the oak woodlands (and grasslands) of the state. More than 75 percent of the oak woodlands biome is grazed, making cattle the most pervasive influence on oak populations by limiting regeneration. One land management practice associated with grazing was the introduction of annual grasses, which have replaced much of the native perennial bunchgrasses found in the Central Valley. These grasses produce many more seeds than the perennials and out-compete the acorns for resources. Grazing animals also eat the acorns and the seedlings and saplings that sprout from them. As a result, very few areas of the oak woodlands biome have shown any regeneration since the early 1900s. In areas where there is no grazing, an overabundance of deer has led to the same result. Inadequate regeneration is affecting the blue oak, valley oak, and Engelmann oak, in particular. Other factors leading to the decline of the oak woodlands biome include fire suppression and clearing for agriculture and urban development.

Sudden Oak Death

Another, more recent threat to the oak woodlands biome comes from the disease "sudden oak death," or SOD. First detected in the state in the mid-1990s, SOD is a fungus-like organism similar to the one that caused the Irish potato famine. The first cases of SOD were reported in Marin and Santa Cruz Counties, and it has spread throughout much of the northern coast range. To date, over one million trees in twelve coastal counties of central and northern California have died from Sudden Oak Death. The trees hardest hit are tanoak (*Lithocarpus densiflorus*), coast live oak (*Quercus agrifolia*), and black oak (*Quercus kelloggii*); the coast redwood (*Sequoia sempervirens*) and Douglas fir (*Pseudotsuga menziesii*) are also susceptible to the disease but appear to suffer foliage damage only, not mortality. The oak woodlands biome is an important part of California's biodiversity with 1,000 native plant species, hundreds of terrestrial vertebrates and thousands of insects as part of this wildlife community. These areas will also have more frequent and intense fires due to buildup of fuels from trees killed by Sudden Oak Death. In addition, the economic impact of Sudden Oak Death is estimated to be in the tens of millions of dollars as a result of the high cost of monitoring and eradication, decreased property values, loss and quarantine of nursery crops that spread SOD, and losses of tourist dollars due to aesthetic impacts on recreational areas.

Grasslands and Marshes Biome

The *grasslands and marshes* biome once covered much of the Central Valley, the surrounding inland valleys of the Coast Ranges, and isolated patches in the Transverse and Peninsular Ranges (Map 8). The climate of this area ranges from hot summer Mediterranean in the north to semiarid and arid in the south, with variable rainfall of 25 to 50 centimeters (10 to 20 inches) per year and hot summers. The grasslands portion of this biome was composed of perennial grasses such as needle grass (*Stipa spp.*), bunch grass (*Poa spp.*), and three-awn (*Aristida spp.*). The freshwater marshes contained cattails (*Typah spp.*), bulrush or tule (*Scirpus spp.*), and sedges (*Carex spp.*).

This biome, with its flat lands and fertile soils, is the most eradicated of all vegetation types found in California (Table 6.6). Today, most of it has disappeared, the marshes drained and the grasslands plowed under and replaced by agriculture and urban development or altered permanently by grazing and introduced grasses. The nonnative grasses are well adapted to both the climate and heavy grazing, and quickly outcompete the native grasses. Only small tracts of the native vegetation still exist, and it is doubtful that even these truly resemble the original species composition of this biome (Figure 6.9).

The wetlands of the Central Valley hosted one of the largest concentrations of migratory birds in the world. Reports from early explorers speak of vast concentrations of waterfowl. Unfortunately, of the 2 million hectares (5 million acres) of original wetlands, only 5 percent remains today. With these wetlands, vast areas of riparian forests grew, particularly along the stream courses of the Central Valley; at present, only about 11 percent of the original riparian forest remains in the Central Valley.

Plant Community	Reduction in Vegetation (%)
Grassland communities	99
Coastal sage–scrub communities	70–90
Vernal pools	91
Wetland areas	91
Coast redwood forests	85

Table 6.6 Human-Caused Reductions in Some California Plant Communities.

Figure 6.9 Consumnes Wildlife Preserve, example of grasslands and marshes biome.
(Image © Terrance Emerson, 2009. Used under license from Shutterstock, Inc.)

Vernal Pools

Unique to the Central Valley are the *vernal pools*, also called hog wallows. These pools form in small depressions, 1 to 2 square meters (11 to 21 square feet) in size. Because of their hardpan floors, these pools hold water during the winter rain; the moisture then evaporates in spring. Very specialized plant communities are found in these pools, making them little islands of unique vegetation. In one study, 101 different plants were identified; 70 percent of these were native annuals, 50 percent were endemic, and only 7 percent were introduced (nonindigenous) annuals. An interesting aspect of vernal pools is that as they dry out, the various annual plants bloom in concentric rings determined by their proximity to the standing water. Common species found in the vernal pools include meadowfoam (*Limnanthes* spp.), downingia (*Downingia* spp.), and goldfields (*Lasthenia* spp). Like the grasslands and marshes biome they are a part of, these too were once widespread in the Central Valley, but human impact has eliminated most of them. Fortunately, some vernal pools have been preserved and can still be seen in the Central Valley (Figure 6.10).

Figure 6.10 Vernal pools in the Central Valley.
(Image © rayvee, 2009. used under license from Shutterstock, Inc.)

Chaparral and Coastal Sage Scrub Biome

The *chaparral and coastal sage scrub* biome covers 12 percent of the state and is California's version of Mediterranean vegetation (Map 8; Table 6.4). This biome is most prevalent in southern California, forming a continuous cover on hillsides and valleys, while more discreet patches are found on dry steep slopes in the foothills surrounding the Central Valley and in the Coast Ranges, Southern Cascades, and Klamath Mountains. In southern California, chaparral and coastal scrub are intermixed, and both are drought and fire adapted (Figure 6.11). The *coastal sage scrub* portion of this biome grows nearest the coast from San Francisco Bay south and extends the farthest inland in southern California, as far east as Riverside County.

The etymology of the word *chaparral* is Spanish, from *chaparro,* meaning "dwarf evergreen oak," although oaks are by no means the dominant species found here. Chaparral vegetation is composed of needle-leafed or broad-leafed sclerophyllous (hard-leafed) shrubs and dwarf trees. These shrubs, with their hard, thick evergreen leaves, stand 0.9 to 1.8 meters (3 to 6 feet) tall and grow close together, branches interwoven, to form impenetrable thickets with virtually no understory. Different chaparral communities are found in California, each defined by its dominant species. Some of the most common chaparral species are the chamise (*Adenostoma fasciculatum*), manzanita

Figure 6.11 Chaparral mosaic in southern California.
(Image © Zack Frank, 2009. Used under license from Shutterstock, Inc.)

(*Arctostaphylos spp.*), California lilac (*Ceanothus spp.*), mountain mahogany (*Cercocarpus betuloides*), Christmas berry or toyon (*Heteromeles arbutifolia*), hollyleaf cherry (*Prunus ilicifolia*), coastal sage scrub oak (*Quercus dumosa*), and the pesky poison oak (*Toxicodendron diversiloba*). Species found only in southern California chaparral communities are the laurel sumac (*Malosma laurina*), the chaparral yucca (*Yucca whipplei*), and the sugar sumac (*Rhus ovata*).

Most chaparral species photosynthesize and grow primarily in the spring, taking advantage of the winter rains penetrating to their roots, as well as the increasing day length. As soil moisture becomes depleted in the summer, chaparral species cope by becoming dormant. During this time period, the vegetation is most susceptible to fire. The evergreen nature of chaparral vegetation allows it to retain its leaves and not expend energy to produce new leaves, as deciduous species do. Thus, when a rare summer storm does happen, they can be photosynthesizing within minutes of water reaching their roots. The leaves of the evergreen sclerophyll do not wilt, so no part of the new water supply is used in rehydrating the leaves to maximize light absorption. Because the leaves are evergreen, they are also drought adapted to inhibit water loss through the leaf pores in various ways, including leaves that are small, thick, and waxy; have their stomata on the underside; or have hairs. All these adaptations help trap water vapor and slow the rate of evapotranspiration (water loss through the leaves).

Coastal sage scrub gets its name both from its location on the coast and the presence of aromatic species. The woody shrubs of this community range in height from 0.6 to 1.8 meters (2 to 6 feet) and are adapted to drought by techniques such as storing water and reducing water loss during the long, hot, dry months. Many of the shrub species are deciduous, or semi-deciduous, and drop their leaves during the summer drought, thus becoming dormant. Southern California's coastline was once covered by coastal sage scrub, but large-scale urban development has left only scattered pockets. Examples of coastal scrub species include California sagebrush (*Artemisia californica*), coyote brush (*Baccharis pilularis var.*

consanguinea), California buckwheat (*Eriogonum fasciculatum*), lemonade berry (*Rhus integrifolia*), poison oak (*Toxicodendron diversiloba*), black sage (*Salvia mellifera*), purple sage (*Salvia leucophylla*), and deerweed (*Lotus scoparius*).

Adaptation to Fire

Fire is an integral part of the chaparral and coastal sage scrub biome and plays an important role in promoting and maintaining species diversity. Not only is the vegetation adapted to recover quickly from wildfires, but some species also actually encourage the spread of fire through the presence of flammable oils in their leaves and woody tissues! In the absence of human fire suppression, wildfires would happen naturally by lightning strikes and would vary in frequency from 10 to 100 years. Fires help clear away dead wood and old, less-productive plants and make space for new and different plant species. The majority of California wildfires happen in this biome, and they occur most often during the driest months, from the end of the summer to the start of the winter rains, and are often aided by the Santa Ana winds.

The vegetation is adapted to regrowth after a fire via (1) producing seeds at an early age, (2) producing fire-resistant seeds that can lie dormant in the soil for decades, and (3) regrowth from the crown or roots. Chamise and manzanita, for example, resprout from their crowns, which contain dormant buds that sprout quickly after a fire. Seeds of other species, like buckbrush (*Ceanothus cuneatus*), can lie dormant in the soil for years until fire-associated temperatures crack the seed coats and allow water penetration and then germination. Unfortunately, chaparral and coastal sage scrub vegetation now forms a mosaic with urban development, and subsequent fires are dangerous to human life and costly in terms of property loss. Current zoning laws and state and federal fire suppression policies have resulted in a buildup of fire fuels (thick underbrush). This has interrupted the natural fire cycle of lower-temperature fires occurring more often and has led to larger, more intense wildfires that occur less often but can be more catastrophic, for both humans and the native plant communities.

Desert Scrublands and Woodlands Biome

The *desert scrublands and woodlands* biome covers about 34 percent of California (all of transmontane California and some of the eastern mountain slopes). Because of the lack of precipitation (average of 25 centimeters [10 inches] or less per year), the desert vegetation is dominated by scrublands, whereas the upper elevations are somewhat wetter and can support the pinyon pine woodlands (Map 8, Figures 6.12 and 6.13).

The desert scrublands and woodlands biome can be divided into three plant communities (though it is, of course, much more varied that this!). The Great Basin floristic province is dominated by the sagebrush scrub community, the

Figure 6.12 A Joshua tree (*Yucca brevifolia*), example of desert scrublands and woodlands biome.

(Image © Katrinaleigh, 2009. Used under license from Shutterstock, Inc.)

Figure 6.13 Stovepipe wells, example of desert scrublands and woodlands biome.

(Image © Mariusz S. Jurgielewicz, 2009. Used under license from Shutterstock, Inc.)

Sonoran floristic province is dominated by the creosote bush scrub community, and the eastern slopes of the Sierra Nevada contain the pinyon-juniper woodland.

The *sagebrush scrub* community is structurally and floristically simple: it is covered with low, grayish shrubs that grow 0.3 to 2 meters (2 to 6 feet) in height and is dominated by Big Basin sagebrush (*Artemisia tridentata*). Sagebrush covers not only the portion of the Great Basin that falls within the borders of California but also vast areas of the Great Basin through Nevada, half of Utah, and parts of Idaho, Oregon, and Wyoming. Winters are cold, often with snow, and summers are very hot and dry. Precipitation is 17 to 37.5 centimeters (8 to 15 inches), much of it falling as snow. The plants are widely spaced as a result of extreme competition for water and nutrients.

The Sonoran floristic province is the southern end of the desert scrublands and woodlands biome and is dominated by the *creosote bush scrub* community. Similar to sagebrush, the creosote bush (*Larrea tridentata*) is quite widespread in the western United States, covering large areas of southwestern United States and northern Mexico. In the Mojave Desert, creosote can be found growing in relatively pure stands or in mixed stands with other shrub species. In the warmer Colorado Desert, the creosote scrub stands are much more diverse and can include ocotillo (*Fouquieria splendens*) and succulents, such as century plant (*Agave deserti*), barrel cactus (*Ferocactus acanthodes*), and jumping cholla (*Opuntia bigelovii*). Both regions display a variety of annuals that flower after the winter rains.

The *pinyon-juniper woodlands* community is found on the eastern slopes of the Sierra Nevada and is a transition zone between the forested and nonforested plant communities. These woodlands are drier than the montane forest of higher elevations. The primary species found are the single-leaf pinyon pine (*P. monophylla*), California juniper (*Juniperus californica*), Utah juniper (*J. osteosperma*), and Sierra juniper (*J. occidentalia*). Other species that can be found there include the scrub oak (*Q. turbinalle*), mountain mahogany (*Ceracarpus ledifolius*), and

bitter brush (*Purshia tridentata*). These woodlands are characterized by scattered individual trees that are usually less than 9 meters (30 feet) tall. Junipers tend to be more drought resistant than pinyon pines, so their distributions are not identical. In some drier areas, the woodlands will consist solely of junipers; in other areas, pinyon pines will be found alone, but most often the two trees are intermixed. The understory in pinyon-juniper woodland is frequently Great Basin sage scrub.

Also within the desert woodlands portion of the desert scrublands and woodlands biome are *fan palm oases* (Figure 6.14).These oases, which contain the California fan palm (*Washingtonia filifera*), are found in areas where springs consistently bring water to the surface. They are typically seen in areas of the Sonoran and Mojave Deserts along fault line springs. The few remaining stands

Figure 6.14 California fan palms at an oasis near Palm Springs.
(Image © Karoline Cullen, 2009. Used under license from Shutterstock, Inc.)

are remnants of a wider distribution that occurred during the wetter climate of the last glacial age. California fan palms and their close relatives, Mexican fan palms, have been planted widely throughout Southern California and have become one of the most distinctive landscape elements in the region. However, the California fan palm is the only palm native to the state of California. Examples of fan palm oases are found in Anza-Borrego Desert State Park and Joshua Tree National Park. The Cahuilla Indians of Southern California used the California fan palm for food, grinding up the fruit to make flour, beverages, and jelly.

Habitat Destruction

California vegetation, especially in the California floristic province, is severely threatened, and many species (28 percent) are considered to be extinct, rare, or limited in their distribution (Table 6.6). Destruction of vegetation is caused by human population growth, which leads to conversion of land for agricultural and urban uses, as well as road construction and dam construction. Increased population also results in increased pollution, groundwater removal, invasion of alien species and suppression of natural fires. Significant losses of the native vegetation of several of the plant communities discussed in this chapter have been well documented and include grasslands, coastal sage scrub, vernal pools, wetlands, riparian woodlands, and the coast redwood forests. The native vegetation has been replaced by agriculture, urban development, and invasive, nonnative plant species.

Introduced Plants

Since the time of European exploration, plants have been both accidentally and deliberately introduced into California. Many weeds, once introduced, have become a well-established part of the landscape, often to the detriment of native species. Introduction began with 16 species as early as 1769 during the establishment of the first mission in San Diego. By the time of the Mexican occupation period (1825–1948), 79 species had been introduced, growing to 134 in the post–gold rush era (1860) (Figure 6.15). Finally, according to

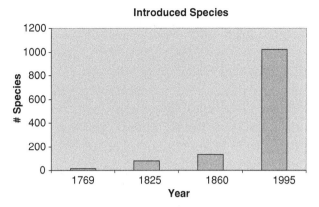

Figure 6.15 Introduced species (1769 to present).

the most recent publication of the *Jepson Manual* (1993), there were 1,023 introduced species, representing 17.4 percent of the current California flora. The vast majority of California's endemic species are out-competed for nutrients and water by these nonnative species.

Summary

This chapter is an overview of the diverse and complex world of California vegetation. The mild Mediterranean climate gives California an extremely high number of species compared to other areas of North America. In addition, because it is separated from the rest of the country by the east–west trending mountains and the desert regions of the state, California also has a very high number of endemic species. On a trip through California's mountains, where orographic precipitation is heaviest, we would see primarily conifers, the most notable being the redwoods of the coastal mountains and the sequoias of the western Sierra Nevada. Coming down into the foothills we would transition into the drier climate of the oak woodlands, ringing the Central Valley and intermixed throughout the many mountain ranges of the state. From here we might travel across the dry Central Valley, once covered by perennial grasslands and marshes but today replaced by agriculture and urban development. Moving into Southern California we would find a mosaic of urban development and fire-adapted chaparral and coastal sage scrub communities. Finally, turning east to travel through the most arid areas of the state, we encounter the highly drought-adapted desert vegetation, such as sagebrush scrub, creosote bush, and pinyon-juniper woodlands. This incredible range of plant life is due to the interwoven factors of topography, geology, and climate; these variables, coupled with soil variation, fire, and human impact, give rise to the most diverse vegetation found anywhere in temperate North America.

Selected Bibliography

Barbour, M., Pavlick, B., Lindstrom, S., and Drysdale, F. (1993). *California's Changing Landscapes: Diversity and Conservation of California Vegetation*. Sacramento: California Native Plant Society Press.

Calflora: *Information on California plants for education, research and conservation*. (2014). Berkeley, California: The Calfora Database [a non-profit organization]. Retrieved [June, 2013], from http://www.calflora.org/

Hickman, J. C. (Ed.). (1993). *The Jepson Manual: Higher Plants of California*. Berkeley and Los Angeles: University of California Press.

Holland, V. L., and Keil, D. J. (1995). *California Vegetation*. Kendall Hunt: Dubuque, Iowa.

Ornduff, R., Faber, P. M., and Keeler-Wolf, T. (2003). *Introduction to California Plant Life*. Berkeley, CA: University of California Press.

Sawyer, J., and Keeler-Wolf, T. (1995). *A Manual of California Vegetation*. Sacramento: California Native Plant Society Press.

Schoenherr, A. (1995). *A Natural History of California*. Berkeley and Los Angeles: University of California Press.

7

Lifestyle Choices: From Urban to Rural

"If they'd lower the taxes and get rid of the smog and clean up the traffic mess, I really believe I'd settle here until the next earthquake."

–Groucho Marx

"The final story, the final chapter of western man, I believe, lies in Los Angeles."

–Phil Ochs

Introduction

Choices of Lifestyles

We all live somewhere, and most of us make various choices about where that will be. In the broad field of geography, researchers have studied and analyzed what goes into individual choice about residential preferences. In turn, other fields have shown a similar interest in the results of these studies, insofar as real estate transactions, commercial development, home building, and similar enterprises want to maximize profits and pursue activities that will have a successful return. Various offshoots of location choices can be found in business site selection, demographic analysis and psychographic profiling, market segmentation studies, and voter districting, to name a few. In short, where people choose to live, why people are attracted to specific sites and the character of different locales are all of interest to policy makers, businesses, and academics.

One of the most applied aspects of location choice is "quality of life." Each year, *Places Rated Almanac* is published, ranking several hundred urban areas in terms of their attractiveness and "livability." In addition, various other publications periodically evaluate places by using whatever criteria are important to their readers (e.g., retirees, those seeking smaller cities, environmentally aware persons). Such efforts reflect the expectation that people do make conscious choices about where they want to live, work, retire, have second homes, and spend time. Of course, this choice making assumes that people are not constrained by other factors, such as economics.

Changing Economic Impacts

Certainly, lifestyle choices are not exclusively a function of personal preferences. A major constraint on such choice is economic realities and changes. The more economically secure a person is, the greater the range of choices. Economic status may change over time as a result of a complex variety of factors. Those factors may be personal and individual, such as stage in life cycle, level of education, unforeseen personal problems, and unexpected windfalls. These factors may be greater than individual factors and may include broader market forces, sociopolitical shifts, global economic forces, and similar systemic upheavals. Increasingly, Californians are experiencing the influences of global changes on individual lifestyles.

Impacts of the "Third Wave"

The noted futurologist Alvin Toffler is known for his creative ideas and books. One of his most interesting works is the book *The Third Wave*, in which he describes a changing world characterized by transitions through various economic and social stages that he calls "waves." These waves certainly are relevant to a discussion of California's varying lifestyles.

Societies in the first wave are characterized by primary occupational activities such as fishing, farming, mining, forestry, ranching, and other activities that extract value directly from the earth. A few world societies are still dominated by these kinds of activities; likewise, even societies that have moved past the first wave retain some pockets of primary, or first-wave, activity. For example, even though California (and the United States) is now dominated by "third wave" forces, portions of the state still engage in ranching, farming, mining, fishing, and logging. These activities are particularly evident in the northern part of the state, the Central and Imperial Valleys, and the desert regions.

As societies evolve, Toffler notes, they typically enter an industrial stage (the second wave). Societies in the second wave are characterized by five factors: synchronization, centralization, standardization, specialization, and size (largeness). These characteristics essentially define an industrial–manufacturing society and how it operates. For example, *synchronization* requires adherence to timetables, schedules, and clocks. Life is constrained by the necessity to "be on time," follow work or school schedules, and assign a time for every purpose. Such behaviors are facilitated by the centralized work and living patterns of an urbanized society—thus helping define lifestyle choices by individuals. *Centralization* of materials and people in second-wave societies is a requirement for greater productivity and control of production of goods and services. Again, this requires greater urbanization. Similarly, *standardization* allows greater interchangeability of components, which also promotes higher and more consistent productivity. This factor finds its way into house and neighborhood design (similar or identical), clothing, consumer products, and tastes, as well as behavioral lifestyles. *Specialization* is a device whereby complex operations (either manufacturing or service) can be divided into discrete pieces for ease of accomplishment. Thus, one function can be learned and done very well without needing to know the rest of the functions that make up the whole. On an assembly line, this means that greater speed and output can be achieved without requiring all operations to be done at the same time and place by the same worker(s). Finally, *size* is generally assumed to mean success: bigger is better; larger is superior to smaller. In terms of lifestyle choices, larger urban areas would be preferred over smaller, less urbanized areas. In turn, choice of living patterns would be biased in favor of bigger-city life.

When societies enter the third wave, however, all assumptions and characteristics of second-wave industrialized societies disappear. Generally, third-wave societies are dominated not by industrial production, but rather by information, services, and high-tech activities. Most observers conclude that the United States has entered the third wave, so Californians are now influenced more by newer characteristics of this wave. In essence, because third-wave societies no longer are dominated by production or manufacturing, they do not need to be synchronized. Tasks and activities may be conducted at any time (and at any place). Because the definitional characteristics of the third wave involve information and services, decentralization becomes the hallmark. In a high-tech society, information can be created and disseminated from and to many places. A worker does not need to be in a large urban setting to conduct business, as electronic communication can occur in even the most rural settings as long as a satellite connection is available. Likewise, services can be scheduled to individual case and need, rather than in a centralized location. Whereas second wave demands standardization, third wave encourages individualism and individuality. Services and information/communication are inherently tailored to specific circumstances and situations in third-wave societies, facilitating creativity in both worker and client. The broader needs of an information society necessitate that workers be more flexible and creative, rather than narrowly specialized as in the second wave. Finally, the compulsion for large scale and size is reduced when production of goods is not the central mission. When size is reduced and activities are more broadly dispersed, the spatial makeup of the society is also modified.

The implication of the socioeconomic shift of American (and Californian) society to the third wave as far as lifestyle choice is concerned is to broaden the options so that people have a greater ability to decide where (and how) they will live. The range of choices varies from the most rural setting to the most highly urbanized, with many variations in between. These choices increasingly will reflect the personalized decisions of the population.

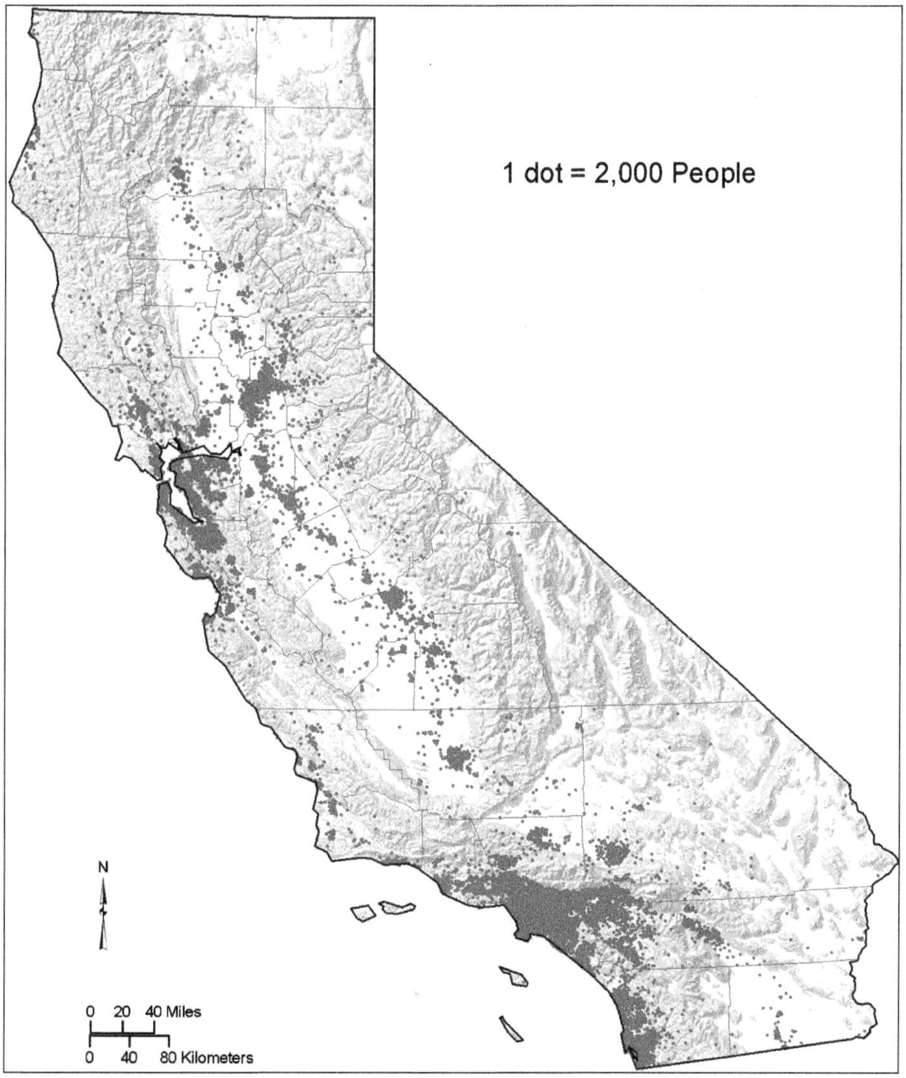

Rural California

Where Is "Rural California"?

If one looks at a map of California, especially one that shows population density (see Map 2), much of the landmass of California has vast areas that would be characterized as rural, or at least sparsely populated. This land includes much of the desert regions, federal and state-owned lands (Bureau of Land Management, National Parks, National Forests, military bases, some reservation lands, state parks, county recreation areas, conservation trust areas, etc.). However, it also includes large private

1 dot = 2,000 People

Map 2 Population distribution (US Census 2007 data).
(Sources: US Geological Survey, ESRI Data and Map, US Census Bureau, Lin Wu)

holdings. Interestingly enough, the rural population of the state is well dispersed throughout almost all counties, ranging from highly rural counties, such as Alpine, Trinity, Sierra, and Mariposa, to rural regions of urbanized counties like San Bernardino and Riverside (Figure 7.1).

What Is "Rural California"?

A search for a widely accepted definition of *rural* would yield a multitude of answers. The one point on which agreement might center is that there is no commonly accepted single or tidy definition. Rather, the definition varies with the agency or expert creating it and reflects, typically, the needs of the moment.

Experts continue to debate the definition of *rural*. Various definitions have been forwarded, with many focusing on what is not rural rather than what is. Frequently, the starting point is the Census Bureau definition of *urban*—with everything else being categorized as "rural." Until around 2000, the Census Bureau used an arbitrary number of 2,500 persons to define "urban," although in subsequent years that definition was abandoned in favor of a more sophisticated methodology that factored in population density and commuter zones. In essence, however, the Census Bureau definition of *rural* continues to be determined by deciding what is urban and then characterizing all that is left over as rural.

A more helpful approach, especially for California, is to look at the characteristics that most experts agree typify rural settings. Although not comprehensive, the following list captures the essence of "rural-ness."

1. *Small size and scale* are hallmarks of the rural experience. Towns are small, government operations are small, and institutions (schools, health facilities, retail, etc.) are small, reflecting the smaller number of persons served.

2. *Part-time services and operations* are frequently the norm, with offices and workers functioning on a reduced-time basis. Often, these workers have other out-of-area jobs to sustain them and can dedicate only limited time to the local rural community. These services may also be provided by volunteers who donate their services for the betterment of the community. In many instances, this is the only way such services can be provided. Thus, libraries, government offices, fire departments, police services, ambulances, and similar activities are provided "as available" (Figure 7.2).

3. *Low population density* (fewer persons per square mile) is a fundamental aspect of rural areas. Insofar as all elements are related, fewer people means lower levels of services and amenities—largely as a function of economics. As a practical matter, this means that much larger landscapes are needed to aggregate enough people to support the amenities—government or private—and that

Figure 7.1 Rural scene in California wine country.
(Image © David M. Schrader, 2009. Used under license from Shutterstock, Inc.)

Figure 7.2 Rural fire stations like this (Running Springs, CA) are typically staffed by volunteers and/or part-time, "paid-call" firefighters.
(Photo: Richard Hyslop)

longer travel times and distances are a part of the rural lifestyle.

4. *Limited resources* typify rural regions. Smaller numbers of people mean both smaller tax bases and smaller customer bases, which in turn means that less variety and fewer choices are available. Limited resources are a normal condition both in terms of social and governmental services and in terms of commercial (retail, entertainment, etc.) opportunities.

5. Rural styles and types are *more complex and diverse* than many believe. Rural areas can include rural agricultural, rural recreational, rural grazing, rural desert, rural forest, and so on.

Rural Economic Phenomena

General Patterns

As might be expected, more rural dwellers are employed in "primary economic" activities than are urban dwellers. Thus, forestry, farming, ranching, mining, and fishing account for a higher percentage of rural employment. Other significant rural occupations include recreation, tourism, government and social services, retail, and food services. None of these options tend to be in the high-pay ranges, that translates to a lower average household income.

Agriculture's Role in the Rural California Environment

Rural and *agricultural* are not synonymous. In many parts of the state, agricultural pursuits are a component of rural life. Although California is the dominant leader in agricultural production and profits in the United States, fewer and fewer people are actually employed in this activity. This, in turn, has had an impact on the rural agricultural population, reducing the number of rural residents who are connected to agriculture or industries that support agriculture. .

Although dealt with in more detail in Chapter Nine a brief mention of the unique place agriculture has in California's rural environment and in the state's economic profile is appropriate.

California remains the number one state in cash farm receipts, posting multibillion-dollar yearly income figures that are almost double those of the next most productive state. Year after year, California also contains the majority of the top 10 agricultural counties in the United States and holds a position as one of the country's top agricultural products exporters. Is it any surprise that agribusiness occupies such an influential role in the state? (Figure 7.3)

The success of agriculture in the state is attributable to several factors unique to California. First, in contrast to many states in the Midwest, California is characterized by almost year-round multicropping. This farming method is facilitated by a wide range of generally mild climates, soils, and terrains. As a result, specialty crops that are not able to be grown elsewhere can be grown in the state. Second, agricultural production in the state is primarily accomplished through heavy and widescale irrigation (and irrigation projects such as the Central Valley Project). Third, most successful crops came about as a result of trial-and-error with introduced crops. Indeed, the choice of what is grown where is not a mere coincidence. Classic theory suggests that available and appropriate land, available water, and distance to markets (or transportation costs) play a large role in decisions about where specific crops are grown.

Nonetheless, many challenges lie ahead. One troubling issue is the rate of conversion of prime agricultural land to urban uses (e.g., farm fields into housing tracts). For example, in

Figure 7.3 Grain silo in Northern California.
(Image © Jennie Book, 2009. Used under license from Shutterstock, Inc.)

the 1990s, Central Valley farm land was being converted to urban uses at a rate of around 8,000 hectares (20,000 acres) per year—largely driven by the state's continued growth in population, the pursuit of reasonably priced homes, and the willingness of commuters to drive longer distances from home to job. By the early 2000s, this number had doubled to around 16,000 hectares (40,000 acres) per year. Another challenge to California agriculture is growing competition from abroad. Environmental problems and the growing awareness of the negative consequences of chemical fertilizers and pesticides (e.g., groundwater contamination and chemical salt buildup in soils) also bring new complications to California agriculture. Finally, the ongoing water shortages in the state do not promise an unencumbered future for agribusiness. With a phase-back of water price subsidies, especially by the federal government, new approaches will be the order of the day.

Population Trends

The current population of California is categorized as 95 percent urban—meaning that approximately 95 percent of the state's population lives in areas designated as urban. Statistically, this means that merely 5 percent of California's population lives in rural areas. These data can be somewhat misleading, insofar as the Census Bureau definition involves calculations of population totals, but the definition also adds considerations of population density (1,000 persons per square mile being the dividing line) and commuter patterns (of at least 25 percent who commute into accessible urban areas) to the formula. Thus, the actual perceived percentage of rural residents may vary significantly depending on an individual's perception and experience. Regardless of the precise numbers, however, population growth statistics still show an increase in California's rural population. Certain indicators point to a steady or modest increase in this trend.

1. The now-acknowledged countermigration from metropolitan areas to rural areas, or at least to less-urban areas, is a real phenomenon—referred to as the "rural rebound" by some experts. Although there is a natural and historical ebb and flow, recent studies have shown that "quality of life" considerations may motivate people to seek less-urbanized regions in pursuit of simpler, more natural, less stressful lifestyles.

2. There are indications that the historic movement away from rural areas may have abated somewhat. Although the outward movement of rural youth may continue, the net out-migration pattern appears to have lessened from the past. An interesting side note is that social scientists have remarked on the "boomerang effect"—that is, some formerly rural youth decide to return to their rural roots to raise their own families.

3. The ability of rural Californians to commute longer distances than rural residents of many other states has facilitated the growth of rural populations. Workers who wish to live in and raise their families in rural areas are not constrained by the lack of jobs in these areas as long as a viable commute is possible. Generally speaking, roads in rural California are of a better caliber and weather is better, which facilitates high-speed traffic. This means that living in a rural area is possible if one is willing to commute longer distances to work (Figure 7.4).

4. The increased mobility and affluence of retirees has helped boost population figures for some rural areas of the state. As people at

Figure 7.4 Many rural roads in California are excellent all-weather, high-speed roads like this one in California ranch county.

(Image © Terrance Emerson, 2009. Used under license from Shutterstock, Inc.)

the leading edge of the so-called baby boom generation enter their retirement years, they often seek either to move to rural areas or to purchase second homes in rural areas. Typically, such moves are selective, focusing on rural regions of a state that has "scenic amenities." Thus, rural areas with trees, mountains, lakes, rivers, and similar attractions prosper, whereas rural areas that are less physically attractive have not benefited from this phenomenon (Figure 7.5). In addition to helping increase the population figures for some rural regions, retirees bring their income into the local communities (what was once termed the "mailbox economy"). They also create an increase in the overall average age of the rural areas, with related demands for services related to this age cohort.

Challenges and Problems in Rural California

As does the state overall, rural California looks to an uncertain future. Projections of modest growth give hope to a more vibrant and prosperous society. As noted earlier, various factors suggest the possibility of a reversal of the economic and population decline that occurred in the recent past, not only in rural California but also in rural regions throughout the country. This is not a certainty, however, nor are all the unique challenges of rural California resolved. In particular, several problems and issues remain

Figure 7.5 Lake Tahoe is a prime example of a rural locale with first-rate scenic amenities.

(Image © Mariusz S. Jurgielewicz, 2009. Used under license from Shutterstock, Inc.)

unresolved. The following are only some of the more prominent difficulties facing rural regions:

1. *Jobs and employment* remain a challenge in rural California. With less economic diversity than urban areas, people in rural areas continue to seek ways to create employment opportunities. Although jobless rates are significantly higher than the overall state figures, the economic impact on rural California is lessened by the fact that the retirement and social security incomes are substantially higher in percentages than in urban areas.

   Certainly, some rural areas of the state are more stressed than others, including the northeast corner, where beef ranching has been on a slow decline, and the northwest corner, where both fishing and lumbering have steadily declined as job sources.

   Likewise, rural agricultural regions such as the Central Valley, have seen a decline in employment, driven in part by intensified competition from places such as Mexico, New Zealand, Western Europe, and Chile. Another factor reducing employment in agricultural California has been the inexorable march of technology and mechanization.

2. *Community services* in rural areas have been notoriously weak or nonexistent, primarily as a result of funding difficulties. Essentially, fewer people per square mile means higher per capita costs and a lower consumer base, which necessarily result in lower levels of services and opportunities. Frequently, this shortfall is partially resolved by higher levels of volunteer service, such as fire departments, ambulance services, and library operations.

3. *Education* in rural regions is again a difficult situation. Because of smaller student populations spread out over much larger geographic areas, one challenge is just to transport these students to school. Likewise, with lower average pay scales, modest funding, and limited resources, educational quality and options vary greatly. Certainly, parent and community involvement tend to be higher in rural school districts, but that may not be adequate to offset reduced opportunities.

4. *Rural health care* is a serious crisis. Rising health care costs is a nationwide issue. The problem is exacerbated by the same economic realities noted earlier for education and community services. Costs of providing health care to a smaller and more dispersed population are higher, and health care professionals can command greater salaries in urban areas. The irony of this situation is that as more retirees look to rural areas for relocation or second homes, they also need to be assured that their health care needs can be met. Lack of rural hospitals, scarce full-time ambulance service, and shortages of doctors and nurses all characterize most rural regions, and the current prospects for the future are not promising.

5. *Recreation* is becoming a greater challenge for rural Californians. Certainly, the scenic rural areas tend to offer many outdoor-related activities, such as hiking, skiing, fishing, camping, and so on. Frequently, however, these activities are undertaken more by urban tourists visiting the regions than by local residents themselves. That is not to suggest that rural residents do not take advantage of the natural environment for recreation but, rather, that they cannot meet all of their recreation desires in the "great outdoors," and they frequently find few alternatives for recreational pursuits. As with services, retail, education, and health care, a relative shortage of population makes it less likely for theaters, restaurants, sporting events, and other spectator activities to be profitable enterprises in rural environs.

6. Finally, the *racial diversity* found in the more urban parts of California is often noticeably lacking. Where there is some racial diversity, it may often be manifested in a bifurcated socioeconomic structure, with the white population occupying the upper tier and nonwhite groups filling out the lower tier.

Regardless of how rural California addresses these challenges, it is clear that the human and physical landscape of the rural regions of the state will continue to evolve and change. Much of the nature of rural California will hinge on how successfully these and other issues are addressed.

Along the Spectrum: Urban, Suburban, and Other Variations

The Context

As noted at the beginning of this chapter, substantial interest, research, and writing have taken place regarding trends, lifestyles, and locational preferences. Traditionally, much of the focus has been on urban models, insofar as the historic pattern has involved people choosing to move from rural to urban areas as a preferred lifestyle. Given this assumption, geographers have developed a whole subspecialty known as *urban geography* in which they analyze both large- and small-scale urban movements. Thus, geographers such as John Borchert and James Vance have looked at the distribution of the US population (and urban clusters) across the country based on a transportation model (Borchert) or on a trade or mercantile model (Vance). Similarly, a popular hypothesis known as the *central place theory* (Christaller) suggests that urban clusters are not mere random accidents but, rather, reflect the need to provide convenient trade, transportation, or administrative centers for the population. These theories help explain the macro-location of cities in various stages of the country's development (e.g., along waterways, railroad lines, convenient surface routes).

On a more micro level, other experts have hypothesized patterns of dispersal *within* existing cities (e.g., John Adam's transportation model of city development and expansion from walking to streetcar to automobile). Even more attention was given to residential and land-use patterns and preferences by urban sociologists and geographers who have studied the urban dynamics of East Coast and midwestern cities and suggested a concentric zone pattern (Burgess), a sector (pie-shaped) model (Hoyt), or a multiple nuclei model (Harris and Ullman). Partly because these models appear to reflect the reality of cities outside California and partly because California's cities seemed to have a unique development pattern, more recent urban geographers have developed an entirely different model that has come to be known as the *Los Angeles School*, in which urban landscapes are characterized by "urban realms" and identity clusters.

Of course, the significance of all of this study is that it reflects the intense level of interest in how people organize their lives on a spatial basis. Adding to the complexity of the study of human living arrangements, increasingly sophisticated data from the US Census Bureau provide a more nuanced picture of the human landscape. For example, recent census data revealed demographic shifts that are influencing where and how Californians choose to live. Factors such as an increase in total households, smaller-sized households, less traditional living arrangements, a widening income gap, a more diverse ethnic mix, a continued shift to the south, a continuation of urban population growth, and an older average age for the population at large all influence the makeup of the state's lifestyles.

The Three Classic Categories

In the past, when geographers discussed living environment choices, they tended to use a broad categorization of rural, suburban, and urban. A brief summary of these categories reveals the following.

Rural

Although the rural category was treated in greater detail earlier, it is worth repeating that the rural population of California still enjoys a modest upward trajectory. The regions that can offer the scenery, calmness, and natural environment along with the prices and services that people want and expect are continuing to grow, partially boosted by retirees, second homes, and "footloose" professions. (A *footloose profession* is a job whose responsibilities do not require that the worker report to a central/specific location. Thus, jobs that can be done by computer or telephone from anywhere [including home], or jobs that focus on a service territory rather

than an office, allow workers to escape from the dominance of urban settings and company or institutional brick-and-mortar work sites.) The attraction for this lifestyle remains the natural setting and the calmer environment (Figure 7.6).

Suburban

The suburban lifestyle continues to hold a strong attraction for Californians. The growth of suburban areas, which spread out from centralized urban settings, continues unabated. Suburb cities or "bedroom communities," such as Irvine or Pleasanton, reflect the attraction of the "Great American Dream" of owning (or at least renting) a home of your own. Although the real estate boom of the late 1990s pushed home prices beyond the means of many Californians, a subsequent readjustment allowed the dream to remain.

Figure 7.6a Mendocino is one rural community that offers a beautiful seashore setting and upscale homes. (Image © Thomas Barrat, 2009. Used under license from Shutterstock, Inc.)

Figure 7.6b Downieville is less scenic and more rural but still connected by satellite dish. (Image © Harris Shiffman, 2009. Used under license from Shutterstock, Inc.)

An interesting irony is found in the fact that some of the attraction of the suburbs is the *perception* that the suburbs are a continuation of what has been called "the American rural ethic." Although it may be a somewhat cruel joke to call Diamond Bar, Murphy Ranch, Philips Ranch, Thousand Oaks, or Moreno Valley rural, the names for these communities reflect the desire of the residents (or developers) to capture an essence of rurality. Clearly, perception is more important than reality, with ranch-style homes, rustic landscaping, and a defined space within which to graze your dogs in the fenced backyard differentiating the suburb from the city (Figure 7.7).

Urban

In a state whose total population shows only growth for the foreseeable future, it is predictable that the urban population of California is also growing, with a few notable exceptions. As noted earlier, the definition of *urban* is a changing concept. Although the exact definition varies from country to country, in the United States (and thus in California) "urban" is now characterized by the US Census Bureau as higher population density augmented by worker commuter zones. Within the broad category of urban are a growing number of subgroups or refinements. These newer concepts are explained more fully below. Categorized or subcategorized, however, the fact is that this lifestyle choice currently is attracting more California residents than it did

during the past few decades. Several factors contribute to this growth.

First, California generally, and Southern California in particular, has been identified as a prime preferred choice for international immigrants. Certainly, the proximity to the Mexican border and the Pacific Rim has contributed to this popularity. However, the traditional factors of economics and climate that have drawn immigrants to California maintain their allure. It is generally accepted wisdom that immigrants have a tendency to migrate to cities, especially where an immigrant community is already established.

Second, a growing trend is to "reurbanize" city centers. Many California cities have undertaken revitalization projects to attract residents back to the cities from the suburbs. Partially driven by longer commuting times, rising gas prices, and desire for convenience, certain groups are finding urban centers to be an attractive choice. Many younger professionals as well as older retirees are recognizing the convenience of centralized, higher-density living. Cities such as Pasadena and San Diego have consciously redeveloped an urban core called "Old Town" to attract such residents to condo, loft, or apartment living in the heart of the area (Figure 7.8). Other inner urban areas, such as the West Adams district of Los Angeles, have seen a

Figure 7.7 A representative suburban neighborhood in eastern Los Angeles County.

(Photo: Richard Hyslop)

Figure 7.8 Restored Victorian neighborhood in San Francisco.

(Image © Michael G Smith, 2009. Used under license from Shutterstock, Inc.)

rebirth through the process of "gentrification," in which deteriorated housing and neighborhoods are revitalized by middle-class professional buyers. These are usually knowledgeable buyers who see great bargains in the underlying "bones" of the homes that they gradually restore to previous attractiveness through the investment of sweat equity and commitment. The result is to acquire a (potentially) nicer home, at a relatively lower price, in a location more convenient to work or entertainment or services. In all fairness, not all California cities have had success with this reurbanization effort.

Third, some cities have experimented with the so-called transportation-core high-density development. The theory here is that placing townhouse or loft living arrangements convenient to rail lines will draw residents who wish to escape their individual cars and commute to work via convenient access to Metrolink, Amtrak, or other rail services. Downtown Pomona and the Soco development of Fullerton represent this approach that, to date, has not enjoyed a high "buy-in" from potential renters or buyers.

A New Jargon of Living Places

Within the broad urban category, experts are increasingly refining the specifics of choice. Recognizing that "urban" covers a wide sweep, experts are now pointing out the variations in urban-ness.

Megalopolis

At the highest level of development (and density) is the urban agglomeration that has come to be known as *megalopolis*. This term was popularized in recent times by French geographer Jean Gottmann, who studied the northeast United States and called the urban corridor from Boston to Washington, DC, a *megalopolis*. Several regions in California could arguably fit into this definition of a large, urbanized region consisting of multiple cities conjoined together into one continuous, higher-density, urban complex (a process referred to as conurbation). Certainly, the greater Los Angeles region seems to fit

comfortably into this category. Thus, whether one lives in San Fernando or West Covina or Anaheim or Pasadena, home is the megalopolis of Los Angeles (Figure 7.9).

Within the large urban agglomerations, others have seen a variation that has been called a *galactic metropolis*. In this configuration, an established, traditional large city is the core of a "galaxy" of related and surrounding urban places. Each of the surrounding urban "planets" exists and prospers as an entity in itself but is loosely tied to the central city by varying factors such as transportation, entertainment and media, commuting, economics, politics, and similar phenomena. In many respects, the central core city acts as an identifier as much as anything else. The example often given is that of the San Francisco Bay area. Certainly, many of the cities of the South Bay area or the East Bay area or the peninsula are dynamic and self-contained. However, the glamour or charm or appeal of "the city by the bay" leads many Bay Area residents and businesses to identify themselves as not from Concord or Pleasanton but, rather, from San Francisco.

In an effort to retain connections to a real or imagined past, and to try to capture a sense of

Figure 7.9 Higher-density neighborhoods characterize the urban residential environment as with this Southern California region.

(Image © iofoto, 2009. Used under license from Shutterstock, Inc.)

urban intimacy, some neighborhoods or developments within the dense urban or suburban core try to create *urban villages* (sometimes called *enclave villages*). The desire underlying this variation is to engender a sense of unique identity within the broader anonymous urban environment. Thus, the area may be retrofitted to close off access streets (Figure 7.10) or rebuilt from the start to include gated and guarded entry points. Occasionally, the enclave may be physically separated from the surrounding area by natural boundaries, such as freeways, flood control channels, green spaces, or similar obstacles to free outside access. The key factor here is to create a sense of separation from the surrounding community. Frequently, such areas are given rustic names (e.g. Philips Ranch, Rancho Estates, or some similar slightly exclusive name). Whether the goal is a greater sense of security, snob appeal, or enhancement of community connection, the trend reflects an apparent broad-based desire to counter some of the isolation of living in, but not being a part of, the big city.

Yet another more recent phenomenon in the urban landscape is the identification of what Joel Garreau has termed *edge cities*. In its broadest sense, an edge city is a relatively new urban region on the fringe (or freeway bypass) of a major

Figure 7.10 Here is an example of a "retrofitted" neighborhood that created its own enclave or "urban village" by closing off public streets and physically restricting access to the neighborhood (in Hacienda Heights).

(Photo: Richard Hyslop)

city or urban metropolis. The typical development pattern involves a fairly new and predominantly residential suburb that quickly evolves into a residential-retail-business-entertainment region at the edge of the more traditional city sprawl. Examples that have been used include Irvine, South Coast Metro, and Warner Center (Los Angeles area); Cupertino, Mountain View, and Pleasanton (San Francisco area); and Mission Valley and San Marcos (San Diego area), to name a few.

Micropolitan

With the desire of many people to escape from the perceived problems and frustrations of the big cities, alternative destinations became more significant. For those who wanted to maintain most of the advantages of urban life (e.g., medical care, entertainment options, diversified retail) with fewer of the problems (e.g., crime, pollution, crowding), smaller cities became a more popular relocation destination. In 2003, the US Census Bureau recognized these cities and began using the term *micropolitan* to categorize them. In its simplest manifestation, such a city consists of an urbanized population of 10,000 to 50,000 people but can also include surrounding scattered populations. It is important to point out that such cities are not connected to or a part of a larger megalopolis but, rather, are stand-alone cities. Some of the Census Bureau–identified micropolitan cities in California are Truckee-Grass Valley, Bishop, Clear Lake, Crescent City, Eureka-Arcata-Fortuna, Red Bluff, Susanville, and Ukiah. The Census Bureau's definitional boundaries, though useful for statistical purposes, do not necessarily capture the essence of the concept, however. Certainly, someone moving to San Luis Obispo or Santa Barbara from the urban megalopolis of Los Angeles or the San Francisco Bay area would see a big difference. The experience remains that of a city but one of more manageable scale and style.

Penturbia

A downsize from the micropolitan setting has been called *penturbia*. The concept was created and named by Professor Jack Lessingner in the early 1990s to explain the outward flow of urban residents to smaller urbanized areas of rural regions beyond the commuter zone of metropolitan areas. Such smaller towns as Mariposa, Sonora, Bishop, Nevada City, Morro Bay, and Sonoma might fit into this category (Figure 7.11). Once again, the goal for moving to such locales is to enjoy the pleasures of small town life and open land. Penturbian communities still offer advantages, although not as abundant as in larger places, such as medical facilities, shopping, entertainment, dining out options, and perhaps a community college. Thus, the former urban/suburban residents can capture some of the sense or essence of a "kinder" rural life without having to endure the inconveniences of true rural life with its far more limited options. In addition, lower housing prices and cost of living are significant attractions. In population numbers, such towns would undoubtedly fall in a range of around 5,000 to 15,000 people, although these numbers could be open to debate depending on various individual factors.

The significance of the expansion of lifestyle categories is to reflect an increasing sophistication in interpretation of the various options available. The rural/suburban/urban troika is insufficient to explain the choices people seek to make. As with California itself, diversity of choice will continue to be a reality for its residents. Furthermore, as the members of the baby boom generation (those born between 1946 and 1964) begin to move into retirement years, they may well opt simultaneously to pursue more than one lifestyle choice, maintaining residences in *both* urban and rural or penturbian settings. Indeed, the future may bring even greater variety and choice.

The Geopolitical Landscape of California

Essentially, the state government of California has broad and comprehensive powers and responsibilities. However, to carry out many of the day-to-day functions that relate to citizens' lives, the state has, through the state constitution, delegated many responsibilities to political subdivisions. The geographic partitioning of California into these localized governing units is relatively straightforward. Unlike many other states, there are no political subdivisions named "villages" or "townships" or "parishes" or "boroughs." Rather, all local government subdivisions fall into one of four broad categories: county, city (or municipal corporation), school district, and special district. Typically, each has its own separate governing body, its own separate taxing authority, its own separate physical boundaries, and its own specific governing task(s).

Counties

California is divided into 58 separate geographic political subunits called counties (although, technically, San Francisco is a blended city-county; see Map 1). Thus, every location in the state is also within a specific county.

Figure 7.11 Frequently, part of the appeal of penturbian cities is their historic character and old-fashioned charm, as seen in this old restored firehouse in Nevada City.

(Image © Peter Weber, 2009. Used under license from Shutterstock, Inc.)

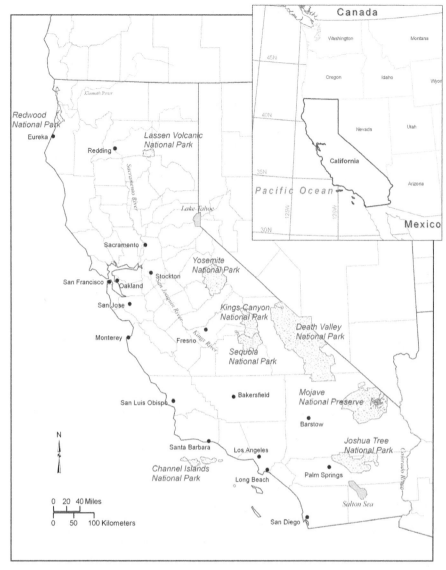

Map 1 Locations of selected geographic features in California.
(Sources: US Geological Survey, ESRI Data and Map. Lin Wu)

form of local government with its own taxing and local governing authority. These subdivisions are called municipal corporations (popular reference to them is ordinarily as a "city," although state law does permit municipal corporations to refer to themselves as "towns" if they choose). State law provides a process for forming such municipal corporations, and, if all requisite elements are present and a majority vote of the residents in the proposed city limits approves, the city will be created. At that point, the city assumes most of the functions that the county had previously provided (e.g., planning, zoning, police, fire, refuse collection, roads). These services are provided through the city's newly created taxing authority and through state-shared revenues that are now redirected from the county government to the city. In turn, the city may create its own departments to provide services or contract with other governments or private entities for such services.

The division and boundaries reflect as much historical happenstance as any coherent planned process. Fundamentally, counties were created to provide service and connection for residents at a more localized level than provided by the state itself. The counties, as subdivisions of the state, provide an administrative structure to implement statewide services (e.g., health services) to local areas. In addition, individual counties can provide certain unique services that may not be needed in other counties (e.g., snow removal).

Cities

Typically, when a county does not or cannot provide its residents adequate service or attention, those residents may choose to create another

According to the California State Department of Finance, as of 2013, there were 482 incorporated municipalities in California ranging in population size from close to 4 million (Los Angeles) to under 125 (Vernon). These data underscore the fact that there are no predictable size constraints on California cities and that little Amador, with a population of slightly under 200, is every bit as much of a city as Los Angeles, with over 3.8 million people. It may be noted that Alpine, Mariposa, and Trinity counties have no incorporated cities, although they do have communities that resemble cities but have no local governing powers. At the other end of the population spectrum, California's largest cities in descending order are Los Angeles, San Diego, San Jose, San Francisco, Fresno, Long Beach, Sacramento, Oakland, Santa Ana, and Anaheim. In terms of land area, California cities range in size from tiny Amador City with 0.78 square kilometers (0.3 square miles) up to Los Angeles with 1,296 square kilometers (498.3 square miles). Arguably, the population and physical size of the largest California cities would place them within the conurbanized category of megalopolis. In contrast, many of the smaller cities of the state, especially those far away from and not connected to the large urban centers, might appear very "rural" to many Californians. Clearly, the idea of "city" in California encompasses a very wide range.

School Districts

California's constitution requires that the legislature provide for public education. Historically, state law established separate governing structures for schools so that they would be managed for educational purposes and not as an afterthought by other local/municipal government bodies. Thus, citizens were empowered to create separate local government entities called *school districts*. Each school district has its own geographic boundary, its own elected governing board (typically titled Board of Education or Board of Trustees), its own funding base, and its own responsibilities for educating the children within its district boundaries. The district boundaries are not required to correspond to any other local government boundaries, insofar as each district is not subordinate to cities or counties other than to the extent that state law delegates some general guideline authority to a county board of education. However, the district governing boards are the primary authority for the districts and their educational responsibilities.

The school districts themselves have partitioned responsibilities. Some districts are *elementary school districts*, which manage the process of kindergarten through eighth grade. *Secondary school districts* manage education from ninth through twelfth grade. If elementary and secondary grade management is merged, the district is known as a *unified school district* (e.g. Glendora Unified School District). Finally, *community college districts* cover lower-division college offerings as well as certain vocational and lifelong learning programs. Each district has its own geographic boundaries that can include all or parts of a city or county or they can cross boundaries and include more than one city or county area. When more than one county is included, it is also known as a *joint school district* (e.g., Lowell Joint School District includes portions of both Orange and Los Angeles counties).

Special Districts

The last unit of local government is known as a *special district*. The primary purpose for the creation of such a local government subdivision is to develop and fund a service that other local governments are not providing. Those services could include cemetery services, parks, libraries, pest abatement, fire protection, water, and hospitals. Currently, the state has approximately 3,400 separate special districts. Their geographic boundaries and sizes vary tremendously. The district may be as small as a neighborhood or as large as several counties. The overwhelming majority of special districts in California serve only a single function, although a few "community services districts" provide multiple services to the residents of their geographic areas. Such districts may be governed by a separately elected governing board (currently about 65 percent of special districts) or may be overseen by the county or city governing structure. As with many issues in our society, economics plays a major role in the creation of a special district. If a county or city is unable to find the resources to provide a needed service, the separate taxing authority of a special district may be the answer, as once formed the district may raise revenues through fees or taxes.

Summary: Patterns for the Future

What general conclusions can be drawn from an analysis of California's lifestyle choices? Although not exhaustive, the following suggestions reflect common observations and predictions.

Population Growth and Challenges

Few experts predict anything except continued growth for the Golden State. As noted earlier, such growth may be more noticeable in the urban areas of California. All along the spectrum and down to the rural regions, growth will be a reality. For the state at large, this may mean increased economic power from sheer numbers alone. Certainly, California's location on the Pacific Rim places the state in an advantageous trade situation, and its coastal setting facilitates extensive interaction with other parts of the world.

With numbers, however, come challenges. Much of the population increase will come from international immigration (both documented and undocumented). Immigration will continue to place strains on the ability of the state to provide such services as health care and education. Population growth also implies more stresses to the environmental fabric of California. More people mean more cars, more pollution, more demands for water and space, and more frustrations driven by an ever-expanding demand for a declining supply of virtually everything.

Sprawl and Space

A constant complaint regarding the human imprint on the natural landscape of California has involved urban sprawl and intrusion into agricultural and natural lands. The state's historic development pattern has been outward rather than upward, disbursement rather than consolidation. Some experts predict that the ever-increasing costs of commuting will drive Californians to return to the center cities and mark a decline of the suburbs. Other experts suggest that telecommuting and a growth of the at-home workplace will ensure that the outward progression of population will continue. Concerns over the conversion of agricultural lands and intrusions into the wild and natural places are real issues, however.

Gridlock

When asked to list some of the most frustrating aspects of life in the state, Californians regularly list traffic congestion and gridlock.

With most of the population clustered into relatively few urban areas, it is not surprising that California has some of the nation's worst congestion (Figure 7.12). This issue has an impact on most Californians every day of the week. In particular, Los Angeles, San Diego, and San Francisco roads and freeways are full day and night, and little relief appears to be in sight.

Budget-Related Issues

Finally, events of the new millennium underscored the interrelated economic problems of California within the world economy. Budget shortfalls plagued the world, the country, and the state. President Barack Obama's first days in office were ushered in with serious economic challenges, including housing and banking crashes, reorganization and bankruptcy of major automobile companies, and an ongoing balance-of-trade problem. These issues were particularly acute in a state where housing costs were already among the highest in the nation, the automobile industry had pulled back from its historic production, and international trade was a mainstay of many local companies. The shortfall of tax dollars worsened a weakened state infrastructure, including higher education, public safety, and social and health programs. Yet through it all, Californians continued to seek new solutions, new approaches, and new

Figure 7.12 Traffic congestion is one of the most common sources of daily frustration for the majority of Californians, as this typical Los Angeles freeway scene indicates.
(Photo: Richard Hyslop)

innovations. That, in essence, captures the spirit of the Golden State—a search for new lifestyles that will suit the challenges of tomorrow!

Selected Bibliography

California Department of Finance website (www.dof.ca.gov).

California State Government website (www.ca.gov).

Garreau, J. (1992). *Edge City: Life on the New Frontier*. New York: Knopf.

Gottman, J. and Harper, R. (Eds.). (1990). *Since Megalopolis: The Urban Writings of Jean Gottman*. Baltimore, MD: Johns Hopkins University Press.

Lessinger, J. (1991). *Penturbia: Where Real Estate Will Boom After the Crash of Suburbia*. Seattle, WA: Socio-Economics, Inc.

Los Angeles Times. Various daily issues.

Rieff, D. (1991). *Los Angeles: Capital of the Third World*. New York: Simon and Schuster.

Savageau, D. *Places Rated Almanac*. IDG Books Worldwide (a new edition is published every two years) The author also has a webpage which outlines various places (http://placesrated.expertchoice.com/)

Toffler, A. *The Third Wave*. (1980). New York: William Morrow & Co.

US Census Bureau website (www.census.gov).

California's Water Resources

"Whiskey is for drinkin', water is for fightin' over."

–Mark Twain

Introduction

Earth is the "water planet," with that precious resource we cannot live without. People living in California read with special interest about water issues such as droughts, water shortages, water conservation, and rate increases. But, how well does the state manage freshwater resources? Where does California's drinkable water come from? Is there have enough water for the 38 million (and growing!) inhabitants, the farms, and the natural environments? The answer to this last question is yes, but ultimately no. Although enough water falls as precipitation within California's borders to supply everyone's needs, the issue is a spatial and temporal one of supply and demand. Most of the precipitation and the resulting surface flow are in the northern part of the state, but the majority of the population resides in the southern part of the state. In addition to this spatial issue is a temporal one: it does not rain or snow year-round in California, as it does in much of the rest of the United States. Instead, there two distinct seasons, wet (winter) and dry (summer), a fact that further complicates water management considerations. But, an even bigger issue than the limits of geographic and seasonal variability is climatic variability. Periods of extended drought

have a tremendous impact on California's water resources and underscore the need for a reliable and sustainable water supply. Thus, reservoirs have been built to store seasonal precipitation and to cope with flooding, and conveyance systems have been built to move vast amounts of stored water throughout the state to urban population centers and agricultural regions.

To examine California's water issues, we must first step back and think about water in general terms. In this chapter, we look at the earth's general hydrologic cycle and then focus on the state's hydrology. Next, we look at how much precipitation California gets each year, what the resulting surface and groundwater reserves are, and how that water has been used historically and examine the water storage and conveyance facilities that have been built in California. We conclude the chapter with a look at the impacts and issues of water use and how to sustain California's water supply into the future.

The Hydrologic Cycle

The term *hydrology* (*hydro-* means "water") refers to the science that encompasses the occurrence, distribution, movement, and properties of the earth's waters, and their relationship with the environment within each phase of the *hydrologic cycle*

(Figure 8.1). The hydrologic cycle takes place in the *hydrosphere*, the watery "envelope" that contains all the water vapor in the atmosphere plus the liquid and solid water on and under the land surfaces and in streams, rivers, lakes, and oceans. Thus, the hydrologic cycle is the movement of water throughout the hydrosphere.

The oceans contain 97.5 percent of the earth's water; freshwater, which is found both on and under the land surface, represents only 2.4 percent, and the atmosphere holds less than 0.001 percent. The small amount contained by the atmosphere may seem surprising because water plays such an important role in the weather. The annual precipitation for the earth is more than 30 times the atmosphere's total capacity to hold water; this fact demonstrates the incredibly rapid recycling of water that must occur between the earth's surface and the atmosphere. Water is also unique in its ability to change state very easily; it can be found readily on the planet in all three forms—solid (ice), liquid (freshwater and salt water), and gas (water vapor).

We can look at the hydrologic cycle in more detail by dividing it into its following parts:

Evaporation—The phase change from liquid water to water vapor.

Condensation—The phase change from water vapor to liquid water.

Transport—The movement of water throughout the atmosphere.

Precipitation—The transfer of water from the atmosphere to the surface as rain, snow, hail, or sleet.

Groundwater—The water located below the ground.

Transpiration—The transfer of water to the atmosphere from plants (through the stomata of leaves).

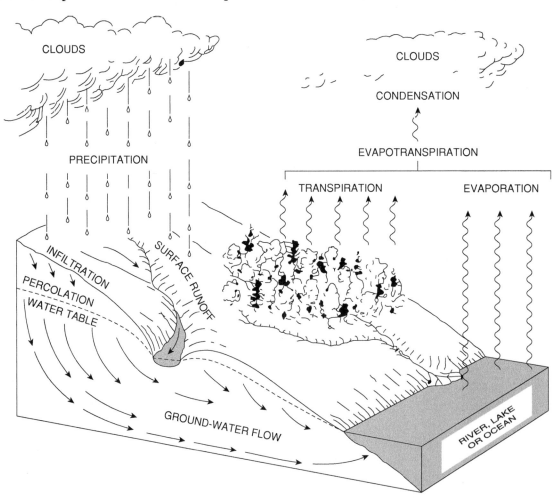

Figure 8.1 The hydrologic cycle.

Runoff—The surface water flowing to, or in, streams and rivers to lakes and oceans.

Percolation—The downward movement of water from the surface through soil and rock material to the water table.

Let's follow the hydrologic cycle in Figure 8.1 beginning with the *evaporation* of liquid water from land surfaces, rivers, streams, lakes, and oceans (caused by the sun's energy). As the air laden with this water vapor rises, it cools and *condensation* occurs: the water vapor gathers into tiny liquid water droplets that form clouds (the droplets are fresh, not salty, water), the prevailing winds *transport* these droplets around the globe until eventually, somewhere, the liquid water falls to the surface as *precipitation*. Then the water may (1) *evaporate* back into the atmosphere, (2) *percolate* down into the ground and become *groundwater*, (3) be taken up by plant roots and be released back into the atmosphere by *transpiration* through the plant leaves, or (4) flow along the surface as *runoff*. Groundwater either seeps its way into the oceans, lakes, rivers, and streams or is released back into the atmosphere through *transpiration* by plants. *Runoff* remains on the earth's surface, where it can flow into lakes, rivers, and streams towards the oceans. That water may *evaporate* somewhere along the way, and the cycle begins again.

How much precipitation becomes runoff and how much percolates into the ground is dependent on the permeability of the ground materials. *Permeability* is the measure of how easily something (like water) can flow through a substance (like soil and rock materials). The more permeable the land surface, the more easily water can infiltrate down into the water table and become stored in areas called *aquifers* (underground layers of porous water-bearing stone, gravel, and earth) (Figure 8.2). Compare, for example, a paved parking lot with a grassy park or area of bare soil. Which surfaces would allow rainwater to infiltrate into it, and which would not? When precipitation occurs faster than it can infiltrate the ground, it becomes runoff.

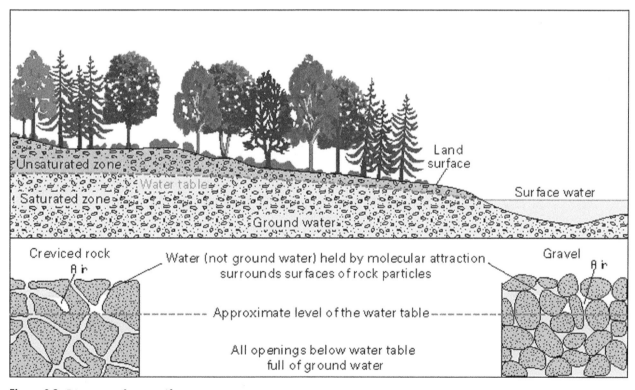

Figure 8.2 Diagram of an aquifer.
(Source: US Geological Survey)

California's State Hydrology

Next, let's examine the hydrologic cycle (Table 8.1; Figure 8.3, Figure 8.4) as it pertains to California. The average annual statewide precipitation is approximately 58 centimeters (23 inches), which equals 200 million acre-feet of water (*maf*: one acre-foot is the amount of water that covers one acre to a depth of one foot). However, this amount of precipitation, and the resulting runoff, is not evenly distributed throughout the state. Approximately 75 percent of the natural runoff occurs north of Sacramento, 40 percent (29 maf) of it from the Klamath Mountains region alone. But much of the demand for that runoff is in southern California, where half the population resides, and in the agricultural areas of the Central Valley.

Of the 200 maf of average annual precipitation, about 65 percent is consumed through evaporation and transpiration by vegetation. The other 35 percent is the state's average annual runoff of about 71 maf; this is the freshwater available to Californians for use. Some of this 71 maf is used for urban and agricultural purposes, and some must be devoted to the natural environment to maintain healthy ecosystems in rivers, estuarine systems, and wetlands. Out-of-state supplies from both the Colorado and Klamath Rivers total 7 maf, boosting the available surface water supply total to 78 maf.

According to the California Department of Water Resources, at a 1995 level of development, water use in California can be broken down into three primary categories of use:

Environmental:	**46 percent**
Agricultural:	**43 percent**
Urban:	**11 percent**
	100 percent

Keep in mind, 78 maf is a yearly average, not something that can be counted on each year to devote to these three general areas of use. Droughts and floods occur often in California, sometimes in the same year. When we examine the yearly runoff records for California, it's easy to see that many years are anything but average. The drought year of 1977, for example, resulted in a total runoff of only 15 million acre-feet, in contrast with the record-setting high of 135 million acre-feet recorded for the winter of 1983 (Table 8.1). Evidence of this type of variability can be seen in the time series of runoff recorded for the Sacramento and San Joaquin River basins. These two basins provide much of the state's water supply, thus their year-to-year hydrology is often used as indices of climatic variability (Figure 8.4, Table 8.2)

Droughts

Droughts are a common and natural occurrence in California. They occur slowly, often over several

Average annual precipitation	23 inches*
Average annual runoff	71 maf
Lowest annual runoff (1977)	15 maf
Highest annual runoff (1983)	135 maf
Total capacity of 1,350 surface reservoirs	41 maf
Imported from out of state supplies	7 maf
2000 average applied urban water use	8.7 maf
2000 average applied agricultural water use	33.7 maf
2000 average applied environmental water use (includes 23.1 maf wild and scenic river flows)	39.6 maf

Table 8.1 California hydrologic data.
*average annual precipitation is calculated on a 30-year average.
(Source: California Department of Water Resources.)

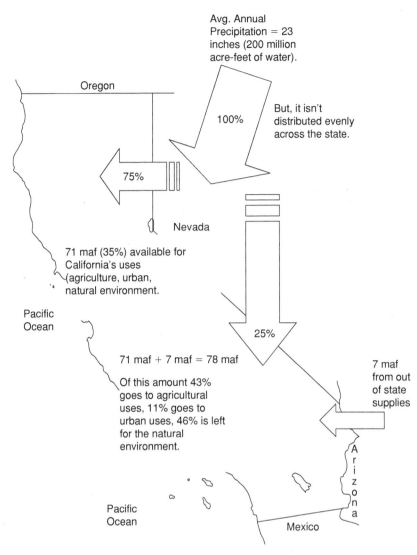

Avg. Annual Precipitation = 23 inches (200 million acre-feet of water).

Oregon

100%

But, it isn't distributed evenly across the state.

75%

Nevada

71 maf (35%) available for California's uses (agriculture, urban, natural environment.

Pacific Ocean

25%

71 maf + 7 maf = 78 maf

Of this amount 43% goes to agricultural uses, 11% goes to urban uses, 46% is left for the natural environment.

7 maf from out of state supplies

Arizona

Pacific Ocean

Mexico

Figure 8.3 California's hydrologic cycle.
(Source: S. Garver)

years, and there is no exact time when a particular drought begins or ends. A *drought* is typically defined as two or three consecutive years of below-average rainfall for the period November to March, when about 75 percent of the average annual precipitation falls. Droughts can be localized to a particular region and may affect only that part of the state, or they can affect the entire state, depending on the precipitation patterns. Those most reliant on annual precipitation, such as ranchers grazing cattle, rural residents relying on wells, or communities with small water systems are the first to feel the impacts of a developing drought. Drought impacts increase with the length of a drought, as reservoirs are depleted and water levels in groundwater basins decline. The

development of California's complex water supply infrastructure, comprised of reservoirs, groundwater basins, and interregional conveyance facilities, helps mitigate the effect of drought periods for most water users.

Measured hydrologic data, primarily precipitation, stream flow, and snowpack measurements, are used in studying the frequency and occurrence of droughts and have been recorded in California since the early 1900s (Table 8.2, Figure 8.5). In addition, some of these records have been extended back in time several hundred more years through the science of dendrochronology. *Dendrochronology* (*dendro*- means "trees") uses tree ring records as a proxy of precipitation and stream flow data to model climate variability on a several-hundred-year timescale. This type of hydrologic record identifies drought periods from 1850 to the present (Figure 8.6). The 1929–1934 drought was a particularly significant event in that it affected not only California but much of North America as well and corresponded with the Great Depression. In California, the reduced runoff measured in the Sacramento Valley during this time period was used to establish the criteria for designing the supply and yield of many large Northern California reservoirs. During this drought period, runoff was only 55 percent (9.8 maf) of the average runoff calculated for the time period from 1901 to 1996.

The 1975–1977 drought, though of shorter duration, was more severe, with the Sacramento Valley runoff at only 37 percent (6.6 maf) of average and the San Joaquin Valley at a mere 26 percent (1.5 maf) of average. The winter of 1976–1977 is the driest year recorded in the state's hydrologic history, and the previous winter is the

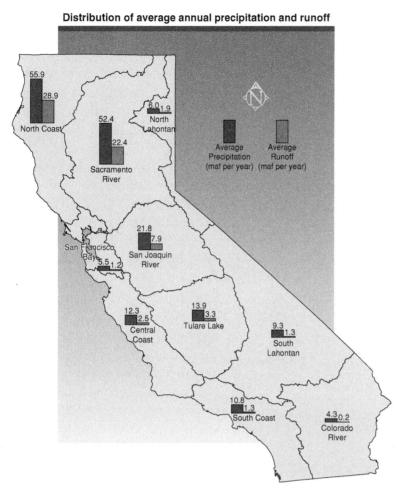

Distribution of average annual precipitation and runoff

Figure 8.4 The ten hydrologic regions of California shown with average annual precipitation and average annual runoff (maf) for each region. (Source: California Department of Water Resources)

fourth driest. These two very dry winters contributed to widespread water shortages and severe water conservation measures throughout California. Agriculturally, serious crop losses were recorded in 31 counties, and a Federal Disaster Declaration was declared in Placer County and surrounding counties.

The drought of 1987–1992 was similar in length and severity to the 1929–1934 drought with precipitation well below average for four consecutive years. Areas of the state most affected were the Central Coast, the northern Sierra Nevada and the Central Valley counties, including Placer County. The Sacramento Valley received only 56 percent (10 maf) of its average runoff and the San Joaquin Valley recorded only 47 percent (2.8 maf). By 1991 the entire state, including urban, rural and agricultural areas, were all suffering from drought conditions.

More current drought conditions are seen in the years 2007—2009 (Figure 8.7), California's first drought for which a statewide

Drought Period	Sacramento Valley Runoff		San Joaquin Valley Runoff	
	maf/yr	Percent of Average (1901–1996)	maf/yr	Percent of Average (1901–1996)
1929–34	9.8	55	3.3	57
1975–77	6.6	37	1.5	26
1987–93	10.0	56	2.8	47
2007–09	11.2	64	3.7	63

Table 8.2 Drought time periods in the Sacramento and San Joaquin River basins.
(Source: California Department of Water Resources)

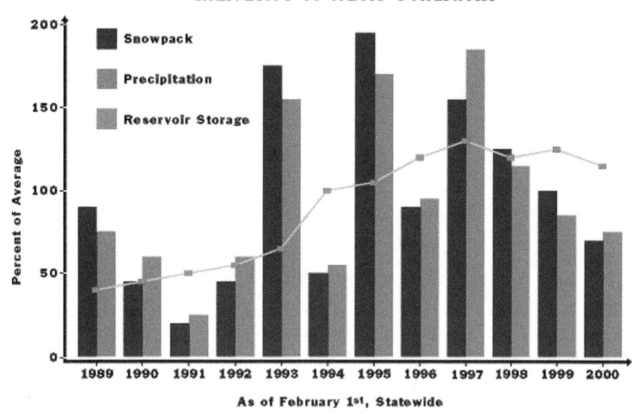

Figure 8.5 Snowpack, precipitation, and reservoir storage as indicators of drought.
(Source: California Department of Water Resources)

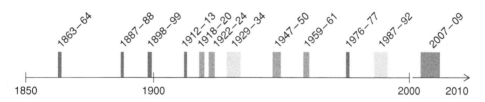

Figure 8.6 Drought record 1850–2009.
(Source: California Department of Water Resources)

drought emergency was declared, and the twelfth driest three-year time period on record. The major difference between this and prior droughts was the severity of State Water Project (SWP) and federal Central Valley Project (CVP) delivery reductions, which began immediately in 2007. The impacts of this drought were most severe on the west side of the San Joaquin Valley, and the resulting water shortages caused significant economic problems for agriculture and rural Californians dependent on agriculture for employment. Particularly hard hit was Fresno County, where unprecedented demands for

social services, such as food banks and unemployment assistance programs, overwhelmed local agencies and resulted in the first state emergency proclamation linking drought with the provision of social services.

Floods

Time periods of above-average precipitation and increased runoff can cause flooding, the other extreme of California's climate variability. Historically, the Central Valley has flooded when the Sacramento and San Joaquin Rivers were inundated with snowmelt from the Sierra Nevada.

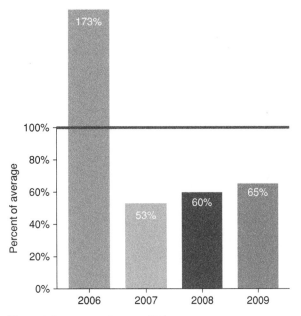

Figure 8.7 Statewide runoff for water years 2006, 2007, 2008, and 2009.

(Source: California Department of Water Resources)

The positive aspect of this history is that, over time, flooding has helped spread the fertile soils that make the Central Valley such a prime agricultural region. Once Europeans settled in the Central Valley and established farms and towns on its floodplains, however, they began to control the rivers with dams, levees, and dikes. As with droughts, California has a recorded history of flood data going back to the 1800s, and severe floods have been recorded in 1850, 1862, 1955, 1964, 1995, and 1997. The 1997 flood in Northern California was the largest in California's history and was caused by a series of storms with heavy, prolonged, and unusually warm precipitation. This warmth caused increased and rapid snowmelt from an above-normal snowpack and resulted in widespread minor to record-breaking floods from central California to Oregon.

Above-average precipitation in California is often the result of an ocean-atmosphere phenomenon known as the *El Niño southern oscillation (ENSO)*, or El Niño for short. During an El Niño, a rise in sea surface temperature in the western equatorial Pacific (2°C or greater) occurs. (Each degree rise in temperature on the Celsius scale corresponds to a 1.8° in temperature on the Fahrenheit scale.) The name El Niño was given because this phenomenon usually peaks in December (the name refers to Christmas: El Niño = "the Christ Child"). These warm water episodes can occur rather frequently (about once every five to seven years) as part of the *southern oscillation*, a cyclical change in the pressure and temperature patterns of the atmosphere and the Pacific Ocean along the equator. El Niño events don't just change the weather patterns in the area of the western equatorial Pacific; these anomalies temporarily alter weather worldwide. Some areas, such as the Pacific Northwest, become drier than usual; some areas, such as California (especially southern California), Arizona, southern Nevada and Utah, New Mexico, and parts of Texas, become wetter than usual. A recent example is the El Niño event of winter 1997–1998, which contributed to the severe flooding in Northern California.

Snowpack

California's snowpack is a natural reservoir of water, and snowmelt is the source of most runoff in the state and for the Colorado River as well (Figure 8.8). The snowpack slowly melts throughout the spring and into the summer, providing a steady supply of runoff. This slow release of water overlaps with the agricultural irrigation season of May through September and enables much more effective use of water from these watersheds than would be possible if the precipitation fell as rain and the runoff was immediate. The largest snowpack reservoir in California is in the high elevations of the Sierra

Figure 8.8 Lingering Sierra snowpack, July 2006, at Virginia Lakes, elevation 10,000′.

(Source: Kirsten Zecher)

Nevada. To make optimal use of each winter's snowpack, data are collected through a process called snow surveying. In this type of data collection, a snow-sampling tube is driven through the snowpack to measure depth; the snow core is then weighed to determine water content. More than 300 snow survey sites are sampled each winter, and this information is critical in helping water planners forecast the water supply for the upcoming year (Figure 8.9).

Snow surveying began in California in the early 1900s, and the early data collection made clear the connection between the amount of snowpack and its water content, and the observed spring rise in the waters of California's lakes and rivers. Scientists recognized that this information was vital to water resource planning and needed to be accomplished in a centralized manner. In 1929, the state legislature established the California Cooperative Snow Surveys Program. Today in California, more than 50 state, national, and private agencies are involved in collecting data each winter.

Groundwater

So far we have discussed water in terms of the surface water that comes from annual precipitation, be it rain or snow. But both surface water and groundwater together are available for use, and, though we don't see it, California actually stores a lot of water underground. So, while we define surface water and groundwater individually, they are very closely linked, and although the land surface is a convenient division, it is a fairly arbitrary one. For example, on its journey from the snowpack of the mountains to the Pacific Ocean, the snowmelt as water could (1) begin as surface runoff, (2) percolate down into the soil and become groundwater, (3) provide base flow to a stream and then become surface flow again, and (4) later in its journey leave the streambed as recharge back into a groundwater table. Groundwater supplies 30 percent of California's urban and agricultural needs, and during a drought that amount can increase to 40 or even 60 percent. Most communities depend on groundwater for a portion of its drinking water supply, and a few cities, such as Fresno, Davis, and Lodi, are completely reliant on groundwater.

Groundwater basins underlie 40 percent of the state, with an estimated storage capacity of one billion acre-feet. Only 250 maf (several times the amount of water stored in all state surface reservoirs combined), though, is usable (pumpable). Figure 8.10 shows the distribution of groundwater basins in the state. Most groundwater resources are located in the Central Valley, east of the Sierra Nevada in the Owens Valley, the southeastern deserts, and scattered throughout the southern Coast Ranges and the Los Angeles Basin. Also, groundwater basins are not found in the granitic Sierra Nevada range. Most of California's groundwater that is accessible in significant amounts is stored in alluvial groundwater basins. *Alluvium* refers to materials deposited by streams and consists of coarse deposits, such as sand and gravel, and finer-grained deposits such as clay and silt. It is the coarse materials, sand and gravel, that have pore spaces where groundwater is stored; these layers are termed *aquifers* (Figure 8.2). The finer-grained layers, clay and silt, do not store water and are called *aquitards*. A groundwater basin may contain one or more aquifers. A *groundwater basin* is defined as an area underlain by permeable materials capable of storing and furnishing a significant supply of groundwater. Groundwater basins are three-dimensional features and may range in size from less than 2.6 square kilometers (1 square mile) in mountain ranges to 2,600 square kilometers (1,000 square miles) or more in flatlands.

Water gets into aquifers in groundwater basins as it percolates naturally through soil, or it may be put there artificially through human-made percolation ponds or injection wells. Unlike a rushing stream or river, water in an aquifer moves very slowly, and the rate at which it moves depends in part on the permeability of the materials (water moves more quickly through coarse material like gravel than through fine-grained material like sand). Aquifers act as natural conveyance and water purification systems while the land above a groundwater basin is put to other uses. Groundwater is not exposed to the elements and is protected from evaporation

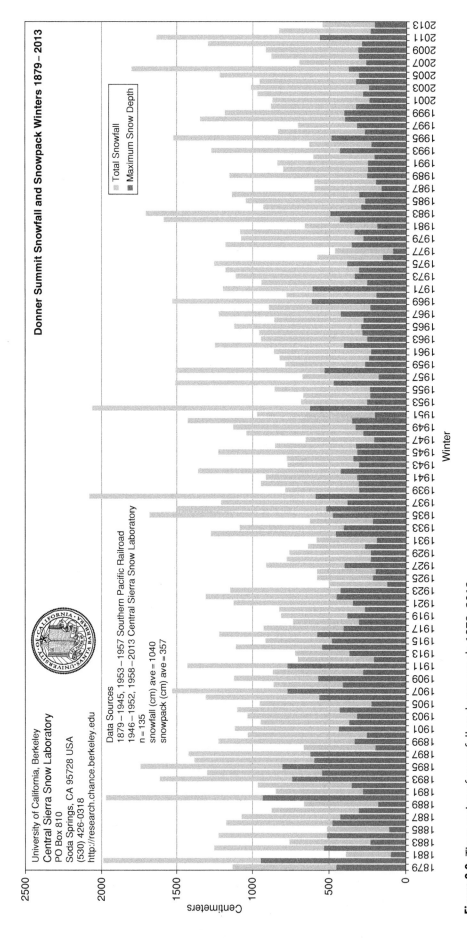

Figure 8.9 Time series of snowfall and snowpack, 1879–2013.
(Reprinted by permission of Central Sierra Snow Laboratory.)

Figure 8.10 Groundwater basin areas of California shown with boundaries of the ten hydrologic regions. (Source: California Department of Water Resources, California Water Plan 2005, Bulletin 160-05)

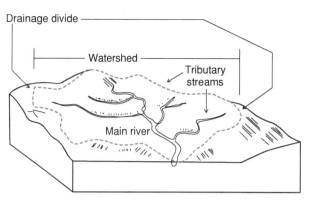

Figure 8.11 Diagram of a watershed and its components.

loss and some forms of pollution. When surface water is depleted during droughts, aquifers can serve as alternative water sources.

It wasn't until the beginning of the 1900s that groundwater came into widespread use for irrigation because advances in drilling and pumping technology made the water easier to access. Groundwater use became so widespread that by the 1930s dropping water tables and escalating pumping costs made necessary the development of surface waters and the construction of extensive storage and conveyance systems such as the SWP and the federal CVP. Even with the creation of these large-scale projects, California's demand on groundwater supplies will continue to increase as the state's population continues to grow.

California's Hydrologic Regions

The 32,000 kilometers (20,000 miles) of rivers and streams in California form sixty major watersheds. A *watershed* (Figure 8.11) is the land area drained by a particular group of rivers, streams, or creeks. A large river, like the Sacramento, drains a huge land area and encompasses a watershed of thousands of square miles. Most large watersheds are made up of several smaller watersheds associated with tributary streams and drainages that contribute flow from smaller sub-basins. The sixty watersheds found in California are divided into ten *hydrologic regions*; these are major drainage areas that share common characteristics of precipitation and runoff, and are also individual water-planning areas. Similarities are observed when comparing the boundaries of these hydrologic regions (Figure 8.4, Figure 8.10) to the geomorphic provinces map (Map 3), as the regions are bounded by the various mountain ranges. The ten hydrologic regions are the Sacramento River, San Joaquin River, Tulare Lake, North Coast, San Francisco Bay, Central Coast, South Coast, North Lahontan, South Lahontan, and Colorado River regions. Each region incorporates the watersheds, the various rivers and streams, and any underlying groundwater basins. Each region differs in its geographic size and shape, amount of precipitation, and runoff and groundwater storage capabilities. These regions also differ in their populations, levels of agricultural activity, urban development, and industrial activities. Thus, each region must have its own carefully thought-out water supply plan.

The Central Valley contains three hydrologic regions. Starting at the northern end of the state, the *Sacramento River Region* drains the northern half of the Central Valley, and although urban development is occurring, it is still a major agricultural center. This region receives most of its

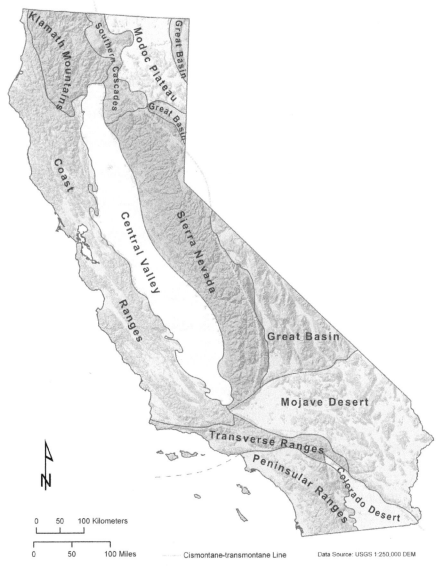

Map 3 Geomorphic regions of California.
(Source: Lin Wu)

0   50   100 Kilometers

0   50   100 Miles   ———— Cismontane-transmontane Line   Data Source: USGS 1:250,000 DEM

developed for agriculture have made the land in this region extremely valuable for developers, and farmland is rapidly being transformed into cities and suburbs. Groundwater is a major water supply source here, as is the Hetch Hetchy Valley Reservoir. The major rivers are the San Joaquin, which carries 9 percent of the state's runoff (6.4 maf), and the Tuolumne, which starts in Yosemite National Park and runs through Hetch Hetchy valley. In the southern third of the central valley is the *Tulare Lake Region*. Climatically this area is a desert, receiving less than 25 centimeters (10 inches) of rain per year. Its rivers include the Kern, Tule, Kings, and the Kaweah, none of which drain into the ocean. The Kern River once ended in the now dry Buena Vista Lake, while the other three drained into what was once Tulare Lake. These lakes, along with extensive wetlands, were drained to make way for the extensive agricultural areas that exist there today.

runoff from the west slope of the Sierra Nevada in the form of snowmelt. The major rivers are the Sacramento, which carries 31 percent of the state's runoff (22.4 maf); the Feather, which is the primary water source for the SWP; and the American, where gold was discovered. This region has been dramatically transformed by numerous dams, levees, flood control facilities, and irrigation facilities. The next hydrologic region to the south is the *San Joaquin River Region*, bounded by the Coast Ranges and the Sierra Nevada crest. It is drier and hotter than the Sacramento River Region and contains 2 million hectares (5 million acres) of irrigated farmlands. The water rights

The *North Coast Region* is in the northwest corner of the state and encompasses the Klamath Mountains and the northern Coast Ranges. This region has the highest precipitation totals in the state, with some areas receiving more than 250 centimeters (100 inches) per year. The streams and rivers in this region carry 40 percent of the state's runoff (29 maf). The primary rivers are the Smith, the only river free of dams; the Klamath, the second largest river in the state; and the Trinity, which drains into the Klamath. But 90 percent of the Trinity River runoff is diverted to the Sacramento Valley for agricultural and urban uses; this has had extremely detrimental impacts on both commercial and Native American salmon fishing in the North Coast Region. Next along the coast is the *San Francisco Bay Region*, which is comprised of watersheds that drain into San Francisco, San Pablo and Suisun Bays. Significant urban and industrial development has transformed this region, and water imports are needed to support the enormous population growth seen in cities such as San Francisco, Oakland, Berkeley, and San Jose. Farther south along the coast from Santa Cruz to Santa Barbara is the *Central Coast Region*. This region covers 2.92 million hectares (~7.22 million acres), much of it agricultural, and contains about 4 percent of the state's population. Major rivers include the Carmel, Big Sur, Salinas, and Santa Ynez. Groundwater is a very important water source in this region, supplying over 80 percent of the annual supply for the region in 1995.

The last hydrologic region along the coast is the *South Coast Region*. This is California's most populated area, containing roughly half the state's population (~19 million people). This area receives a paltry 2 percent of the state's average annual precipitation. Because of the intensive urban development, much of the watershed surface area in this region has been covered in nonpermeable surfaces, like asphalt and concrete. This cover inhibits groundwater recharge and contributes to ocean pollution via storm water runoff during precipitation events. The 82–kilometer (51–mile) Los Angeles River is the water source that originally supported the early inhabitants of Los Angeles, but as a result of flooding it was channelized with concrete as the city grew up around it. Starting in the 1990s, work began to rehabilitate the Los Angeles River and return stretches of it to its natural state, to create parks, walking paths, and bikeways. Numerous reservoirs in the area have been constructed to hold water from the SWP, the Los Angeles Aqueduct, and the Colorado River Aqueduct. These reservoirs include Lake Casitas, Castaic Lake, Big Bear Lake, Lake Mathews, and Morena Lake.

East of the Sierra Nevada are the final three hydrologic regions. The North and South Lahontan Regions are named for Lake Lahontan, an ancient lake that covered more than 22,100 square kilometers (8,500 square miles) of the western Great Basin during the Ice Age. Lake Tahoe is the biggest water body in the *North Lahontan Region*. This small region in the northeast corner of the state has a precipitation range of 20 to 175-plus centimeters (8 to 70-plus inches) from the high desert to the alpine areas. The primary land use is cattle ranching, with pasture and alfalfa as the main crops, along with forest and recreational lands. The *South Lahontan Region* contains both Mount Whitney and Death Valley, the highest and lowest points in California. It also contains Mono Lake, into which mountain streams carry snowmelt, and the Owens River, which carries snowmelt the length of the valley to the now-dry Owens Lake. In the southern end of this region is the Mojave Desert. Desert cities found here are growing rapidly and have to import water; they include Lancaster, Palmdale (Los Angeles County), Victorville, and Apple Valley (San Bernardino County). The *Colorado River Region* is found in the very arid southeastern corner of the state. The Colorado River forms the southeastern border of the state, though minimal runoff from California drains into it. The Colorado River drains a large portion of the western United States including areas of Wyoming, Colorado, Utah, New Mexico, Nevada, and Arizona. The Salton Sea is the other large body of water in this hydrologic region, covering 936 square kilometers (360 square miles) of this desert area. The Salton Sea is situated in the Salton Trough, an area below sea level, and was accidentally created from broken irrigation channels in the early 1900s. It has become an important bird habitat since much of the natural habitat along the coast of southern California has been lost.

The History of Water Use in California

If we could go back to the time before the European settlement of 1769, we would see a very different waterscape in California from what we see today. Rivers flowed uninterrupted, and many lakes and marshes expanded and contracted with the wet–dry seasonal nature of California and with the climate variability of wet and dry years. This waterscape supported abundant wildlife, including fish, birds, and game. The Central Valley, in particular, would look dramatically different during the time of the Native Americans. The Sacramento and San Joaquin Rivers were free to flow through the Central Valley, their banks changing course year to year and overflowing with silt-laden floodwater. The southern end of the Central Valley once contained Tulare Lake, a large lake with four times the surface area of Lake Tahoe, as well as Buena Vista and Kern Lakes. East of the Sierra Nevada, in the Owens Valley, could be found the free-flowing waters of the Owens River, Owens Lake (now dry), and extensive wetlands.

Native American Period

Before European settlement, the population of Native Americans in California is estimated to have been about 275,000 to 300,000 people. Although that number is small compared to the current population, it was one of the most populated areas in North America. Despite this, Native Americans did not negatively affect the environment in any lasting way but instead lived in balance with the natural habitat. Village locations were determined by the available resources (food and water), and for the most part tribes settled near the water sources so that they would not have to divert or store it.

Native Americans developed only two areas in California for agriculture: the Owens Valley and the Colorado River Valley. The Owens Valley is a 120-kilometer-long (75-mile-long) narrow valley between the Sierra Nevada and the White-Inyo Range. It is technically a desert, receiving only 15 or fewer centimeters (6 or fewer inches) of rain per year, but it gets extensive runoff from Sierra Nevada streams that are the result of melting snowpack. The Paiute Indians (around 1000 AD) built irrigation systems here, including a dam on Bishop Creek, that diverted water several miles downstream to two large agricultural plots (3.2 by 3.2 kilometers and 6.4 by 1.6 kilometers [2 × 2 and 4 × 1 miles] in size).

Here they grew native plants, such as yellow nut grass, tubers, and wild hyacinth.

The other area where irrigation took place was along the lower Colorado River. Several tribes (Quechan, Mohave, Halchidhoma, and Kama) practiced agriculture here, again before the Spanish arrived. Crops grown in this area included native vegetation, a type of maize, watermelons, muskmelons and black-eyed peas. These nonnative plants had been introduced by the Spaniards in other areas of North America and had migrated north faster than the European settlements did. These tribes had numerous small plots along the banks of the Colorado River, and depended on the annual floodwater for irrigation instead of building dams and canals. Agricultural production provided tribes in both these regions with 30 to 50 percent of their food supply. In neither area did they abandon their hunting and gathering practices. Although water was obviously important to these tribes, they did little to disturb the land in their irrigation practices. Native American tribes had no concept of owning land or of owning water rights. Thus, because no one owned it, it couldn't be bought or sold. The arrival of Europeans brought with it a very different set of values regarding land, water, and ideas of ownership.

Spanish and Mexican Period

The Spanish and Mexican chapter of California history extends through both the Spanish Mission period of 1769–1821 and the Mexican period of 1821–1846. Although only 77 years in length, this time period brought a dramatic change in the way water was perceived and used in California. The Spanish began their settlement of Alta California in 1769 with the mission system in San Diego. With them came a new culture and value system in which nature was perceived as a divine gift from God to be subdued and exploited for the betterment of humankind—and in the name of Spain. This was a very different value system from that of the Native American peoples already living in California. Unlike the Native Americans, the Spaniards saw water as a symbol of power. Their homeland of Spain was also an arid place, with few rivers, and their customs and laws recognized the importance of controlling this scarce commodity. Under Spanish law, water was not given to the individuals residing in the missions, pueblos, and presidios in California, but instead it was given by the monarch to the entire community. This common right of all the residents to a fair share of the available water also came with the responsibility to help build and maintain the *zanja madre* (main ditch) and support irrigation canals. Residents also had the responsibility to not waste water by letting their ditches overflow or by allowing water to damage roads. Thus, through shared community rights and responsibilities, elaborate water systems were built throughout the mission system, with much of the labor supplied by Native American Indians.

For example, at the Santa Barbara Mission (Figure 3.2), the Franciscan fathers introduced agriculture to the Chumash Indians and directed the planting of corn, wheat, barley, beans, and peas. Orange and olive trees were brought in and planted, and grapevines were cultivated. There were also large numbers of cattle, sheep, goats, pigs, mules, and horses. Water needed to be diverted from the mountain creeks to irrigate the fields and for livestock and domestic use. To impound these waters, the Indian Dam was built in 1807 in the foothills above the mission; the water flowed by gravity to the mission through an aqueduct. Today, the ruins of the aqueduct, dam, and storage reservoir can still be seen.

The rise of the ranchos (1821–1846) and the secularization of the missions (1830s) put new pressure on water resources. The number of ranchos grew to 800, and land grants were given out for thousands of acres of land. Although the rights of rancheros were considered inferior to those of the communities (i.e., the presidios, pueblos, and missions), water rights did start to vary. Some rancheros built dams, for example, even though these structures inhibited the flow of water used by a nearby community.

American Period

The Spanish and Mexican period ended with the sudden influx of Americans and other people from around the world as a result of the discovery of gold. This important event changed California and its use of water resources forever. The changes in water use reflected the increasing population, the introduction of American technology, and a different philosophy. The American philosophy was that nature was an abundant and infinite resource to be exploited for individual gain. Hispanic culture stressed the need, in arid environments, to have water rights belong to the community, but Americans believed in individual rights and minimal governmental interference. This individualist, entrepreneurial nature was strongest in people who chose to come west in the first place, choosing to leave behind the watchful eyes of their families, communities, churches and government. The Indian and Mexican ways of life could not hold up to the onslaught of people coming into the state. The population soared, from 10,000 non-Native Americans in 1846, to 1,000,000 by 1849, and 1.5 million by 1900. Towns, agriculture, and commerce grew, all of which led to huge and irreparable changes to the natural landscapes and waterscapes in California.

Americans coming to California for the gold rush needed water for their mining. The presence of few rivers and minimal precipitation was a big problem as many of the gold deposits were miles away from the nearest water source. Thus, miners diverted water from rivers and streams through flumes, ditches, and pipes to the gold fields. One especially horrific use of water during this time period was for *hydraulic mining* (Figure 8.12). By the 1850s, the easy way to find surface gold was gone, and methods such as hydraulic mining were needed to get to the gold underneath the ground. This technology required massive amounts of water delivered to high-pressure hoses. The hoses were then used to blast away entire hillsides to expose the gold deposits underneath. The resulting erosion caused environmental disaster and delivered millions of cubic feet of dirt and rock to the rivers below. Water companies formed to deliver the needed water became rich, often inheriting mining claims when the miners' water bills could not be paid. Downstream the farms, towns, and rivers were inundated by debris from the water

Figure 8.12 Hydraulic mining.
(Source: Library of Congress)

runoff and silt of the hydraulic mining. An estimated 42 billion cubic meters (1.5 billion cubic feet) of soil was moved during the gold rush. Farmers and landowners downstream took the miners and water companies to court and finally, in 1883, hydraulic mining was outlawed.

The Evolution of Water Rights in California

Water rights law in California differs from that in the rest of the nation because of seasonal, geographic, and quantitative differences in precipitation that resulted in a combination of riparian and appropriative rights. Laws dealing with water in California have reflected changing economic, environmental, technological, and social conditions. Because water has changed much of the face of California, making habitation, farming, and recreation possible, legal control of this resource represents a significant area of legislation and litigation. It is, therefore, important to understand the fundamental legal doctrines that underlie much of the debate. It has been suggested that two realities govern the "law" of water in the arid West: (1) the basic availability of water and (2) the related issue of how the available water is distributed geographically throughout the region. Much of early American water law was based on the premise that water was a ubiquitous resource (available everywhere)—a concept that provided a "poor fit" for California and other western states.

Riparian Rights

Riparian (meaning located by the banks of a river, stream, or other body of water) *water rights* come with property ownership adjacent to a source of water and is the most common water doctrine in the United States. With statehood, California adopted this English common law familiar to the Eastern Seaboard. These water rights are considered to be a part of the ownership interest of the land and a form of real property right. Such rights do not require permits, licenses, or government approval. Riparian rights remain with the property when it changes hands, although parcels severed from the adjacent water source generally lose their right to the water. The extent of this right to use water has been debated for years, with two approaches emerging from the discussion. The "reasonable use" theory limits water use only to the extent that such use does not negatively affect other users. Thus, the premise is that all riparian users have equal rights to the water supply. The alternative theory, which might be called that of "domestic use," is that the riparian owner may take water for domestic purposes only (family, livestock, and gardening) and that other uses of water may be considered infringements of other riparian owners' rights. The doctrine of riparian rights has both advantages and disadvantages. Generally, diversion of water is not permitted under the doctrine, and this limits the use of water by nonriparian landholders. This is a critical problem in the arid western states, where there are few free-flowing streams. Because of the limitations of riparian doctrine, therefore, another major water allocation system was developed.

Appropriative Rights

One year after California adopted riparian water rights laws, its legislature recognized the appropriative rights system as having the force of law. In the absence of abundant free-running water, the doctrine of riparian rights seems to be neither useful nor appropriate. In the dry western states, consequently, the doctrine met several challenges. Because of extensive federal ownership of vast tracts of land, early residents could not claim water rights from land they did not legally possess. A practical solution evolved that created superior rights for those who first put the water to beneficial uses. This came to be known as the doctrine of prior appropriation and permitted the first person to appropriate water from a stream to continue using that amount of water, against all subsequent users, as long as that original use was continuous. Appropriation rights, however, can be lost by failure to continue using the water acquired in this fashion.

In essence, this doctrine established a first come, first served legal process for water in the arid west. Miners and settlers were able to regulate legal control of water in accordance with this rule and establish some order and predictability in the process. It should be noted that the doctrine did not overturn riparian rules or take precedence over the rights of the federal government, which still owned the land, but served only to settle disputes between competing individuals, all of whom were claiming rights to water on land they did not own.

Later federal legislation attempted to clear up the question of riparian versus appropriation rights. Both the state and the federal government desired to respect the developing theory of appropriation while still preserving some measure of respect for the riparian doctrine insofar as it might apply to any particular future case. In its purest application, however, prior appropriation separated water rights entirely from land rights and put the control of water in the hands of the state as a public trust. This often meant, in practice, that each western state could establish its own set of rules governing appropriation of water, thus undercutting riparian doctrine almost totally.

Most western states eventually developed a permit system whereby water users would apply to the state for certain uses and, upon compliance with permit requirements, would receive permission from the state to appropriate water for beneficial purposes. This placed significant power in the hands of the state agency, as well as heavy responsibility for monitoring competing uses and determining overall best interests of the public at large. Competing water uses to be considered included personal, household, livestock, agricultural, recreation, fishing, and many others. Finally, the doctrine created a new approach to water use that tried to establish preferential uses of water and protect as many landowners as possible. The new system was better suited to the western states but needed further refinement to meet the specific needs of each state.

California Doctrine and California Water Code

Water distribution in California created a unique problem that called for a blending of ideas from both riparian and appropriation theories. The modified doctrine needed to address the geographically and temporally imbalanced distribution of water. Given this imbalance of location and occurrence, the state attempted from the outset to provide an orderly and consolidated water plan that would equally serve the needs of all its residents. To accomplish the goal of fair water use, California has created a modified doctrine using aspects of both of the recognized approaches. Through various court cases, and eventually by constitutional enactment, a series of rules on water use has evolved.

1. The state declared that it would recognize and honor the water rights acquired under the prior appropriation process. This protected those water users who had established time priority.
2. The state acknowledged that riparian rights did exist. The California courts noted that the original riparian rights had belonged to the original owner of the land, the US government, which took title from Mexico, and that upon transfer to private parties these riparian rights were also transferred.
3. Riparian rights acquired by a recent transfer would not be absolute. They would be subject to those prior appropriation rights that existed at the time of transfer and would be further limited by a proviso that the riparian rights would extend only to such amounts of water as were reasonably required for beneficial use of the land. This could prevent a riparian owner from selling water above the amount needed for the riparian land alone.

4. The water resources of the state were declared to be so vital as to require such water to be used for the general welfare and for the public interest. This meant that water use would be subject to specific state control and regulation (although riparian and appropriation owners' rights were to be respected).

5. All water above the amount needed for limited riparian and appropriation uses would be considered "excess water" subject to state control. This control would be exercised by a state water board.

6. All uses of water that came about as a result of appropriation would be deemed to be "public uses" and thus subject to state control. In practice, this meant that the bulk of water available in the state was subject to governmental control. Once this fact was established by definition, it remained only for the state to determine priorities of use and provide a permit system for use and distribution.

The philosophical position of the state is reflected in the *California Water Code*. This places the highest priority on domestic use, followed by irrigation. Furthermore, applications by a municipality for use of water by its residents are given priority over most other competing uses. Beyond the priorities thus outlined, it is left to the judgment of the water board to determine allocations in such a way as to serve the broad public interest. All uses are handled by a permit application process, and the board acts on a case-by-case basis. Finally, the board is enjoined to give constant attention to state water plans for coordinated water use (see the discussions on the SWP and the CVP earlier in this chapter).

The efforts of the state of California to develop a comprehensive approach to the use of water were not quickly or easily accomplished. The California Doctrine reflected an uneasy compromise of various rights and arrived at a position that would benefit the most people to the least detriment of any individual. Obviously, some individuals and localities were necessarily subordinated to the general welfare, a circumstance that created bitter and longstanding controversies, the conflict between Los Angeles and Owens Valley being a case in point.

Groundwater Law

California groundwater law is also complex. Groundwater exists in one of two ways: (1) percolating groundwater or (2) subterranean stream. If the flow of groundwater is confined to a known and defined subsurface channel, it is a subterranean stream. Groundwater not flowing as a subterranean stream is classified as percolating groundwater. In 1903, the California Supreme Court established the *doctrine of correlative water rights*. This doctrine states that overlying users of percolating groundwater and riparian users of subterranean streams must share the available supply. If a shortage exists, each overlying or riparian rights holder must cut their use to some degree. Overlying and riparian users have priority over appropriators who may take only surplus water. Groundwater also can be appropriated and diverted outside of groundwater basins by cities, water districts, and other users whose lands do not overlie a groundwater basin. In 1914, California created a water right permit process governing the appropriation of surface water and subterranean streams. Appropriations of subterranean streams require a permit from the State Water Resources Control Board.

Development of California's Water Distribution System

Between 1860 and 1930, local groups of farmers and communities across the state began banding together to undertake small water projects such as draining wetlands; building levees, small dams, and irrigation systems; and digging wells for groundwater. Cooperatives and development companies formed to finance and construct larger local water projects to support growing towns and agricultural areas in places like the San Joaquin Valley and southern California. But people worried about giving control of such a public resource, like water, to private concerns. In a move toward increased public control, the first irrigation district, the Turlock Irrigation District, was formed under the Wright Irrigation District Act of 1887. The act evolved into the California Irrigation District Act of 1917 and paved the way for other types of water development and delivery districts, such as county water districts and special services districts. California's two major urban areas, Los Angeles and San Francisco Bay, recognized the limitations of their local water supplies and were the first to look toward remote water sources. The federal government also played a major role in the development of the West's water resources. As early as 1875, the US Army Corps of Engineers began work on the Sacramento and Feather Rivers to improve navigation. In 1920, the US Geological Survey proposed a comprehensive, statewide plan for conveyance and storage of California's water supplies. This plan served as the framework for an eventual State Water Plan, which later formed the basis for the federal Central Valley Project (Figure 8.13).

The present-day water distribution system has been put into place over the last 150 years and includes water projects at the local, regional, state, and federal levels. These water projects each store and deliver water to consumers, sometimes moving that water several hundred miles from its source. The major systems of aqueducts and supporting infrastructure that deliver water throughout the state include the (1) State Water Project, (2) Central Valley Project, (3) Colorado River Aqueduct, (4) Los Angeles Aqueduct, and (5) Tuolumne River and Hetch Hetchy delivery systems. Each of these is operated by a federal, state, or regional agency that wholesales water supplies to local water districts for delivery to consumers (Figure 8.14, Map 9).

The State Water Project

California's Department of Water Resources operates the enormous California State Water Project, commonly known as SWP, which averages an annual delivery of 2.3 maf and delivers water up to 965 kilometers (600 miles). This conveyance system begins in northern California at the Lake Davis, Frenchmen Lake, and Antelope Lake reservoirs on the upper Feather River. Water is moved over half the length of the state, much of it along the California aqueduct through the Central Valley to its southern terminus at Lake Perris, located in Riverside County.

Urban and industrial consumers use 70 percent of the water supply with the remaining 30 percent going to agricultural areas in the San Joaquin Valley. This project is the largest state-built multiuse water storage and conveyance system in the United States and provides water supplies for 25 million Californians and 305,500 hectares (755,000 acres) of irrigated farmland. In addition, there are hydroelectric power plants associated with it, and much of the electricity generated is used for pumping stations to lift the water over the mountain barriers along its route. Initial planning for the project started in the mid-1940s, and construction began in 1957. Various phases were built in the early 1970s, and additional work on new facilities continues today.

Figure 8.13 Irrigation of new crops, southern San Joaquin Valley.
(Image © Richard Thornton, 2009. Used under license from Shutterstock, Inc.)

Figure 8.14 Aerial view of aqueduct carrying water.
(Image © iofoto, 2009. Under license from Shutterstock, Inc.)

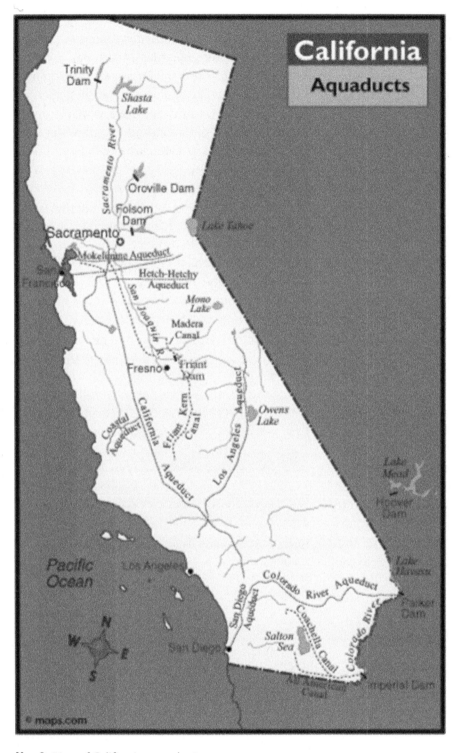

Map 9 Map of California aqueducts.

Central Valley Project

The federally funded CVP, one of the largest water systems in the world, began in 1935 and is operated by the US Bureau of Reclamation. It extends from the Cascade Range in the north to the semiarid but fertile agricultural areas in the southern Central Valley. The original goal was to control seasonal flooding in the Central Valley and shift water supplies south to irrigate 1.2 million hectares (3 million acres) of drier farmland, though it also supplies domestic and industrial water for California's Central Valley communities and the San Francisco Bay Area, and is the primary source of water for much of California's Central Valley wetlands. The Central Valley farmland that the CVP irrigates is some of the most productive in the state and it has been estimated that the value of the crops and related service industries in this region has returned Congress's original $3 billion investment 100 times over.

The CVP consists of 20 dams and reservoirs, 11 power plants, and 800 kilometers (500 miles) of major canals and related facilities. It stores 7 maf (~17 percent of the state's developed water) and delivers water to 139 landowners and 8 water districts, irrigating approximately one-third of the state's agricultural land. Although the SWP gets most of its water from the Feather River watershed, the CVP diverts water from five major rivers and dams: Trinity River and Trinity Dam, Sacramento River and Shasta Dam, American River and Folsom Dam, Stanislaus River and New Melones Dam, and the San Joaquin River and Friant Dam. It was hoped that diverting water supplies south to the San Joaquin

Valley would help decrease the over pumping of groundwater in that area. Instead, more acreage was put into farmland production, and groundwater pumping actually increased. The CVP supplies roughly one million households with their water needs each year, along with enough electricity for two million people annually. Pursuant to the Central Valley Project Improvement Act, over 1.2 maf of water is dedicated each year to wildlife refuges and wetlands to support the habitats of fish and other wildlife.

Colorado River Aqueduct and Canals

The Colorado River Aqueduct (CRA) was built between 1933 and 1941 by the metropolitan water district (MWD) of southern California to ensure a steady supply of drinking water to Los Angeles, though it now supplies water to southern California communities from Ventura County to San Diego County. Created by the famous Hoover Dam, Lake Mead is the primary reservoir for storing water along the lower Colorado River basin. The Colorado River aqueduct begins downstream at Parker Dam and can carry 0.5 maf per year to southern California—3.8 billion liters (1 billion gallons) per day! The idea of building an aqueduct to transport Colorado River water to southern California was conceived by William Mulholland and was the largest public works project undertaken in southern California during the Great Depression. Southern California shares the water resources of the Colorado River with six other states and Mexico. All available water from the Colorado River has been allocated to the various water rights holders; in fact, it is overallocated. The reason is that the original allocations were based on overestimates of the annual runoff of the Colorado River. California's current allotment from the CRA is 4.4 maf per year (it supplies up to 12 percent annually of southern California's water). Of that amount, 75 percent is used to irrigate 364,000 hectares (900,000 acres) of farmland in the southeast corner of the state, primarily Imperial County. The remaining portion travels to urban areas, 390 kilometers (242 miles) across the desert, in the Colorado River Aqueduct.

The Metropolitan Water District is the agency that handles this water and distributes it to fourteen cities, twelve municipal water districts, and one county water authority in the southern California region. For years, California has been able to take more than its allocated 4.4 maf because other states were not yet ready to divert their full share. For example, in the year 2000, the CRA delivered 5.3 maf to southern California. By 2002, however, Arizona and Nevada (with a growing Las Vegas) began to put pressure on California to cut back on the extra 8,000,000 acre-feet of water. Southern California water planners, unable to pull more water from northern California, had to work on a reallocation of water within the southern California region. This led to a struggle between urban water users living in coastal southern California counties and agricultural users in inland Imperial County. The result was that the Imperial and Coachella Valley agricultural areas had to give up water they would have used for agriculture in order to supply the water needs of urban southern California.

Los Angeles Aqueduct

The Los Angeles Aqueduct was the first major long-range water delivery project ever built in California. A near drought in 1904, coupled with the inability of the Los Angeles River and local wells to meet the needs of a growing population, had made Los Angeles keenly aware of the need for long-term planning of reliable water sources. William Mulholland, superintendent of the Los Angeles water district, and Fred Eaton, a former mayor and engineer for the city, conceived of and planned the project. Mulholland, who had emigrated from Ireland and taught himself engineering at night, began working for the Los Angeles City Water Company as a ditch tender and had risen quickly through the ranks to become the superintendent of the now public city of Los Angeles water system. Their idea was to build an aqueduct from the Owens Valley, east of the Sierra Nevada, to the city of Los Angeles. The water source for the Owens Valley and Owens River is snowmelt from the Sierra Nevada, which feeds the mountain streams that in turn flow into the Owens River in the valley itself. In addition, this area also had large groundwater resources (Figure 8.10). To undertake this $25 million project, the city of Los Angeles had to first purchase 120,000 hectares (300,000 acres), 98 percent, of the Owens Valley to secure the water rights. However, the Owens Valley was a very active region of farming, ranching, and small towns, and the residents were not going to be willing to sell off their water rights to Los Angeles. Fred Eaton, who also worked as an engineer for the Federal Bureau of Reclamation, let the Owens Valley residents think that the interest in their water was a result of the federal government wanting to undertake a large irrigation project in the area. The actual purchase of land and water rights was done in a very underhanded way, effectively bilking the residents of the Owens Valley out of their water rights, and is an unfortunate chapter in California history. The aqueduct was completed in 1913, over the protests of the Owens Valley residents, and to this day transports water 400 kilometers (250 miles) to the city of Los Angeles. Currently, the Los Angeles Department of Water and Power (LADWP) transports 400,000 acre-feet from the Owens Valley to Los Angeles each year, serving 3.2 million people. A more complete history of the building of the Los Angeles Aqueduct can be found in Mark Reisner's book *Cadillac Desert*. The movie *Chinatown* presents a fictional version of this rather dark story of the farmers of the Owens Valley fighting for their very existence against the mighty city to the south (Figure 8.15a-c).

Mono Lake Extension of the Los Angeles Aqueduct

By the 1930s, demand for water in Los Angeles was once again outpacing available supplies. The LADWP started buying more water rights, this time in the north end of the Owens Valley, near Mono Lake. An extension from the north end of the Owens Valley to the original Los Angeles Aqueduct was completed in 1941, and the streams that fed Mono Lake were diverted into the new aqueduct for transport to the ever-growing city of Los Angeles. Unfortunately, this had dire

effects on the Mono Lake ecosystem. Mono Lake is an inland body of salt water with no outlet, and it serves as a critical nesting spot for migratory birds, including almost the entire population of California gulls (*Larus Californicus*). Because the creeks were being diverted, the water levels in the lake began to fall and the decreased inputs of freshwater caused the water to become even more saline and alkaline than it already was (it is two to three times saltier than the ocean, with a pH of ~10, similar to glass cleaner). This change in the lake's chemistry disrupted the food chain and sharply decreased the population of brine shrimp that the migrating birds depended on for their food source. Falling water levels also created land bridges to the islands where the birds nested, allowing predators to feed on the eggs. Even the air quality of the region was being affected because of the caustic dust from the exposed lake bed (Figure 8.16).

In the mid-1970s, after several decades of falling lake levels, a group of biologists began a study of Mono Lake to determine the effects of this significant decrease in freshwater inflow. In 1978 the Mono Lake Committee was formed to advocate for the environmental health of Mono Lake and, along with the National Audubon Society, sued the LADWP, arguing that the stream diversions violated the public trust doctrine (which states that navigable bodies of water must be managed for the benefit of all people). In 1983, the case went to the California Supreme Court, which ruled in favor of the committee. Finally, in 1994, after exhausting all its legal appeals, the LADWP was required to begin letting enough stream water flow back into Mono Lake to raise the lake level to within 7.5 meters (25 feet) of the 1941 level (1,948 meters [6,392 feet] above sea level). As of 2007, Mono Lake had reached its highest water level since the 1970s (1,946 meters [6,385 feet] above sea level), with 2.1 meters (7 feet) left to go to reach the court ordered level. Hydrologists say the lake is on track to rise to the height prescribed by the Department of Water Resources by 2014. The added water is making Mono Lake less salty, which is increasing the population of brine shrimp. Other signs that the ecosystem is

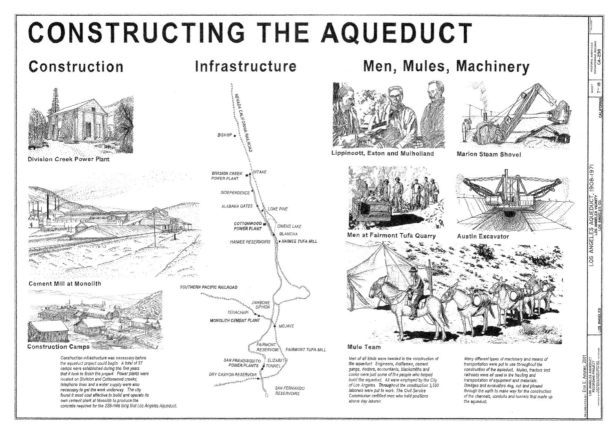

Figure 8.15a The aqueduct construction information.
(Source: Library of Congress)

Figure 8.15b The "Cascades," terminus of the aqueduct, located in present day Sylmar at the southern end of the Newhall Pass.
(Source: Library of Congress)

Figure 8.15c November 5, 1913, opening day of the aqueduct bringing water 250 miles from the Owens River.
(Source: Library of Congress)

Figure 8.16 Present-day Mono Lake.
(Photo: Lin Wu)

beginning to repair itself is the presence of the willow flycatcher *(Empidonax traillii)*, an endangered species not seen at Mono Lake for generations. Today, the Mono Lake Committee continues in its mission as a watchdog agency for the restoration of the habitats in and around Mono Lake.

Historic Lake Levels

1941 - (before diversions) lake level 1,956 meters (6,417 feet) above sea level
1982 - (lowest recorded level) lake level 1,942 meters (6,372 feet) above sea level
2007 - (highest water level since the 1970s) 1,946 meters (6,385 feet) above sea level
2013 - (present day) lake level 1,944 meters (6,381 feet) above sea level
2014 - (stabilization level) 1,948 meters (6,392 feet) above sea level

Tuolumne River and Hetch Hetchy Delivery System

The Tuolumne River and Hetch Hetchy water delivery system begins inside Yosemite National Park and transports Tuolumne River water through the Hetch Hetchy Aqueduct to supply 2.3 million people in San Francisco and surrounding areas. The O'Shaughnessy Dam, completed in 1923, dams the Tuolumne River and creates the Hetch Hetchy Reservoir. This system is owned and operated by the San Francisco Public Utilities Commission and also generates electricity, a significant revenue source for San Francisco.

The history of Hetch Hetchy is a sad one. John Muir was opposed to its being built, but Congress authorized the city of San Francisco, in the Raker Act of 1913, to construct a dam and reservoir on the Tuolumne River in Hetch Hetchy Valley, inside Yosemite National Park. The first phase of the project was completed in 1923, with electrical power being generated from the dam. Ten years later the pipelines were completed, and San Francisco began receiving water supplies. In the 1980s, President Reagan's secretary of the interior proposed the restoration of Hetch Hetchy Valley, and the Bureau of Reclamation completed a preliminary study of this plan for the National Park Service. Since that time, studies on the restoration on Hetch Hetchy have been undertaken. In 2006, the Department of Water Resources released its own study, which stated that the restoration of Hetch Hetchy is feasible and would cost $3–10 billion. To date, no concrete plans are in place to begin the restoration of this beautiful, glacier-carved valley, though the nonprofit organization, *Restore Hetch Hetchy*, continues its work on the goal of restoring this beautiful valley to its natural state (Figure 8.17).

Impacts and Issues of Water Use

As we have discussed, water use in California is divided into three general categories: environmental, agricultural, and urban. Each of these has its own particular water needs and uses, and each suffers from and/or contributes to the problems of water shortages and water degradation. The natural environment, which once received all the available water to sustain its diverse habitats, is now left with only 46 percent of California's water. Although it is admirable that we as humans have the engineering abilities to build giant water systems that store and move water to where we need it, there are serious environmental consequences of taking that much water away from the natural environment and "giving" it to agricultural and urban users.

A staggering 95 percent of California's wetlands are gone forever because of both the lack of water and the transformation of those areas for agricultural uses. Gone, too, are 90 percent of California's riparian woodlands, the plant communities that grow near streams, rivers, wetlands and lakes. Many of the state's rivers and streams simply have too little water to support the riparian vegetation or the fish and other organisms that make up the aquatic ecosystems. Numerous species of fish, such as salmon and trout, are born in freshwater, migrate to the ocean, and return to freshwater rivers and streams to spawn. The many dams built on rivers throughout the state kill fish inside spinning turbines, and keep them from swimming upstream to spawn. In 1883, more than 700,000 salmon were caught by commercial fishermen. Since that time so many dams have been built and so much stream habitat destroyed that only 10 percent

Figure 8.17a Present-day Hetch Hetchy.
(Source: Image © Bryan Brazil, 2009. Used under license from Shutterstock, Inc.)

of the necessary stream habitat remains accessible to salmon and trout today. Sadly, the great amount of water taken from California's natural environment each year and the resulting habitat loss have made California the state with the most endangered and threatened species.

Agricultural uses receive 43 percent of California's available water. According to the California Farm Bureau, the food that each of us consumes represents a staggering 17,100 liters (4,500 gallons) of water per day! About 3.6 million hectares (9 million acres) of farmland use irrigated water, much of it going to alfalfa, cotton, rice, grapes, and pasture. These crops use different amounts of water; cotton, for example, uses significantly more than grapes. Overirrigation causes runoff that carries pesticides, fertilizers, and salts, thereby polluting both surface and groundwater. Agriculture has also historically been a driving force behind habitat loss, draining the state's wetlands, for example. Despite the problems of agricultural water pollution, farmers were exempt until 2002 from the federal Clean Water Act of 1977, which put controls on pesticides and herbicides entering the nation's groundwater and surface water supplies.

Salt build up in soils is an ongoing problem many California farmers faces as much of California's soil is derived in part from marine sediments, which are naturally high in salt. Salts are present in the water used for irrigation (Figure 8.18), but crops take up the water through their roots, leaving the salt minerals behind to become concentrated in the soil. Over time, agricultural soils become too high in salts

Figure 8.17b Hetch Hetchy Valley circa 1911.
(Source: Library of Congress)

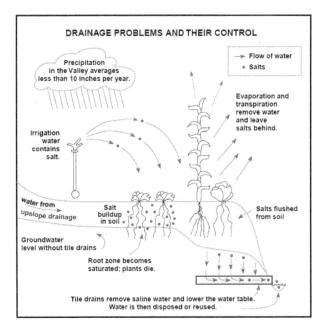

Figure 8.18 Issues with salts in Central Valley agriculture. (Source: California Department of Water Resources)

to grow crops, so farmers need to flush the salts from the soil by applying even more water than the plants need. But where do the salts go when flushed out of the fields? They either percolate down into the groundwater, or they are flushed through pipes and drainage ditches into a body of water, such as a river, stream, lake, or wetland. Wherever the salts go, they are a major source of environmental concern. The Salton Sea, located in Imperial County, is an example of where the salts flushed from agricultural fields can end up. The 200,000 irrigated hectares (500,000 acres) in this area contribute roughly four million tons of dissolved salts and tens of thousands of tons of fertilizers each year into the Salton Sea, creating a huge evaporation pond that is 25 percent saltier than the ocean. Although the Salton Sea suffers from serious water degradation, it is still a productive fishery and an important stopover for roughly 250 different species of migrating birds. Salton Sea restoration efforts to stabilize salinity and create healthy fish and wildlife habitats are ongoing.

Urban users live in communities, big or small, and use water in homes, businesses, schools, and factories. There are 38 million such users who account for 11 percent of the statewide water use. Depending on what part of the state the urban users live in, the water they use might come from groundwater, a reservoir, an aqueduct, or most likely a combination of those sources. Those users living in California can save water and ease the burden on the state's water supplies in many ways (Table 8.3), including turning off the tap while brushing teeth, installing low flow toilets, waiting for a full load before running the dishwasher or washing

Conservation Measure	Estimated Water Savings
Turn off the tap while brushing teeth	2 gallons/minute
Shorten showers by 1–2 minutes	2.5 gallons/minute
Turn off tap while washing dishes	2 gallons/minute
Fix leaky faucets	20 gallons/day
Fix leaky toilets	30–50 gallons/day
Run dishwasher only when full	2–5 gallons/load
Run clothes washer only when full	15–20 gallons/load
Don't use toilet as a trash can	1.6 gallons/flush
Replace old, high volume flushing toilets	30–50 gallons/day
Earn rebate for new high-efficiency clothes washer	20–30 gallons/load
Water yard before 8 AM to reduce evaporation	Up to 25 gallons/day
Install smart sprinkler controls	Up to 40 gallons/day
Sweep driveway and sidewalks, don't hose	Up to 150 gallons/time
Mulch around plants to reduce evaporation	Hundreds gallons/year

Table 8.3 Ways to be water wise.

machine, landscaping with drought-tolerant species, using mulch around plants, and limiting outdoor sprinkler usage.

Summary: Sustaining California's Water Supply into the Future

Californians have faced, and will continue to face, ongoing challenges from the state's natural hydrology—namely, the spatial and temporal issue of supply (precipitation) in the northern third of the state and demand (population) in the southern end. This dichotomy has led to the development of an intricate system of reservoirs, dams, and aqueducts built and maintained by a variety of agencies on the federal, state, and regional levels. Couple this already precarious balance with issues of drought, growing population, and environmental degradation, and the need for serious long-term planning to ensure reliable water supplies becomes obvious. An additional challenge that Californians will face is that of climate change. The design and operation of California's current water systems were based on a hydrologic record that is not reflective of the new precipitation patterns that climatic change will bring. Scientists predict that climate changes will significantly alter California's precipitation, both the patterns and the amounts, from that shown by recorded hydrologic data. For example, snowpack has already been decreasing in the Sierra Nevada, which will result in a seriously diminished natural water storage capacity. Precipitation is also forecasted to become more variable, making it more difficult to rely on the flood and water supply management systems currently in place. The citizens of California need to continue their individual efforts to learn to use water efficiently in their homes, schools, businesses, and farms. Municipalities need to increase their use of recycled water, and water facilities need to continue to improve their operation and efficiency.

Selected Bibliography

Carle, D. (2000). *Water and the California Dream.* Berkeley and Los Angeles: University of California Press.

Carle, D. (2004). *Introduction to Water in California* (California Natural History Guides, 76). Berkeley and Los Angeles: University of California Press.

Department of Water Resources. (2000). *Preparing for California's Next Drought: Changes Since 1987–92.* Sacramento, CA: Author.

Department of Water Resources. (2003). *California's Ground Water*, Bulletin 118. Sacramento, CA: Author.

Department of Water Resources. (2005). *California Water Plan 2005*, Bulletin 160–05. Sacramento, CA: Author.

Erie, S. P. (2006). *Beyond Chinatown: The Metropolitan Water District, Growth, and the Environment in Southern California.* Stanford, CA: Stanford University Press.

Hanak, E., Lund, J., Dinar, A., Gray, B., Howitt, R., Mount, J., Moyle, P., and Thompson, B. (2010). Myths of California water—implications and reality. *West-Northwest Journal of Environmental Law and Policy,* 16(1).

Hanak, E., Lund, J., Dinar, A., Gray, B., Howitt, R., Mount, J., Moyle, P., and Thompson, B. (2011). *Managing California's Water: From Conflict to Reconciliation.* San Francisco, CA: Public Policy Institute of California.

Hundley, N. (2001). *The Great Thirst: Californians and Water: A History.* Berkeley and Los Angeles: University of California Press.

Reisner, M. (1987). *Cadillac Desert: The American West and Its Disappearing Water.* New York: Penguin Books.

US Geological Survey. (2013). Diagram of an Aquifer. Retrieved [June, 2013] from http://water.usgs.gov/ogw/aquiferbasics/

Economic Geography

California is still the land of dreams…. It's the place where Apple, Intel, Hewlett-Packard, Oracle, QUALCOMM, Twitter, Facebook and countless other creative companies all began. It's home to more Nobel Laureates and venture capital investment than any other state.

–Jerry Brown

Introduction

As introduced at beginning of the book, California, if it were an independent nation, would rank between the sixth and tenth (depending on the source) most powerful economy in the world (ahead of Spain, Mexico, and South Korea). New York, the next largest state economy, falls short of California by more than a third. What are the factors that contribute to the economic power of the state? In this chapter, we are not attempting to give a complete answer to this question; rather, we follow the question to explore various aspects of the state's economy from geographic perspectives. We will start with an overview of the basic components, move on to explore one of the main foundations (the population), and then look into a few economic sectors in the state—the entertainment industry, the tourism industry, the agricultural industry, the housing market, and the high-tech industry.

Components of the California Economy

Several factors help explain California's unique economic status, with geography playing a significant role. *Location* has boosted various opportunities, such as international trade, fishing, and ocean-oriented activities. *Climate* has encouraged tourism, migration, agriculture, and other activities. *Natural resources* have fostered a tradition of primary economic activities, such as logging, mining, fishing, farming, and ranching. In short, the influence of California's unique geography cannot be ignored.

An enduring strength of California's economy has been its resilience and diversity over time. In terms of employment, the state has demonstrated an ability to capitalize on technological and industrial trends. Whenever a cyclical downturn or change occurs in international politics or national priorities, California has changed its employment and fiscal directions.

Thus, when defense and aerospace industries declined, high tech and information technology surged to the fore. When high tech suffered setbacks, international trade and small entrepreneurial activities picked up. In short, unlike some other regions with limited economic activities, California has been able to reap the harvests of its social-cultural-economic diversity. Currently, approximately three-quarters of the state's population are employed in service-related jobs. This employment profile is only part of the picture; in terms of dollar contributions to the total state economy, certain activities stand out as major factors in California's economy.

One of the most remarkable components of the state's economy is its agricultural sector, with a wide variety of products that include grapes (wine), dairy, citrus, flowers, and exotic crops. Manufacturing, though not the leading sector, still accounts for a substantial output, particularly of electronics, software, and related products. California has also long been known as a major media and entertainment venue, with Hollywood assuming an almost mythical status. Tourism contributes important income to the state with various theme parks (e.g., Knott's Berry Farm, SeaWorld, Disneyland, Magic Mountain), natural attractions (e.g., Yosemite National Park, Devil's Postpile National Monument, Sequoia National Park, Death Valley National Park, miles of beaches), and major urban areas such as San Francisco, Los Angeles, and San Diego.

Among the primary economic activities, mining, fishing, and timber harvesting have been traditional mainstays. California remains one of the country's most productive mining states, producing oil and natural gas, boron (the only source in the United States), gold, sand and gravel, gypsum, and potash. Fishing has also had a long history in California. With its productive ocean waters the state has engaged in harvesting fish of all types including swordfish, tuna, salmon, shark, and halibut. In recent years, however, overfishing of certain species has caused a marked decline in the industry, and environmental concerns are playing a greater part in official decision making about how much regulation of this industry is needed.

With the new century well under way, service industries in California have assumed the greatest position in terms of employment and income. This all-encompassing category includes health care, amusement, recreation, food industries, financial and professional services, information technology, retail, trade, transportation, warehousing, business services, education, broadcasting, motion picture production, and Internet businesses. A detailed description of these activities is beyond the scope of this chapter, but certain key areas unique to the state are worthy of further discussion.

Population

The scale and composition of a population are among the most important control factors of the economy of a state. Population in California has continuously grown since the first population boom in 1850. The chart in Figure 9.1 compares the top three most populated states in the country from 1900 to 2010, and the growth rate of California has outpaced the other two states in most of the decades. Since surpass New York in 1960s to become the most populated state, California's population nearly doubled New York in less than half of a century. The growth though

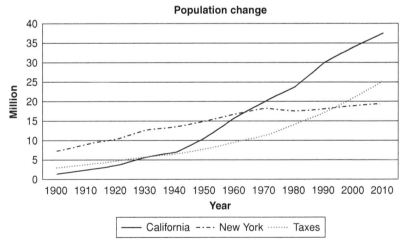

Figure 9.1 Population growth trend from 1900–2010.
(Source: Lin Wu, Data: US Census Bureau)

has slowed down in the last a few decades. Table 9.1 shows that the percentage of population change in the decade dropped below 10 for the first time in 2010. It is projected that this trend will continue for the next a few decades primarily due to the decrease in immigration, following the national trend.

California is among the states with the highest foreign born population. Comparing Tables 9.2 and 9.3, over 27% of California's population are foreign born, whereas only 13% in the country are foreign born in 2010. These characteristics remain consistent in the last a few decades as illustrated by the tables. The percentage of people born in

California has steadily increase from the 1990 census of 46% to 54% in 2010 census; however, for the nation as a whole, this numbers stayed same with a slightly decrease from 62% to 59%.

Based on 2000 census, California for the first time became a state without a majority (over 50%) ethnic group. The diversity of the population continues in the 2010 census as illustrated by Table 9.4, with Hispanic and Latino population (38%) nearly tie with the non-Hispanic White (40%) population. California also has the largest Asian population in the country (percentagewise, second to Hawaii), with majority of them concentrated in the Bay Area and Southern California.

	1960	1970	1980	1990	2000	2010
Total	15,717,204	19,971,069	23,667,902	29,760,021	33,871,648	37,253,956
Change		4,253,865	3,696,833	6,092,119	4,111,627	3,382,308
Percent change (%)		27.07	18.51	25.74	13.82	9.99

Table 9.1 California Total Population by Decade (1960–2010).
(Source: 2000, 2010 US Census Data)

	1990		2000		2010	
	Number	**Percent (%)**	**Number**	**Percent (%)**	**Number**	**Percent (%)**
Total population	29,760,021	100.00	33,871,648	100.00	37,691,912	100
Total native population	23,301,196	78.30	25,007,393	73.83	27,496,855	72.95
Born in United States	22,905,935	76.97	24,633,720	73.00	27,052,171	71.77
Born in same state	13,797,065	46.36	17,019,097	50.25	20,466,513	54.30
Born in different state	9,108,870	30.61	7,614,623	22.48	6,585,658	17.47
Born outside United States	395,261	1.33	373,673	1.10	444,684	1.18
Total foreign-born population	6,458,825	21.70	8,864,255	26.17	10,195,057	27.05
Naturalized	2,017,610	6.78	3,473,266	10.25	4,771,515	12.66
Non-naturalized	4,441,215	14.92	5,390,989	15.92	5,423,542	14.39

Table 9.2 California Citizenship Status (1990–2010).
(Source: 2000, 2010 US Census Data.)

	1990		2000		2010	
	Number	**Percent (%)**	**Number**	**Percent (%)**	**Number**	**Percent (%)**
Total population	248,709,873	100	281,421,906	100	309,349,689	100
Total native population	228,942,557	92.05	250,314,017	88.95	269,393,835	87.08
Born in United States	225,695,826	90.75	246,786,466	87.69	265,166,046	85.72
Born in same state	153,684,685	61.79	168,729,388	59.96	181,747,947	58.75
Born in different state	72,011,141	28.95	78,057,078	27.74	83,418,099	26.97
Born outside United States	3,246,731	1.31	3,527,551	1.25	4,227,789	1.37
Total foreign-born population	19,767,316	7.95	31,107,889	11.05	39,955,854	12.92
Naturalized	7,996,998	3.22	12,542,626	4.46	17,476,082	5.65
Non-naturalized	11,770,318	4.73	18,565,263	6.6	22,479,772	7.27

Table 9.3 United States Citizenship Status (1990–2010). (Source: 2000, 2010 U.S. Census Data.)

The large population has played an important role in California's economy. At one end, the population contributes to consumption and other economic activities that lead to higher GDP (Gross Domestic Product, a measure of economic vitality); at the other end, to support a large population, it requires creating large number of jobs and developing infrastructure and other support system, which could be challenging at the economic down times.

Entertainment Industry

The entertainment industry in California has been a huge part of the state's economy, especially in the Los Angeles region, though economics alone is not the sole reason for the significance of entertainment. Culturally, *Hollywood* has become the defining term worldwide for California's film industry. Various sites in the Los Angeles region represent this symbolic focus, with images of the Hollywood sign, the

Hollywood Walk of Fame, Universal Studios, Grauman's Chinese Theater, the Hollywood Bowl, and the back lot of Paramount Pictures to name a few. It is no coincidence that the yearly Academy Award ceremonies are broadcast from Los Angeles. In short, from its beginnings to the present, the movie and film industry has been a significant presence in California.

As noted in Chapter 3, key factors led to the establishment of the movie industry in Southern California. Although the threat of eastern bankers and the convenience of the Mexican border are no longer underpinnings of the industry, the powerful attraction of climate and abundant natural settings continues to be a strong magnet for the entertainment industry.

In 2011, motion pictures and television productions in California were responsible for close to 0.2 million direct jobs and $17.0 billion in wages, including both production and distribution-related jobs. According to some estimates, between one-half and two-thirds of

		1990	2000	2010	Change 1990–2000	Change 2000–2010
Non-Hispanic	White	17,029,126	15,816,790	14,956,253	−1,212,336	−860,537
		57.22 (%)	46.7 (%)	40.15 (%)	−7.12(%)	−5.44 (%)
	Black or African American	2,092,446	2,181,926	2,163,804	89,480	−18,122
		7.03 (%)	6.44 (%)	5.81 (%)	4.28(%)	−0.83 (%)
	Native American	184,065	178,984	162,250	−5,081	−16,734
		0.62 (%)	0.53 (%)	0.44 (%)	−2.76(%)	−9.35 (%)
	Asian	2,710,353	3,648,860	4,775,070	1,042,243	1,126,210
		9.11 (%)	10.77 (%)	12.82 (%)	38.45(%)	30.86 (%)
	Pacific Islander		103,736	128,577		24,841
			0.31 (%)	0.35 (%)		23.95 (%)
	Other race	56,093	71,681	85,587	15,588	13,906
		0.19 (%)	0.21 (%)	0.23 (%)	27.79(%)	19.40 (%)
	Two or more races	NA	903,115	968,696	NA	65,581
			2.67 (%)	2.60 (%)	NA	7.26 (%)
Hispanic or Latino		7,687,938	10,966,556	14,013,719	3,278,618	3,047,163
		25.83 (%)	32.38 (%)	37.62 (%)	42.65(%)	27.79 (%)
All persons		29,760,021	33,871,648	37,253,956	4,111,627	3,382,308
		100.00 (%)	100.00 (%)	100.00 (%)	13.82(%)	9.99 (%)

Table 9.4 California Ethnic Categories and Change from 1990–2010.
(Source: 2000, 2010 US Census Data.)

all films made in the United States are still produced in California. In recent years, the major economic downturn, as well as growing costs and increased competition from other states and Canada, has somewhat eroded California's firm hold on industry productions. Nonetheless, no other state has such a dominant role in the field, and this preeminent position is not one the state wishes to lose. Ongoing discussions at both the local and state levels are aimed at finding better incentives and inducements for film companies to keep California in the forefront of the entertainment enterprise.

Certainly, new technologies have presented new challenges to record and movie producers in the form of file-sharing, counterfeiting, and piracy. Those same new technologies, however, also provide new ways for consumers to access various forms of entertainment. Downloading music and movies to portable devices merely expands the time and place options and arguably makes music and video more accessible and more in demand.

Similarly, in addition to films, other forms of media and entertainment have had a major presence in the state, with the recording industry, television, and theme parks adding to employment and revenue that is generated. The famous Capitol Records Building, erected in 1956, still houses recording offices and studios and makes a dramatic visual statement to passersby (Figure 9.2). The cultural and economic impact of entertainment is also found in the ancillary businesses that flourish as a result of the presence

Figure 9.2 The capitol records building.
(Source: Image © JustASC, 2009. Used under license from Shutterstock, Inc.)

of "Tinseltown" in Southern California: nightlife and tourism, movie theaters and video retailers, themed restaurants and stores, prop houses, lighting companies, and special effects and editing businesses.

Certainly, the economic challenges and difficulties in California are not unique to the state, nor are they absent in other locales. The oft-cited threat of businesses abandoning the state must be viewed in the context of what competing locations have to offer that is better than California. If California's economy is under stress, so too is the rest of the world's. Although tax incentives and preferred treatment of film industry activities may occur in other states or countries, California leaders are not oblivious to this reality and have attempted to investigate similar incentives. A sometimes unacknowledged advantage California enjoys is its tradition and association

as the creator of dreams ("the dream factory"). This tradition continues to cast a golden glow around the world and has shaped how millions of people in the global community view the entertainment industry known as Hollywood, California.

Tourism

California ranks number one in the nation in tourism, whether it is measured by the number of visitors or by the income generated. In 2012, the total direct travel-related spending in California reached $106.4 billion, which directly supported more than 917,000 jobs. California ranks as the top travel destination of all the states, with more than 300 million domestic visitors and 15 million international visitors each year. California's tourism economy is one and a half times the size of Florida's, which ranks second in the nation, more than double the size of New York's, and about five times the size of Hawaii's.

The great diversity of tourist attractions in California contributes to the economic success of the industry. Tourist destinations and attractions in the state include many striking natural scenes, framed in the national and state parks, prime beaches dashed along 1,760 kilometers (1,100 miles) of California coast, spectacular mountain resorts perched atop mountain slopes, world-renowned amusement parks clustered in Southern California, unique cities and towns located in various historical and cultural areas, and a whole spectrum of museums, historic sites, and other landmarks scattered across the state. Tourism dollars and jobs are not evenly distributed throughout the state, but instead are concentrated in coastal and southern California counties (Figure 9.3). The distribution of travel industry earnings also differs by sector, with the accommodation and food services categories topping the list (Table 9.5).

California has a large share of the nation's national parks and other types of national park units, including national monuments, historical parks, and recreation areas.

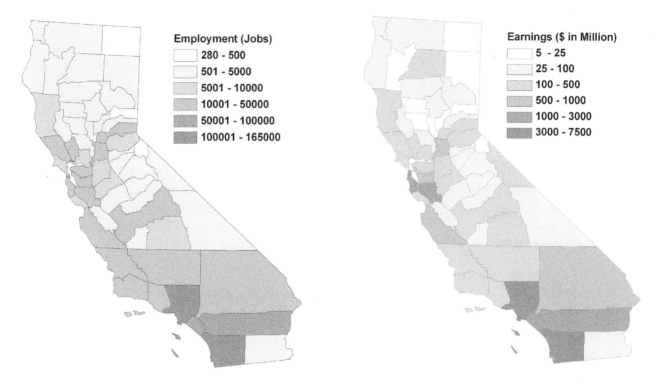

Data Source: Dean Runyan Associates (www.deanrunyan.com) 2007

Figure 9.3 County-based earnings and jobs created by the tourism industry.
(Source: Lin Wu, Data: Dean Runyan Associates)

Travel-Related Industries	Industry Earnings from Travel Spending ($ Billion)		
	1992	**2002**	**2012**
Accommodations & food service	6.9	11.3	17.1
Arts, entertainment, & recreation	3.9	6.0	8.1
Retail (including gasoline)	1.6	2.4	2.7
Auto rental & ground transportation	0.2	0.5	0.6
Air transportation (visitor only)	0.8	1.0	1.6
Other travel related	3.0	3.4	2.2
Total direct earnings	16.4	24.6	32.3

Table 9.5 Travel-Related Income.
(Source: Dean Runyan Associates.)

In addition, 278 state park units are scattered throughout the state (Figure 9.4). The unique and spectacular views and features in these parks draw state, national, and international visitors. The annual number of recreational visitors to California's national parks range from more than 300,000 in Channel Islands to more than 3 million in Yosemite (Table 9.6). Based on 2011–2012 data, the annual number of visits to all the state parks totaled 67.9 million people! From an economic perspective, the parks by themselves may not be the main income generator; however, by drawing hundreds and thousands of people to these areas, the parks become the major economic propellers for the local and regional areas in which they are located.

Another major driving force of the tourism industry is the state's amusement parks. Disneyland, the second most attended theme park in the world, attracts approximately 15 million visitors annually (Figure 9.5). Of the top 13 most attended theme parks in North America, all are located in either Florida or California (Table 9.7). Disneyland, Disney's California Adventure, Universal Studios, and SeaWorld California, all in Southern California, are among

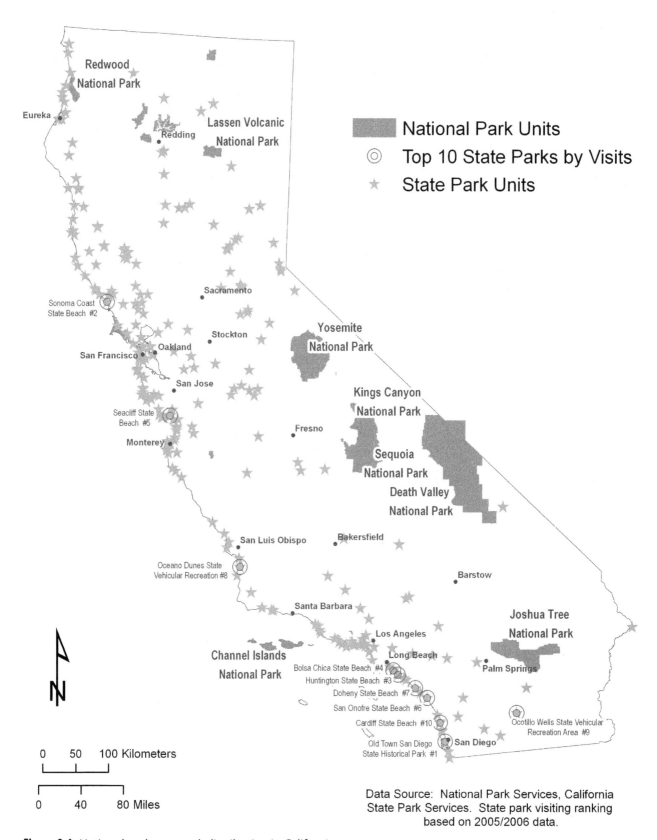

National Park Units

◎ Top 10 State Parks by Visits

★ State Park Units

Redwood National Park

Eureka

Lassen Volcanic National Park

Redding

Sacramento

Sonoma Coast State Beach #2

Stockton

Oakland

San Francisco

Yosemite National Park

San Jose

Seacliff State Beach #5

Monterey

Fresno

Kings Canyon National Park

Sequoia National Park

Death Valley National Park

San Luis Obispo

Bakersfield

Oceano Dunes State Vehicular Recreation #8

Barstow

Santa Barbara

Joshua Tree National Park

Channel Islands National Park

Los Angeles

Long Beach

Palm Springs

Bolsa Chica State Beach #4

Huntington State Beach #3

Doheny State Beach #7

San Onofre State Beach #6

Cardiff State Beach #10

Ocotillo Wells State Vehicular Recreation Area #9

Old Town San Diego State Historical Park #1

San Diego

0   50   100 Kilometers

0   40   80 Miles

Data Source: National Park Services, California State Park Services. State park visiting ranking based on 2005/2006 data.

Figure 9.4 National and state park distribution in California.

(Source: Lin Wu, Data: National Park Service; State Park Service)

National Parks	1999	2008	2012
San Francisco Maritime NHP	3,535,315	4,086,211	4,144,912
Yosemite	3,493,607	3,431,514	3,775,038
Joshua Tree	1,316,340	1,392,446	1,384,849
Sequoia	873,229	930,011	1,002,265
Death Valley	1,227,583	871,938	923,344
Kings Canyon	559,534	574,870	588,043
Channel Islands	607,057	332,177	290,157
Redwood	369,726	396,899	398,566
Lassen Volcanic	353,756	377,361	377,298
Pinnacles NM/NP	-	166,988	240,474

Table 9.6 Annual Recreational Visits to National Parks in California.

(Source: National Park Service.)

Figure 9.5 Mickey's neighborhood in Disneyland. With approximately 15 million visitors annually, the park is ranked the second most visited amusement park in the world.

(Photo: Lin Wu)

the top 25 most visited theme parks in the world (based on 2007 data).

In addition to the parks, many unique cultural landscapes and landmarks in California add colorful attractions to the tourism industry. One example is Solvang, a charter city with a population slightly over 5,000, nestled in the wine country of the Santa Ynez Valley, about two hours' drive from the population centers of Southern California. This so-called Danish capital of America, Solvang, with its European village-like shops lining the streets and Danish-style buildings decorated with signature windmills, attracts more than one million visitors each year (Figure 9.6). Retail, accommodation, and food services provide most of the jobs and income for the city.

Not only places like Solvang, with a long history of tourism, serve as tourist attractions in the state; new tourist sites are being established. Figure 9.7 is a photograph of the Sundial Bridge across the northern end of the Sacramento River at Turtle Bay in Redding, completed in 2004.

During the day, the 66-meter (217-foot) pylon, made of 580 tons of steel, casts a rotating shadow over the tile-covered garden at the north side of the bridge, telling time for the visitors; at night, the entire bridge floor, made of glass, lights up. The unique architecture of the bridge (designed by well-known architect Santiago Calatrava), combined with its unique physical geographic location (across the Sacramento River, surrounded by hills, framed by the higher mountains in the background), has attracted many visitors.

There is no doubt that the diverse physical and cultural geographic characteristics of the state contribute to the creation of the many tourist attractions; however, we should not overlook other geographic factors that have helped California stay on top of the tourism industry. One of these factors is the state's large and diverse geographic areas and large population. The tourism industry could be sustainable with the state's own population alone. In fact, over

Rank	Park and Location	Attendance (2007)
1	MAGIC KINGDOM at Walt Disney World, Lake Buena Vista, FL	17,060,000
2	DISNEYLAND, Anaheim, CA	14,870,000
3	EPCOT at Walt Disney World, Lake Buena Vista, FL	10,930,000
4	STUDIOS at Walt Disney World, Lake Buena Vista, FL	9,510,000
5	KINGDOM at Walt Disney World, Lake Buena Vista, FL	9,490,000
6	UNIVERSAL STUDIOS at Universal, Orlando, FL	6,200,000
7	SEAWORLD FLORIDA, Orlando, FL	5,800,000
8	DISNEY'S CALIFORNIA ADVENTURES, Anaheim, CA	5,680,000
9	ISLANDS OF ADVENTURE at Universal, Orlando, FL	5,430,000
10	UNIVERSAL STUDIOS HOLLYWOOD, Universal City, CA	4,700,000
11	BUSCH GARDENS TAMPA BAY, Tampa, FL	4,400,000
12	SEAWORLD CALIFORNIA, San Diego, CA	4,260,000
13	KNOTT'S BERRY FARM, Buena Park, CA	3,630,000
14	CANADA'S WONDERLAND, Maple, ON	3,250,000
15	BUSCH GARDENS EUROPE, Williamsburg, VA	3,157,000

Table 9.7 Top 15 Amusement/Theme Parks in North America (Ranked by Attendance).
(Source: Themed Entertainment Association and Economics Research Associates' Attraction Attendance Report 2007.)

Figure 9.6 A European village-style shopping center in Solvang.
(Photo: Lin Wu)

Figure 9.7 Sundial Bridge at Turtle Bay in Redding.
(Photo: Lin Wu)

80 percent of domestic travelers in California are Californians. Beautiful Yosemite National Park is equally as attractive to Californians as it is to people from all over the world. A San Diegan in the southwest coast area of the state would have to travel more than 1,280 kilometers (800 miles) to visit Redwood National Park on the northwest coast; that distance would cross many state lines on the East Coast of the country.

Another factor is the "Hollywood effect." Many movies made in the state captured various local scenes, famous or ordinary, and brought them to the world in a dramatized fashion. The 2004 movie *Sideways* may not be among the most well-known movies, but after the movie aired, it brought an influx of visitors to the local winery areas in the Santa Ynez Valley, which provided the backdrop for the movie. Finally, the unique geographic location of the state—its isolation and connectivity, as discussed in Chapter 1—makes it a prime destination for international travelers.

These combined factors make the tourism industry in the state more resilient compared with other states and other industries. The war on terror in the early 2000s and the economic downturn in the late 2000s did dent the tourism industry in the state; however, the impact was limited and short-lived. In many cases, the drop in international and out-of-state travelers was compensated for by the increase of in-state travelers who used primarily ground transportation to avoid the hassle and high cost of air travel. It is safe to say that, without considering other economic factors, the geographic factors alone are enough to conclude that the tourism industry will remain one of the vital economic components of the state.

California's Agriculture

Although agriculture may not be the first thing that comes to mind when people think about California, it is, in fact, one of the most significant economic activities in the state and one of the main contributors to the state's annual gross domestic product (GDP). California is the largest agriculture economy in United States and has been for more than 50 years. In fact, California's agricultural production is so large that it produces 30 percent more than Iowa, the country's second-largest agriculture producer (Table 9.8). Third-place Texas, with a land area 68 percent greater than that of California, produces only a little more than half of what California does. At the county level, the top most productive agricultural counties in the country are often found in California.

No other regions in the United States, or even in the world, with similar agricultural land areas can match the amount and diversity of agricultural commodities that California produces. California grows a wider variety of

commodities than any other state; it is the leading producer of more than 50 commodities, some of which are produced only in California (Table 9.9). California also leads the United States in exports of agricultural products—primarily beef, cotton, grape products, almonds, fish, and oranges. Pacific Rim countries—in particular, Japan—receive 55 percent of these exports, followed by Canada (18 percent), Europe (9 percent), and Mexico (5 percent). The acreage under agricultural production occupies one-third of California's 100 million acres, and agriculture and related industries provide 1 of every 10 jobs in the state. On the downside, agriculture consumes the biggest share of the state's precarious water supplies and poses some of the biggest environmental challenges.

State	Rank	Billon Dollars
California	1	43.5
Iowa	2	29.5
Texas	3	22.7
Nebraska	4	21.8
Illinois	5	19.8

Table 9.8 Top Five Agricultural States in Cash Receipts, 2011.
(Source: California Department of Food and Agriculture.)

Solely Produced in California (99 Percent or More)	California Leads the Nation in Production
Almonds, artichokes, dates, figs, raisins, kiwifruit, olives, clingstone peaches, pistachios, dried plums, pomegranates, sweet rice, ladino clover seed, walnuts	Avocados, broccoli, carrots, cantaloupe, table and wine grapes, cut flowers, potted plants flowers, alfalfa hay, honeydew, lettuce head and leaf, lemons, limes, Bartlett pears, chili peppers, bell peppers, vegetable and flower seeds, spinach, strawberries, watermelon, …

Table 9.9 Selected Commodities in which California Leads the Nation.
(Source: California Department of Food and Agriculture, 2011.)

Natural Resources

California's position as national leader in agriculture is a direct result of its physical resource base: prime valley flatlands, fertile soils, the ocean's moderating effect on climate, a lengthy growing season, and a varied topography creating many unique microclimates. California's Mediterranean climate and fertile soils allow for year-round production of many commodities, including lemons, artichokes, avocados, broccoli, cabbage, carrots, cauliflower, celery, lettuce, mushrooms, potatoes, spinach, and squash. California's varied topography creates many diverse microclimates that allow for a wide variety of crops grown commercially only in California; these include almonds, artichokes, figs, olives, walnuts, persimmons, pomegranates, prunes, and raisins.

California is home to many fertile soils. Deep alluvial deposits in the Central Valley, Imperial Valley, Napa Valley, Sonoma Valley, and other lowlands laid the foundation upon which these soils developed. It is no surprise that most agriculture areas in the state are located in these valleys, basins, and plains. The fertile San Joaquin series, found in the Central Valley's prime agricultural area, is the state soil. The limiting factor to agriculture in California is water. The uneven seasonal and spatial distribution of precipitation poses great challenges to agricultural activities and has necessitated the creation of some of the largest water storage and conveyance infrastructure in the country (see Chapter 8).

Farms and Farmers

If you ask anyone who does not live in a California agricultural area to describe a "typical California farmer," you probably would get a pause or a description of someone wearing a straw hat and checked shirt out working in the fields (Figure 9.8). In reality, there is no single description to describe a "typical" farmer in California. A farmer could be a field worker who works in the field year-round or a college professor who teaches during the off seasons and works in the field during the busy seasons. A farmer could be a multimillionaire entrepreneur or someone with an income that barely supports her family.

Figure 9.8 A grape grower and his grandson at work in the Napa Valley.
(Source: Kristin Surber)

Likewise, no single depiction could effectively describe a typical farm in the state. California has approximately 77,000 farms and ranches covering about one-third of the state. The average farm size is roughly 139 hectares (347 acres), which is smaller than the national average of 177 hectares (443 acres). The average farm size in California is smaller than the national average because 60 percent of the state's farms are less than 20 hectares (50 acres) in size; this is also an indicator of specialty crop operations. Agriculture provides a range of employment opportunities for millions of Californians, including field workers, factory workers, managers, researchers, and educators. In the heart of the state's agricultural regions, most economic activities are tied to agriculture.

Major Products and Distribution

A complete survey of the commodities produced in California is not possible and is beyond the scope of this book; however, a zoomed-out overview and a zoomed-in look at a few examples help us see the geographic characteristics of agricultural production and distribution in the state. In general, California is the leading US producer of milk and cream, grapes, nursery products, almonds, lettuce, hay, flowers, and foliage. California is also the exclusive US producer of almonds, artichokes, dates, figs, kiwifruit, olives, persimmons, pistachios, prunes, raisins, clovers, and walnuts.

Livestock and Poultry

Livestock and poultry account for about a quarter of California's gross cash income (Figure 9.9). Major products of livestock and poultry include dairy products, beef cattle and calves, chickens and eggs, turkeys, and sheep and wool products (Table 9.10).

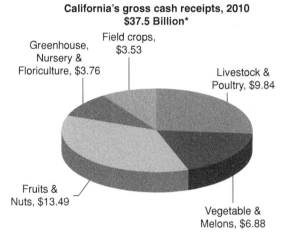

California's gross cash receipts, 2010 $37.5 Billion*

Field crops, $3.53

Greenhouse, Nursery & Floriculture, $3.76

Livestock & Poultry, $9.84

Fruits & Nuts, $13.49

Vegetable & Melons, $6.88

*Totals may not add due to rounding.
Agricultural statistical overview

Figure 9.9 Major agriculture sectors by gross cash receipts in billions of dollars in California. (*Gross receipts* are the total annual income without subtracting any costs or expenses.)
(Source: California Department of Food and Agriculture)

Livestock, Poultry, and Products	Cash Value ($1,000) 2010	Cash Value ($1,000) 2011
Cattle and calves	2,068,412	2,825,125
Milk and other dairy products from cows	5,928,150	7,680,566
Hogs and pigs	36,063	39,196
Sheep and lambs	66,060	-
Poultry and eggs	367,788	391,578
Aquaculture	58,200	64,036

Table 9.10 Selected Values of Livestock, Poultry, and Products (2010 and 2011).
(Source: California Department of Food and Agriculture)

Except for a few locations with extreme conditions, such as cold mountaintops, frigid terrains, and dry deserts, cattle and other livestock are ranched everywhere in the state (San Francisco is the only county out of the 58 counties in the state that does not have a recognizable cattle population). The density of the production, however, varies significantly. Factors behind the distribution patterns include distance to market, available land, and different management and production practices. For example, in the northeast region of the state, cattle are raised in open ranges and called range cattle. It requires large areas to support these types of cattle populations, and productivity is usually low. By comparison, in areas such as the Central Valley and the Imperial Valley, cattle are held in lots and fed processed food; these are called feedlot cattle. It takes considerably less land area to raise feedlot cattle, and productivity is significantly higher than that of range cattle.

Field Crops

Field crops account for $3.53 billion of the state's agricultural economy (Figure 9.9). Among the major field crops, hay has highest acreage in the state. It is the most tolerant of environmental conditions and therefore can grow well in most areas in the state. For hay products, however, environmental factors become important at the drying stage, after the hay is harvested, when dry conditions must prevail. Because of California's sunny climate, most of the hay products in the state are dried naturally.

Cotton and rice are the second- and third-ranked field crops produced in the state and have differing environmental requirements (Figures 9.10 and 9.11). Cotton grows well in loamy soils and needs warm temperatures, relatively dry conditions (to curb the insect problem often associated with cotton production), and a rain-free harvest season. These conditions are best met in the San Joaquin Valley and the Imperial Valley. These areas contribute to the high cotton production in California (10 percent to 14 percent of the total US yearly production).

Rice, unlike cotton, requires water-preserving soils (such as clay), warm temperatures, and

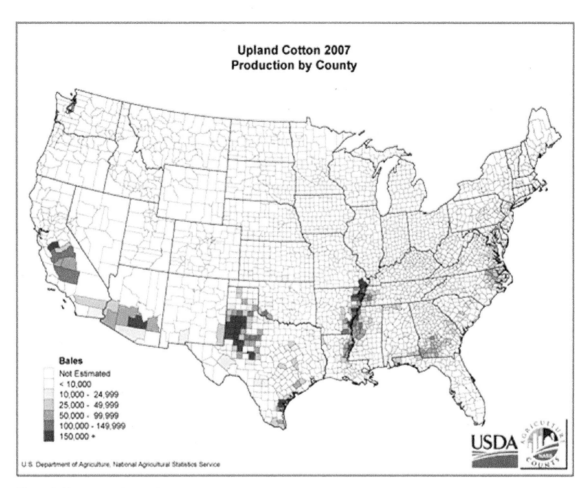

Figure 9.10 Cotton field distribution pattern.
(Source: US Department of Agriculture)

abundant water, especially in the initial planting season, when the rice fields must be completely flooded. These conditions are best met in the Sacramento Valley, where most of California's rice products are produced. Although rice was introduced to the state by Asian immigrants in the 1800s, today various rice products produced in California are exported to Asian counties.

Vegetables, Fruits, and Nuts

Over 50 percent of the nation's vegetables, fruits, and nuts are produced in California. Some, like citrus or peaches, are commonplace and are grown on both large and small farms. Other are specialty crops, such as figs, olives, macadamia nuts, passion fruit, and papaya; these tend to be produced on small farms and cater to niche markets, such as the diverse immigrant communities found across the state. It is probably not an exaggeration to say, "Name a vegetable, any vegetable, and you can find it grown in California."

Californians enjoy not only endless varieties of produce found in the market but also the year-round supply of them as a result of the long growing seasons.

In terms of citrus, California ranks first in lemon production and second in orange production (Florida produces 73 percent of all the oranges in the nation; California produces about 25 percent of the total). As with all citrus products, orange production requires relatively warm temperatures year-round, and production is significantly affected if winter temperatures drop below the freezing point. To avoid the cold winter, most of the orange and other citrus products are planted in the southern part of the state and in the thermal belts (discussed in Chapter 4) in the Central Valley and other locations where winter temperatures usually do not drop below freezing.

In addition to year-round warm temperatures, constant moisture is required for growing juicy, flavorful citrus. Such conditions are met

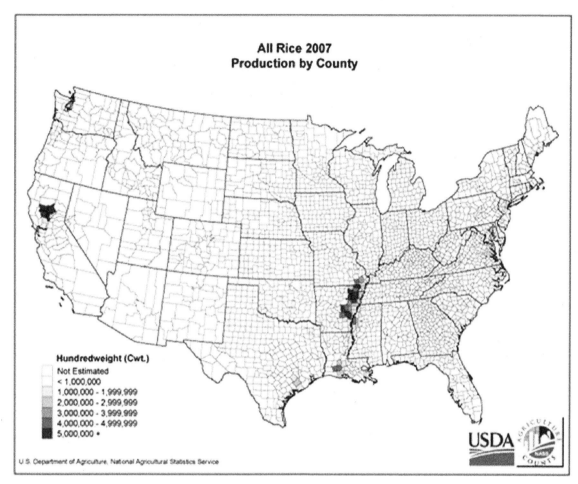

Figure 9.11 Rice field distribution pattern.
(Source: US Department of Agriculture)

only along the southern and central coastal areas where summer fog supplies moisture during the rain-free summer season. As a result, citrus production is concentrated in the central coast, with Ventura County leading the state (and the nation) in lemon production. California's citrus products are world-renowned and are exported to many other states and countries. Unlike citrus crops, apples require relatively cold winter conditions; therefore, apple production is limited to the northern part of the state.

California also leads the nation in many nut crops, some of which are produced exclusively in California, such as English walnuts and almonds. Perhaps the most well-known agricultural products in the state are grapes and products made from grapes—namely, wine and raisins. Grapes grow in very specific types of microenvironments, and because of that specific products are often grown at different locations.

Most grape production in the state is located in the San Joaquin Valley in Fresno, Tulare, Kings, and Kern Counties. Microclimate and soil conditions play key roles in grape production, especially in wine grape production. Slight differences in temperature, moisture, and soil conditions have a significant effect on the grapes and the wine produced from them. In 2011 the grapes crushed to produce variety wine products totaled 3.9 tons, and red wine accounts for almost half of the grapes for wine production.

Opportunities and Challenges

Favorable environmental factors and a diverse population make agriculture a major contributor to the economic viability of the state. Agriculture accounts for a large portion of the state's exports; it provides employment for millions in the state's labor force, and it provides food security for the state's large population. The state's multibillion-dollar agriculture industry, however, has faced increasing challenges in recent years, including extreme weather conditions affecting production, prolonged droughts threatening the water supply, increased labor costs making competition with other markets difficult, and shrinking size of prime farm land because of increasing population and urbanization. Based on the 1997 and 2002 US agriculture census data, California lost 1 percent of its farmland during this five-year period, which translates to more than 400,000 hectares (1 million acres).

Perhaps the biggest challenge the industry will face in the decades to come is the move toward sustainable practices to reduce the impact of agricultural activities on the environment (discussed in Chapter 11). The agriculture industry is one of the largest air, water, and soil pollution generators in the state. To reduce the environmental impact in future practices and to clean up the mess accumulated over the last century will require enormous efforts in education, research, and technology development as well as changes in policy making and management practices.

California's Housing Market

Present-day California is known for its extremely expensive housing markets (some of the highest in the country!), with coastal and urban areas such as Los Angeles and San Francisco even more expensive than inland and rural areas. But California's housing markets have not always been so expensive, nor have they always differed so markedly from those in the rest of the United States (Figure 9.12). Nationwide, median home values have risen steadily from roughly $30,000 since the first housing census in 1940 to over $211,000 in 2012 (these prices are adjusted for inflation). This generally upward national trend has differed from decade to decade, with the greatest change being the 30 percent

increase in median prices from the beginning to the end of the 1970s. Throughout the country, different states have had median housing prices above or below the national average, as well as differing patterns of decade-to-decade changes.

California's median house price has always been above the national average; in fact, the state is second only to Hawaii with the highest median housing prices in the country. This price did not begin to diverge dramatically from the national average until the 1970s, in great part as a result of the dramatic increase in population. Another factor contributing to this short-supply and high-demand real estate market was that, starting in the 1960s, California communities began passing laws regulating the development of land to control urban growth patterns and to set aside protected areas as off limits to development. These two key factors caused prices to increase markedly in comparison with the national average. Whereas nationwide, since World War II, median house prices have been roughly three times or less the median household income, in California prices in some parts of the state have climbed to as much as 10 times the median income, making housing unaffordable to many and creating opportunity for huge profits for those who could afford to invest in real estate. As of 2012, California ranked one of the most severely unaffordable housing markets in the country; also in this decade, California

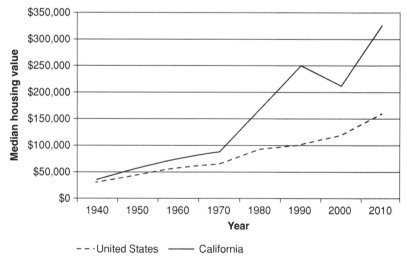

Figure 9.12 A comparison of median housing values for the United Sates and California from 1940 to 2000 (adjusted for 2000 dollars). (Source: US Census Data)

has experienced some of the steepest declines in housing values in the country.

When we look in more detail at annual changes in California housing prices since 1970, several upward and downward trends are observed (Figure 9.13). An above-average and rapid increase in property values in a housing market is sometimes referred to as a *housing bubble*; it continues until the real estate market reaches unsustainable levels and then, like a bubble, it bursts and housing prices decline, sometimes quite severely. As shown in Figure 9.13, housing values in California increased markedly throughout the 1970s and then decreased in the early 1980s. This housing bubble was followed by a bubble that began in the mid-1980s and ended in a steep decrease in housing values in the early 1990s. Another housing bubble began at the end of the 1990s and continued into the 2000s. In the late 2000s, the bursting of this most recent bubble created some of the severest price decreases in the real estate history of the state and caused wide spread foreclosures. While the housing markets are still sluggish in 2013, some economists predict that they will rebound soon.

Coastal versus Inland Markets

Just as housing markets differ from state to state, housing markets can differ from area to area within a state. In general, the urban and coastal housing markets have behaved differently from the inland or more rural housing markets. As shown in Figure 9.14, the urban and coastal markets, such as the San Francisco and Los Angeles regions, have had consistently higher median prices, along with steeper price increases and decreases during the various real estate bubbles. By comparison, the inland markets of Sacramento, in the Central Valley, and

the Inland Empire markets of Riverside and San Bernardino have had lower median housing values overall, and although these markets have experienced the same housing bubble time periods, these markets do not peak and decline with the same intensity as seen in the coastal markets.

The history of these different housing markets differs as well in terms of when they were developed. From post–World War II to the 1960s, housing development was focused in the more desirable coastal areas because land was available and houses were still affordable. The greatest population growth seen during this time period was in the coastal counties of Southern California, such as Santa Barbara, Ventura, and Orange, and in the San Francisco region and its surrounding counties. But, by the end of the 1970s, housing prices in these coastal markets had increased significantly, and homes were no longer affordable to many Californians. The increase in the median values of these markets was the result of the decrease in available land for building; simply no more land was available for building new houses, so the prices of existing housing stock increased. Thus, by the end of the 1970s, the new population growth centers and subsequent housing development had shifted to inland markets such as Riverside and San Bernardino counties in Southern California and in the Central Valley in areas such as Sacramento. Housing development could occur here because land was still available and more reasonably priced; thus, population growth shifted to these inland areas.

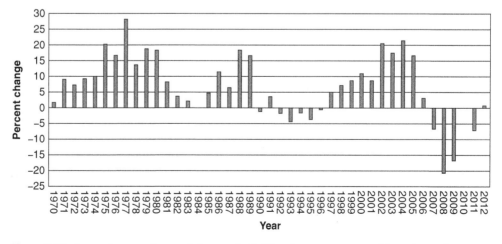

Figure 9.13 Annual percentage of change in California average annual housing prices. (Source: Lin Wu, Data: California Association of Realtors)

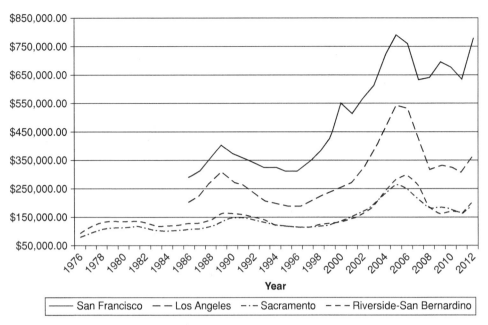

Figure 9.14 Median home price comparison of coastal versus inland markets.
(Source: Lin Wu, Data: National Association of Realtors)

The Silicon Valley

The name Silicon Valley originally came from the computer chip production industry located in the South Bay area in the 1970s. Today, the name has become synonymous with the high-tech industry. The regional association with this name varies depending on the context. Within California, in most of the cases, it refers to the high-tech industry region in the Bay Area, especially in the South Bay area, with San Jose claiming the title "Capital of Silicon Valley." From within the country, the term is often used in reference to the high-tech industry in California, and sometimes, when the term is used in an international context, it refers to the high-tech industry in the United States. The name itself suggests the importance of the high-tech industry in California and the United States.

From an employment perspective, the high-tech industry accounts for approximately 4.3% of the total employment in California based on 2010 statistics. Nationally, only 2.9% of all jobs are in the high-tech industry. The statistics often includes sectors in computer industry, electronic product manufacturing, telecommunications, Internet service providers, Web search portals, data processing, and computer systems design

and related services. In recent years, there has been a trend for the high-tech industry to shift from manufacturing to service. In 1990, 66% of the high-tech employment in California was in the manufacturing sector; by 2010, that number declined to 46%, and the trend continues. This is a reflection of changes in the high-tech industry as service (e.g., companies that provide social network services) becomes more and more important and the manufacturing component decreases accordingly. The shifting of the international labor market also has an impact on the numbers. Although there was a trend to move high-tech manufacturing overseas because of lower costs, a recent phenomenon of "onshoring," or "reshoring," has been observed; however, "offshoring" still outpaces "reshoring" in the current economy.

Labor and the economy are not the only significant aspects of the high-tech industry in California. Silicon Valley in many ways has a profound influence on the culture of California and beyond. Think of our everyday life: beyond fulfilling our physical needs, we rely on many of the products provided by the California technology companies, for example, Apple, Facebook, Google, and Intel. The cultural landscape in the Silicon Valley has also changed drastically in recent years. In its early years, Silicon Valley was characterized by small, low-key business buildings located alongside the streets and sometimes mixed with residential areas. The only thing that drew your attention to an Apple facility was a small wall on the street that bore an Apple company logo, and these facilities were scattered. Today, the high-tech companies are competing to design and

Figure 9.15 A scene of the Silicon Valley.
(Photo: Lin Wu)

build major business building groups (often called "campuses") with unique architectural designs that are changing the skyline of the Silicon Valley—in this context, the South Bay area (Figure 9.15).

Summary

California has always been, and most likely will remain, a vibrant, multifaceted, and risk-taking economy that is a leader in many realms. Its location, climate, and natural resources have created an economy that exploded with the gold rush and expanded to include international trade, fishing, oil and natural gas industries, tourism, entertainment, high-tech industries, and agriculture, to name a few. This large and diverse economy is obvious when we note that California, if it were a country, would rank somewhere between the sixth and tenth largest economies in the world!

In this chapter, we attempt to briefly examine some of California's key economic components and how they relate to the geography of the state. The state's wonderful Mediterranean climate and great physical diversity contributes greatly to the tourism and entertainment sectors of the economy; those same factors also contribute to the incredibly diverse agricultural economy of the state. The abundant natural resources contribute to a plethora of primary economic activities, such as mining, fishing, oil and natural gas extraction, and timber production. With the start of the new century, service industries have become the largest employer in the state and cover a wide range of areas including health care, food, retail, recreation, financial and professional services, information technology, trade and transportation, education, motion picture production, and Internet businesses.

Selected Bibliography

California Department of Food and Agriculture (CDFA). (2013). *California Agriculture Production Statistics.* Retrieved August 2013 from http://www.cdfa.ca.gov/statistics/

California Department of Parks and Recreation. (2013). *Park Systems Statistic Reports and Various Sources.* Retrieved August 2013 from http://www.ca.gov/

Coe, N., Kelly, P., and Yeung, H. W. C. (2013). *Economic Geography: A Contemporary Introduction* (2nd ed.). Hoboken, NJ: Wiley-Blackwell.

McKnight, T. (2003). *Regional Geography of the United States and Canada.* Englewood Cliffs, NJ: Prentice Hall.

National Park Services.(2013). *Visitor Use Statistics.* Retrieved August 2013 from https://irma.nps.gov/Stats/

State of California. (2013). *Various Information on Economy.* Retrieved August 2013 from http://ca.gov/

US Census Bureau website. (2013). Most important source of data and information on population and economic status and trend in the United States (http://www.census.gov/data).

California's Coastal Ocean Region

"The sea, once it casts its spell, holds one in its net of wonder forever."

–Jacques Cousteau

Introduction

This chapter is an introduction to the coastline, the nearshore and offshore ocean waters, and the marine resources of the eastern Pacific along the western coast of California, focusing on both the biological and physical environments. For simplicity, we refer to these areas collectively as California's coastal ocean region. The coastal ocean region hosts a wide diversity of habitats and contains some of the most biologically diverse natural communities in the world, with a vast array of plant life, invertebrates, amphibians, reptiles, fish, birds, and marine mammals. This region is a vital resource for California, and it is a major aspect of the state's economy through such activities as tourism, fishing, and marine transportation.

From its wild rocky shores in the northern part of the state to the wide sandy beaches in the south, the beautiful and varied California coast is 1,770 kilometers (1,100 miles) long. It is constantly being reshaped by the effects of tectonic uplift, water, and wind to form a variety of landforms, including mountain ranges, marine terraces, headlands, sea cliffs, dunes, beaches, estuaries, and bays (see Chapter 2 for more information on coastal landforms). The northern coastline differs in appearance from the southern coastline in part because it trends in a northwest–southeast direction, whereas at Point Conception the southern coastline turns sharply to the east (Map 3). This eastward turn creates the Southern California Bight (a *bight* is a wide open-mouthed bay), which extends past the southern border of California into Baja, Mexico, and includes the Channel Islands. The orientation of the southern coastline and the presence of the Channel Islands protect the southern shoreline from the rougher storm-driven waves of the north, helping to create the wide sandy beaches that southern California is known for. The more north–south trend in the northern coastline causes it to be hit head-on by winter storms; thus, it has a more rugged appearance and lacks the wide sandy beaches of the south (Figures 10.1 and 10.2).

The Pacific Ocean has had, and continues to have, a strong influence on California in some important ways. The movement of the ocean waters in waves and tides continually shapes the various landforms found along the coast, and the moderating effect of the ocean is key in

Figure 10.1 Rugged coastline of Northern California.
(Image © Keith McIntyre, 2009. Used under license from Shutterstock, Inc.)

Figure 10.2 Wide sandy beaches of Southern California.
(Image © V.J. Mathew, 2009. Used under license from Shutterstock, Inc.)

Figure 10.3 California surfer catches a wave.
(Image © cassiede alain, 2009. Used under license from Shutterstock, Inc.)

creating California's wonderful Mediterranean climate. California's location on the edge of the Pacific Ocean has shaped the way it was explored in that, unlike the landlocked western states,

California was explored predominantly from the sea inward, both by the Spanish and then by the Americans. The ocean is also an important factor in California's economy, particularly through tourism, fishing, and marine transportation. The beautiful coastal ocean region of the state certainly has a strong influence on where Californians want to live. Approximately 75 percent of the population lives at or near the coast, in heavily urbanized regions in the San Francisco Bay and South Coast areas, a mixture of both rural and semi-urban regions along the Central Coast, and primarily rural regions along the North Coast (Map 2). Culturally, the ocean waters, beaches, and shorelines of California have become an iconic symbol that defines the state. Images of a surfer on a wave or people having a bonfire on the beach have been shown in countless television shows, movies, and advertisements and bring the Golden State to mind (Figure 10.3).

The Chumash

Some of California's early inhabitants, such as the Chumash Indians, lived along the coast as early as 13,000 YBP (years before present). The Chumash had numerous villages along the coast from Paso Robles south to Malibu, and they also occupied some of the Channel Islands, including San Miguel, Santa Rosa, Santa Cruz, and Anacapa Islands (Figure 10.4). The waters of the eastern Pacific Ocean were an important resource for the Chumash, who collected shellfish, hunted marine mammals, and fished in the streams, estuaries, and nearshore waters. To hunt and gather fish and shellfish, they built several types of canoes and made a variety of fishing gear, harpoons, fish spears, and nets. Because the protected waters of the Santa Barbara Channel are relatively calm, they were able to fish throughout the year for annual runs of albacore, tuna, and sardines. The early Native American tribes had minimal environmental impacts on the coastal ocean region, especially in comparison with those caused by later explorers and settlers, such as the Spanish. Starting with the arrival of the Spanish in the late 1700s, settlers began decimating seal and sea lion populations for oil and

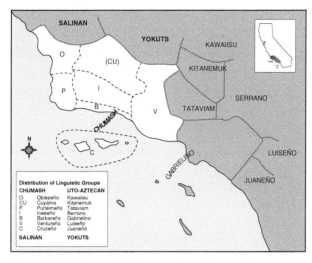

Figure 10.4 Chumash villages along the southern coast of California before European settlement.

(Source: Santa Barbara Natural History Museum)

skins, sea otter populations for fur, and whales for oil; damming streams for irrigation; filling in wetlands for agriculture; and generally destroying much of the coastal habitats through pollution and habitat destruction.

Ocean Economy

Today, California's coastal ocean region is an important component of the state's overall economy. In fact, California boasts the largest ocean economy in the United States both in terms of employment and gross state product (GSP). In terms of the overall US ocean economy (2005), California contributed the largest share of any state, with a GSP of ~$42.9 billion (about 19 percent of the US total) and roughly 400,000 jobs. The highly varied sectors of the state's ocean economy include such activities as commercial fishing, kelp harvesting, offshore oil and natural gas extraction, tourism and recreation, maritime transportation, and port development and maintenance. Particularly strong sectors of the ocean economy are tourism and recreation, with beaches a top destination for tourists, and the marine transportation sector, with the ports of the Los Angeles and San Francisco regions among the largest and busiest in the nation.

These ports are an essential aspect of California's international trade; in 2007, for example, 40 percent of the total containerized

Figure 10.5 Sea lions sunbathing at Pier 39, Fishermen's Wharf, San Francisco.

(Image © RJR, 2009. Used under license from Shutterstock, Inc.)

cargo entering the United States and 30 percent leaving it passed through California ports. In addition, the ports have a variety of nonshipping-related activities and facilities, such as restaurants, hotels, services for passenger ships, and a variety of tourist attractions, such as the ever-present sea lions lounging on the docks at Fishermen's Wharf in San Francisco (Figure 10.5).

Environmental Protection

Fortunately, Californians recognize the intrinsic and economic value of the coastal ocean region and have passed legislation to protect this very special resource. Starting in 1972, California voters passed an initiative aimed at the preservation and restoration of the coastal ocean region for current and future generations to enjoy. This initiative also created the California Coastal Commission, whose job is to work with both the federal and local governments in the planning and regulation of urban development and natural resource use in the coastal ocean region.

In 1976, the California Coastal Commission was made permanent when the California State Legislature enacted the California Coastal Act. Although much progress has been made in environmental protection and habitat restoration in the coastal ocean region, much work remains to be done. This region is home to numerous threatened species, including fish, such as the endemic tidewater goby (*Eucyclogobius newberryi*), and birds, such as the California brown pelican (*Pelecanus occidentalis californicus*) (Figure 10.6). Species such as these are often at risk as a result of the destruction of their native habitats. For example, 90 percent of the original wetlands along the coast of California are gone, and most major rivers that drain to the ocean have been dammed and channelized with concrete; this keeps coastal estuaries from receiving their supply of freshwater, deprives the beaches of sand, and inhibits fish, such as salmon (*Salmonidae*) from swimming upstream to spawn.

Figure 10.6 California brown pelicans.
(Image © vm, 2009. Used under license from Shutterstock, Inc.)

The Pacific Ocean and The California Current

To put California's coastal waters in perspective, we need to look at the eastern Pacific Ocean from a global perspective. The Pacific Ocean is the largest of the world's oceans. Covering about one-third of the earth's surface, it is larger than all of the landmasses combined. It extends all the way from the Arctic to the Antarctic and from Asia and Australia to North and South America (Figure 10.7). At the equator, the Pacific is divided into north and south. Thus, California's western border is along the eastern edge of the North Pacific, an area referred to as the North Pacific gyre. The North Pacific gyre (Figure 10.8) is made up of four currents, which combine to move water in a huge clockwise pattern—the

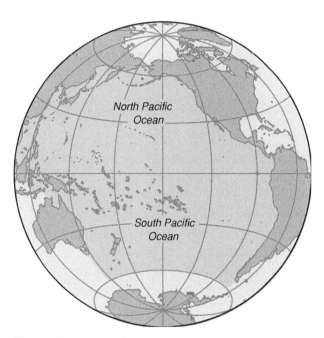

Figure 10.7 The Pacific Ocean.

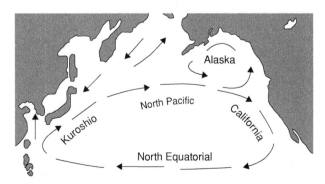

Figure 10.8 Currents of the the North Pacific gyre.

North Pacific Current to the north, the California Current to the east (along the western coast of California), the North Equatorial Current to the south, and the Kuroshio Current to the west (along the eastern coast of Japan).

The California Current flows north–south along the western coast of North America from British Columbia in Canada; along the coasts of Washington, Oregon, and California; and as far south as the tip of Baja California. This movement of water from north to south means that the waters along the California coast are cooler than coastal areas of similar latitude on the eastern coast of the United States (such as North and South Carolina), which have the warmer gulf coast waters flowing from south to north along the Atlantic coast. In addition to the north–south flow of the California Current, another important aspect of these coastal waters is the presence of extensive upwelling of subsurface waters (Figure 10.9). *Upwelling* is the vertical upward flow of water at depth to the ocean surface. The upwelling of water along the coastal waters of California is the result of the prevailing atmospheric flow of northwest winds; as the winds blow from the northwest, surface water is driven away from the coast and is replaced by colder, denser water from below. The important result of the upwelling of ocean waters is that nutrient-rich material (dissolved nutrients from the decay of organic material) is brought to the ocean surface. This regular supply of nutrients by coastal upwelling means that the California Current region has an incredibly high degree of biologi-

cal productivity. These nutrients are essential for phytoplankton growth, the base of the ocean's food chain, which in turn supports a complex ecosystem of invertebrates, fish, birds, and marine mammals. This is the reason why upwelling areas along the west coasts of landmasses, such as California, are where the world's best fisheries are found. Upwelling also varies seasonally; with changes in the prevailing wind patterns, the strongest period of upwelling in California's coastal waters is from March through September.

Along the north coast, the California Current, with its southward flow of cold water, stays close to the shoreline until, at Point Conception, it begins to veer offshore (Figure 10.10). Along the south coast, in the Southern California Bight region, a countercurrent brings warmer water north from the semitropical region of the Pacific Ocean along the coast of southern California. The temperatures of these two currents create different biological regions in northern and southern California. The Southern California Bight region, with its relatively warmer waters, supports temperate and warm water fish and other organisms, whereas the northern ocean region supports species adapted to its colder waters.

The marine environments of the eastern Pacific Ocean are also divided into the nearshore waters, which are the fertile shallower waters overlying the continental shelf (to a depth of 100 meters [330 feet]), and the less productive offshore waters,

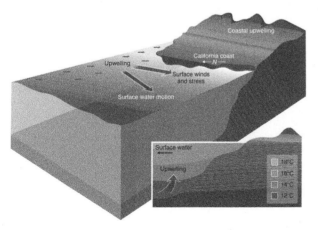

Figure 10.9 Coastal upwelling.
(Image © Kendall Hunt Publishing Company.)

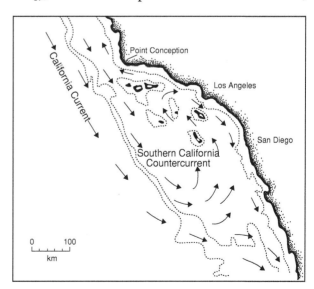

Figure 10.10 Location of California current and Southern California countercurrent.

where the continental shelf drops off to the deep sea floor (extending 320 kilometers [200 miles] offshore). The ocean basin offshore is tectonically active, the result being that the *bathymetry* (the depth of the water relative to sea level) of the waters off California is as highly varied as the land surfaces of the state. Like land surfaces, where we see the variety of the landscapes, the bathymetry of the oceans contains its own plains, canyons, faults, and mountain ranges. The *continental shelf*, the underwater extension of the continent, is quite narrow along the western coast of the United States (6.5 to 8 kilometers [4 to 5 miles] wide) in comparison to the broader shelf of the Eastern Seaboard. In addition, the shelf is very rugged, with elongated basins bounded by faults and deep underwater canyons. The most dramatic example is found in Monterey Bay, whose underwater canyon is deeper than the Grand Canyon.

Nearshore Waters

California's nearshore waters benefit greatly from seasonal upwelling. In the spring, upwelled nutrients and increased sunlight cause phytoplankton, tiny free-floating plants that are at the base of all food webs in the ocean, to bloom in large quantities. This in turn causes large populations of zooplankton, tiny aquatic animals such as copepods and krill, to increase in number. The zooplankton communities provide a food supply for fish that in turn are a source of food for birds and marine mammals.

Kelp Forests

The primary habitat found in the nearshore waters along most of California's coast is the kelp forest. Kelp, which are large, swaying underwater plants, provide food and shelter for a wide variety of marine organisms. Kelp forests are made up of multiple species, with the predominant one being the giant kelp (*Macrocystis pyrifera*). The giant kelp is incredibly fast growing—more than 25 centimeters (10 inches) per day. It anchors itself to the rocky ocean floor, and its gas-filled floats keep it upright. Kelp forests provide habitat for invertebrates, such as snails, crabs, and brittle stars, which live on the

kelp blades, and bottom-dwelling organisms, such as anemones, sea cucumbers, abalones, sea stars, and urchins. Fish that inhabit the kelp forests include surfperch (*Brachyistius frenatus*); rockfish (*Sebastes* spp.); kelp bass (*Paralabrax clathratus*), a popular recreational species; and the state marine fish, the Garibaldi (*Hypsypops rubicundus*) (Figure 10.11).

Marine mammals, such as sea otters (*Enhydra lutris*), also live within the kelp beds and can be seen floating on their backs feeding on abalone or sea urchins (Figure 10.12). Harbor seals (*Phoca vitulina*) might be observed swimming through the kelp forests hunting for fish. The giant kelp has a commercial value and over time has been harvested for a number of different uses. During World War I, it was harvested and burned and the ashes used to produce gunpowder. Today, kelp is primarily used to produce algin, an emulsifier used in the production of paints, cosmetics,

Figure 10.11 Kelp forest with starfish, bright orange garibaldi, swaying kelp, and brown sea fans.
(Image © Kelpfish, 2009. Used under license from Shutterstock, Inc.)

pharmaceuticals, and processed foods, such as ice cream. Birds found in the nearshore water and on the beaches include numerous species of gulls, the brown pelican, and shorebirds, such as sandpipers, godwits, and curlews (Figures 10.13 and 10.14).

Figure 10.12 A sea otter wraps itself in kelp.
(Image © A. Craig, 2009. Used under license from Shutterstock, Inc.)

Figure 10.13 Seagulls flying on Drakes Beach, California.
(Image © Cate Frost, 2009. Used under license from Shutterstock, Inc.)

Figure 10.14 A godwit hunts for food.
(Image © Stubblefield Photography, 2009. Used under license from Shutterstock, Inc.)

El Niño

A phenomenon called *El Niño*, which is an oscillation of the ocean-atmosphere system in the tropical Pacific, can have serious effects on the marine life in California's nearshore waters. El Niño has long been observed by fishermen living in Peru along the western coast of South America. For centuries, these fishermen have noticed that during some years the waters of the eastern Pacific are markedly warmer than other years and that the temperature change is associated with a reduction in fish yields. This event was dubbed El Niño, or Christ child, as its effects were most noticeable at Christmastime.

Today scientists know that an El Niño event is a change in the way the ocean and atmosphere interact along the equatorial Pacific (Figure 10.7). During a normal (non–El Niño) year, the trade winds along the equator blow from east to west and keep a warm pool of water "piled up" in the western Pacific. Then every five to seven years, for reasons that aren't completely understood, this situation changes: the trade winds slacken or even reverse themselves, and the warm pool of water flows east toward the western coast of South America. When the warm water hits the coast of South America, it turns south and north, extending along the coasts of both North and South America. The result for California is that the nearshore waters are inundated with the very warm and nutrient-poor El Niño waters and the usual upwelling of cold water along the coast stops. The effects on the nearshore marine communities can be quite severe with a domino-like effect of decreasing phytoplankton production, declining fish populations, and starving seabirds and marine mammals.

The Southern California Bight

The waters off Southern California are referred to as the Southern California Bight (SCB) (Figure 10.10). This region is created by the long eastward bend in the coastline that begins at Point Conception, in Santa Barbara County, and extends all the way past the southern border of California into Baja Mexico. Although a bight is defined as "a bend in a coastline forming a wide open-mouthed bay," the SCB is more than just that. The presence of the bight creates a gyre that mixes the warm waters of the California Countercurrent from the south with the cool waters of the California Current from the north. This mixing of waters of different temperatures helps create a unique marine ecosystem that extends from Point Conception into Baja and that is home to a wide diversity of marine life, supporting both temperate and warm water fish and other organisms.

Today, the SCB area is one of the most highly urbanized areas in the country, with a population of roughly 16 million people. A population of this size has enormous impact on the waters of the Southern California Bight, causing pollution from a variety of sources, including municipal and industrial wastewater discharge, stormwater runoff, oil and natural gas extraction, commercial and recreational fisheries, ship traffic, and recreational activities. There is great concern about the environmental degradation of the SCB, as there have been extremely detrimental effects observed on the kelp forests and associated fish and invertebrate communities. Currently, over $17 million per year is spent by public utilities, private industry, and government agencies on marine monitoring efforts in the SCB.

Municipal Wastewater and Stormwater Runoff

It is important to recognize that 80 percent of the pollution entering the SCB originates on land, so even southern Californians who don't live right on or near the coast affect the SCB in profound ways through their daily activities—namely, the production of municipal wastewater and stormwater runoff. Municipal wastewater is a form of point source pollution, meaning that it comes from an identifiable source—in this case, a sewage treatment plant—and after treatment it is discharged via a pipe into the ocean. Stormwater runoff, in contrast, is considered nonpoint source pollution, meaning that rainfall and irrigated water run over the land surfaces and pick up pollutants such as pesticides, oil, fertilizers, and pet waste. It is not treated at all in a wastewater treatment plant; instead, it flows directly through the storm drain system and empties into the ocean at the shoreline. This *toxic soup* of wastewater is referred to as nonpoint source pollution because its sources are highly varied and nearly impossible to trace back to their original source. Because of the Mediterranean climate of the region, with its wet winters and dry summers, the systems designed to handle stormwater runoff and municipal wastewater flows are separate, as stormwater flow is much more variable than wastewater flow.

Roughly 22,600 square kilometers (8,700 square miles) of highly urbanized southern California drains into the SCB, and much of this huge land area is covered with impermeable materials such as concrete, asphalt, and roofs. This increases the amount of stormwater runoff during rainfall events because there is less soil and vegetation to absorb the rainwater. Even major rivers such as the Los Angeles and the San Gabriel have been lined with concrete for much of their length in a process known as channelization. This stormwater soup of chemicals, bacteria, oil, fast-food wrappers, animal waste, and cigarette butts flows through the vast system of drains, pipes, and channels right into the ocean at the shoreline, where high bacterial levels can lead to beach closures (Figure 10.15).

Sewage is sent to municipal treatment plants before reaching the ocean. These plants receive sewage from domestic, commercial, and industrial

Figure 10.15 Stream carrying stormwater runoff into the ocean at Arroyo Burro Beach, Santa Barbara. (Source: S. Garver)

sources. In general, each person generates 100 gallons of wastewater per day. Throughout the SCB region, roughly 1.5 billion gallons of municipal wastewater is released into the ocean every day. Most of this sewage is treated to primary and advanced primary levels, with some secondary treatment, before being discharged into the ocean through outfall pipes. Primary sewage treatment removes solids by first passing it through screens and grit chambers to remove solid material; then suspended solids are settled to the bottom of tanks as sludge, and grease floats to the top and is skimmed off. The resulting *effluent* (water portion of sewage) is then chlorinated to kill any remaining pathogens. Primary sewage treatment alone is no longer considered sufficient, and the Federal Clean Water Act of 1972 mandates that all wastewater treatment plants should provide secondary sewage treatment that adds a biological process using bacteria to remove dissolved organic matter. Currently, only a small amount of sewage (570 million liters/day [150 million gal/day]) is treated at a tertiary level, which includes primary and secondary treatment plus removal of nutrients (nitrogen and phosphorus) and toxic chemicals. Effluent treated to this level can then be used as reclaimed water for landscape irrigation.

There are numerous wastewater treatment plants in the SCB area; the largest and oldest is the Hyperion Treatment Plant in Los Angeles. When Hyperion opened in 1894 it discharged raw sewage into Santa Monica Bay, with detrimental effects for many years on the ecosystem of the bay. The enormous amount of suspended solids buried kelp beds and reduced light penetration, inhibiting new growth. Improvements in wastewater treatment and the replanting of kelp beds has helped revive the kelp forests in the SCB. In the 1950s, Hyperion was upgraded to partial secondary treatment, and more recently it was upgraded to full secondary treatment. Today, Hyperion averages 1,285 million liters per day (340 million gallons per day) of wastewater and has a wet weather capacity of 3,215 million liters per day (850 million gallons per day). As individuals, the best way that we can cut down on the wastewater we produce

each day is to conserve water; the less water we use, the less we send down the drain to our already overburdened wastewater treatment plants (see Table 8.3: Ways to Be Water Wise).

Offshore Waters

The *offshore waters* are defined as beginning at a depth of 100 meters (328 feet) and continuing out into the open ocean to a distance of 325 kilometers (200 miles); most of California's offshore ocean region is beyond the continental shelf. As the continental shelf drops off to the deep sea floor, the open ocean waters become much less fertile as a result of the diminishment of upwelling of nutrients to the surface. Because less food is available, pelagic fish (fish that live in the open ocean) tend to be nomadic and travel great distances to find food sources. Some of these fish have great commercial value, such as anchovies, sardines, and herrings, along with larger predators such as tuna, salmon, and shark. Pelagic birds that travel great distances to feed on crustaceans and small fishes include species of petrels, gulls, and frigate birds.

Sardine Fisheries

If you have ever visited Monterey Bay and its historic Cannery Row, you have been to the heart of what was once the world's largest sardine fishery. This time period in Monterey Bay is the setting for John Steinbeck's novel *Cannery Row* (1945). Sardine fishing began here in the early 1900s, and by the 1940s there were more than 100 canneries and processing plants, not just in Monterey Bay but all along the coast from San Diego to San Francisco. By the 1950s, however, after only about 50 years of fishing, the schools of sardines suddenly disappeared. This led to a heated debate between scientists and the state of California about the causes of this collapse: Was it overfishing, natural fluctuations in sardine populations, or a combination of both? Years later marine biologists measuring fish scale deposits in deep ocean sediments off Southern California found alternating layers of sardine and anchovy scales, indicating major recoveries and subsequent collapses of these fishes over a 1,700-year period. What scientists concluded from this evidence was that both sardines and anchovies vary in abundance over periods of approximately 60 years and that the rise and fall in their populations is tied to cyclical changes in ocean temperature, with cold water ocean cycles favoring anchovy populations, and warm water cycles favoring sardines. Researchers determined that the current sardine recovery seen since the 1950s is similar to those of the past and that sardines disappear periodically, even without fishing pressures. After the sardines vanished in the 1950s, the fishing industry turned to other fish, including anchovies and squid.

Marine Mammals

The marine mammals found in California's coastal ocean region comprise about 35 species in all and are made up of whales, dolphins, porpoises, seals, sea lions, and sea otters. Like all mammals, they give birth to live young, have mammary glands, and have hair on their bodies at some stage of development. The whales are divided into baleen whales and toothed whales. Baleen is a brushlike strainer in the whale's mouth that allows it to feed by filtering large amounts of water for zooplankton, such as krill and small fish (Figure 10.16). Baleen whales commonly found in California are the blue whale (*Balaenoptera musculus*), the largest mammal on earth (up to 30 meters [100 feet] in length); the acrobatic humpback whale (*Megaptera novaeangliae*), which can throw itself completely out of the water (Figure 10.17); and the state marine mammal, the gray whale (*Eschrichtius robustus*), which has one the longest migrations of any mammal (approximately 16,000 to 22,400 kilometers [10,000 to 14,000 miles] round trip), traveling annually from its feeding grounds in the Bering Sea to mating grounds in Baja California, Mexico.

Toothed whales found in California include sperm whales, dolphins, and porpoises. Sperm whales (*Physeter macrocephalus*) are the largest of the toothed whales and can grow to lengths of 18 meters (60 feet). Porpoises and dolphins are smaller members of the toothed whales; dolphins have longer beaks than porpoises, and typical species you might see along the coast of California include the common dolphin (*Delphinus delphi*) and the Pacific white-sided dolphin (*Lagenorhynchus obliquidens*) (Figure 10.18). Porpoises commonly found in California include the black and white Dall's porpoise (*Phocoenoides dalli*), the fastest swimmer of the dolphins and porpoises, and the harbor porpoise (*Phocoena phocoena*), which stays near the shore and thus has been hunted for centuries.

Sea lions and seals differ from whales, porpoises, and dolphins in that they give birth and molt (shed fur) on land but find their food in the ocean. Sea lions (Figure 10.5), elephant seals (Figure 10. 19), and harbor seals (Figure 10.20) are the most commonly observed of all the marine mammals along the coast of California.

Figure 10.16 Baleen in the mouth of a gray whale.
(Image © jocrebbin, 2009. Used under license from Shutterstock, Inc.)

Figure 10.17 A humpback whale breaches out of the water.
(Image © Richard Fitzer, 2009. Used under license from Shutterstock, Inc.)

Figure 10.18 Pacific white-sided dolphin leaping out of the water.

(Image © Suzi Logan, 2009. Used under license from Shutterstock, Inc.)

Figure 10.19 A family of elephant seals on the beach.

(Image © Galina Barshaya, 2009. Used under license from Shutterstock, Inc.)

Figure 10.20 Harbor seal pup at on the beach at La Jolla.

(Image © Jose Gil, 2009. Used under license from Shutterstock, Inc.)

The California sea lion (*Zalophius californicus*) can grow to 2.1 meters (7 feet) long and weigh 225 to 337 kilograms (500 to 750 pounds). Its main rookeries (breeding grounds) are on the Channel Islands and along the central coast.

Steller sea lions (*Eumetopias jubatus*) are concentrated north of Point Conception with rookeries offshore in the San Francisco Bay area on Año Nuevo Island, the Farallon Islands, and Seal Rocks. Seals differ from sea lions in that they lack external ears and can't turn their hind flippers forward so they are less mobile on land. The smaller harbor seal (*Phoca vitulina*) grows 1.2 to 1.8 meters (4 to 6 feet) long and is commonly seen in bays and harbors. Harbor seals do not form organized rookeries on the coast or islands like other seals and sea lions. The northern elephant seal (*Mirounga angustirostris*), with its big snout, is the largest of the seals and can weigh up to 2,250 kilograms (5,000 pounds). Elephant seal rookeries are located on the Channel Islands, Año Nuevo Island, and the Farallon Islands.

Sea otters (*Enhydra lutris*) are the smallest of the marine mammals, weighing about 40 kilograms (90 pounds). They have no fat or blubber, so they depend on their thick coats to keep them warm. This is why they are so often observed incessantly cleaning their coats—to maintain its insulating quality. They live in the kelp beds along the coast from Point Conception to Monterey Bay and eat sea urchins, abalones, and other invertebrates (Figure 10.12).

All California marine mammals are protected under the Marine Mammal Protection Act (MMPA), which was enacted in 1972 and prohibits the hunting of marine mammals in US waters. Some of the species, such as the gray whale, have been successfully repopulated, and, although they remain on the protected list, they are no longer considered endangered. Other species, such as the sea otter, Steller sea lion, and the blue, humpback, and sperm whales remain listed as endangered and depleted, meaning that these species are considered below their sustainable population levels and under threat of extinction (Table 10.1).

Coastal Landforms

A *coastline,* the interface between land and ocean, is a very active area. The waves, currents, tides, and winds supply energy that is continually

Species	Status under MMPA	Notes
Blue whale (*Balaenoptera musculus*)	Endangered/depleted	Once sought after by whalers for their large amounts of blubber. Currently, California has the largest concentration of blue whales in the world (~2,000).
Humpback whale (*Megaptera novaeangliae*)	Endangered/depleted	One of the most endangered whales, only 10 percent of population remains. Currently, ~800 humpbacks feed along the coast of California.
Gray whale (*Eschrichtius robustus*)	Delisted/recovered	Intensive hunting caused population to drop to ~2,000 whales. In 1946 an international agreement was reached to stop hunting. Currently, the population has grown to over 26,000, similar to populations before whaling. Removed from the endangered species list in 1994.
Sperm whale (*Physeter macrocephalus*)	Endangered/depleted	Hunted in the past for oil and ambergris (a waxy substance in the digestive system used in perfume). Current West Coast estimated population is ~1,200.
Steller sea lion (*Eumetopias jubatus*)	Endangered/depleted	Current California population ~500. Still great concern because population has dropped 80 percent in the last 30 years. Researchers believe that a decline in food sources could be the cause.
Sea otter (*Enhydra lutris*)	Endangered/depleted	California population once over 250,000, then hunted to the verge of extinction for its fur. Current population ~2000. Although protected from hunting, sea otters are very vulnerable to oil spills.

(*Endangered*—any species that is in danger of extinction throughout all or a significant portion of its range; *Depleted*—any species below its optimal sustainable population; *Recovered/Delisted*—a species is "recovered" when it no longer requires protection and thus is also delisted).

Table 10.1 Status of Selected California Marine Mammals.

eroding and depositing sediments to produce a variety of landforms. The Pacific coast, with its rugged, high-relief appearance, is on an active continental margin and thus is categorized as an erosional coast (see Chapter 2 for more information on coastal landforms). By comparison, the Atlantic coast, with its gentle terrain, is a depositional coast, with many sources of sediments and very different types of landforms being created from the action of water and wind.

Sea Cliffs, Marine Terraces, and Headlands

Characteristic landforms seen on erosional coasts, like California's, include sea cliffs (Figure 10.21), marine terraces (Figure 10.22), and headlands (Figure 10.23). These landforms bear testament to the active tectonic processes at work that result in faulting and uplift. In addition to tectonic processes, all these landforms are shaped by water and wave action and are also the product of their geologic makeup. Sea cliffs are typically composed of sedimentary rocks, such as sandstone and shale, which are easily eroded by wind and wave action. Sea cliffs are created from the undercutting action of waves at the base of a cliff, which forms a notch. Slowly, the notch gets larger, until the cliff above eventually collapses, causing it to "retreat" back. In a similar process of erosion, marine cliffs can also be carved into sea caves and sea arches; when an arch eventually collapses; the sides left standing in the water are called sea stacks.

Marine terraces can be seen along much of the coastline, rising like stair steps from the sea cliffs inland. Each of the stairs is actually a sea cliff that formed at some point in time, and then, as the movement of the plates pushed the coastline upward, or as climate change caused a rise or fall in sea level, a new terrace step is created. California has extensive areas of marine terraces from Ft. Bragg in Mendocino County to Point Buchon in San Luis Obispo County and Dana Point in Orange County. Marine terraces tend to be unstable and subject to slope failures. By contrast, headlands are composed of harder material that is not so easily erodible, such as granites and basalts. Morro Rock in San Luis Obispo County is an example of a headland. It is actually an extinct volcano and was named by Juan Rodriguez Cabrillo when he first saw it in the sixteenth century (*morro* means "crown" in Spanish).

Beaches and Dunes

In contrast, beaches and dunes are depositional features, formed by the action of wind and water moving and depositing sediments. Beaches are features that are always in flux; in fact, they are defined as the place along a coast where sediment is in motion. Sediment is brought to the coast by streams and rivers. Once at the beach, this material is constantly being moved along the shore by the action of the waves and currents in

Figure 10.21 Sea Cliff at Bodega Bay.
(Image © Andy Z., 2009. Used under license from Shutterstock, Inc.)

Figure 10.22 Marine terrace at Dana Point showing extensive erosion as a result of the 1997–1998 El Niño.
(US Geological Survey)

Figure 10.23 Morro Rock.
(Image © Leifr, 2009. Used under license from Shutterstock, Inc.)

a process called *littoral drift;* in California, this river of sand moves south along the coastline. Beaches also change dramatically in appearance on an annual basis. In winter, storms erode material and deposit it in offshore sand bars, creating winter beaches that are narrow and steep; in summer, the gentle waves return the sand to shore and the same beach becomes wider and more gently sloping.

In general, differences in appearance between the beaches of northern and southern California are the result of the prevailing wind direction and the geology of the area. In northern California, cove or pocket beaches are most common. Here, sea cliffs composed of harder igneous rock, granites and basalts, have been sculpted by the prevailing northwesterly winds and high-energy waves hitting the coast head-on over millions of years, forming coves in the coastline where small beaches form (Figure 10.24). In Southern California, the sea cliffs are composed of shale and sandstone, which are easily erodible, and the coastline is more protected by its eastward trend and the Channel Islands; thus, beaches tend to be much wider and longer, and are broken up by widely separated rocky points (Figure 10.2). The extensive urban development seen along much of California's coastline has had a detrimental effect on the processes involved in beach formation. The damming of so many rivers has dramatically reduced the supply of sediments to California's beaches, and the littoral drift of sediments has been interrupted by the placement of structures such as breakwaters and jetties, which are designed to inhibit the natural transport of sand along the coast.

Dunes can also be seen along the coast and, like beaches, are dynamic, constantly changing landforms. Dunes are formed when winds blow dry sand particles inland, away from the ocean, where they slowly accumulate around objects such as vegetation or driftwood. Over time, the drifted sand forms a barrier that will continue to shift but becomes more stabilized as vegetation gets established. Coastal dune fields form into different sizes and shapes, such as parallel ridges or U-shaped patterns, depending on the amount and type of sand available and the prevailing wind patterns (Figure 10.25).

Over 25 dune areas are found along the California coast. One of the largest is the Nipomo Dunes Complex located along the central coast. This dune complex is quite old, dating from 18,000 YBP, and has the tallest dunes in the western United States. More than 200 birds, including the endangered California least tern (*Sternula antillarum browni*), are found at the Nipomo Dune Complex, and it contains 18 species of rare, endangered, or sparsely distributed plants including surf thistle (*C. rhothophilum*) and beach spectacle pod (*Dithyrea maritima*). Invasive plant species are an issue at this and other dune fields, including European beachgrass (*Ammophila arenaria*), which was planted more than100 years ago to help stabilize dunes and keep sand from blowing onto areas such as agricultural fields. Today,

Figure 10.24 Pocket cove beach at Big Sur.
(Image © Andy Z., 2009. Used under license from Shutterstock, Inc.)

Figure 10.25 Coastal dunes.
(Image © Linda Armstrong., 2009. Used under license from Shutterstock, Inc.)

it is recognized that European beachgrass crowds out native plants and interrupts the natural processes of the dune fields. Currently, the Nipomo Dune Complex is recognized as a National Natural Landmark and has been set aside for conservation.

Coastal development has disturbed dunes at many points along the coast, and, as at the beaches, dune erosion is happening at a greater rate than sand deposition. Dams trap river sediments, depleting the sand supply, and coastal protective structures, such as jetties, disrupt the natural recycling of sand from sandbar to beach. In addition, recreational activities such as walking and off-road vehicle driving disturb vegetation and leave the dunes vulnerable to wind erosion.

Wetlands

Coastal wetlands are transition zones between land and ocean where freshwater and salt water mix. California's coastal wetlands, such as estuarine salt marshes and mudflats, support an abundant and diverse assemblage of plants and animals existing in aquatic, semiaquatic, and terrestrial habitats. Wetlands are also an important interface between land and ocean, supplying nutrients and organic materials to ocean waters and giving shelter to juvenile fish of numerous species. They also help buffer the land from storms and filter runoff to the ocean that is potentially laden with pesticides, fertilizers, or other pollutants.

Plants that grow in coastal wetlands are halophytic, meaning they are able to tolerate salt. Typical plant species include eelgrass (*Zostera marina*), found in the subtidal mudflats; cordgrass (*Spartina foliosa*), growing in the lowest zone of the saltmarshes; pickleweed (*Salicornia* spp.), growing in the middle marsh; and saltgrass (*Distchlis spicata*), growing in the highest zone where the wetlands are transitioning to solid ground. Numerous fish species spend at least part of their life cycle in wetland areas, including the California white sturgeon (*Acipenser transmontanus*), Pacific herring (*Clupea pallasii*), California killifish (*Fundulus parvipinnis*), and the endangered

Delta smelt (*Hypomesus transpacificus*), which is found only in the San Francisco Bay Estuary. The coastal wetlands of California are also part of the Pacific Flyway, one of the four primary bird migration routes in North America. During annual migrations, flocks of waterfowl, such as kingfishers, brants, pintails, and canvasbacks, can be observed as they stop over to rest and feed. One type of birds that nests in the coastal wetlands is the California clapper rail (*Rallus longirostris obsoletus*). This shy little water bird was once found in wetlands throughout northern and central California but currently is found only in the San Francisco Bay Estuary. It has been on the endangered species list since 1970. It was also found on the menus of many San Francisco restaurants until 1915, when hunting the clapper rail became illegal.

The San Francisco Bay estuary is the largest wetland region on the West Coast, accounting for 90 percent of California's remaining wetlands, and it is the confluence of the San Joaquin and Sacramento Rivers. As a result of increasing urbanization over the last 150 years, the bay itself has shrunk by one-third, and a mere 130 square kilometers (50 square miles) remains of the original 2,210 square kilometers (850 square miles) of freshwater wetlands and saltmarshes that once surrounded the bay. In addition to loss of wetlands, the San Francisco Bay estuary system suffers from depletion of freshwater inflows, various forms of urban and agricultural pollution, introduction of nonnative species, and a decline in both terrestrial and aquatic species. Currently, the US Geological Survey and other agencies are conducting research in the San Francisco Bay estuary to monitor the extent and impact of these effects on this important estuary.

Although it is tragic that so much of California's coastal wetlands have disappeared, there has been a growing awareness on the part of Californians that coastal wetlands have great economic, environmental, and recreational importance, and as a result much work is being done to protect and restore wetland habitats. This is evident in the state's 1993 adoption of a policy of no net loss of wetland habitats. In addition, numerous restoration projects throughout the state are ongoing in several areas, including Humboldt Bay in northern California, the San Francisco Bay Estuary, and the Bolsa Chica and Ballona Wetlands in the southern part of the state.

Islands

The islands and rocks found off California's coastline provide critical breeding grounds and sanctuaries for a wide variety of birds and marine mammals. Some also contain rare and endemic species of land plants and animals that have evolved in isolation from the mainland. The Farallon Islands, off the coast near San Francisco, is a group of seven sparsely vegetated islands that were named by a priest on Spanish explorer Sebastian Vizcaino's expedition of 1603 (Figure 10.26). The word *farallon* in Spanish means "rocky promontory rising from the ocean." The Farallons are the largest seabird nesting colony in the contiguous United States, providing critical nesting sites for hundreds of species of resident and migratory birds, including half of the world's population of ashy storm petrel (*Oceanodroma homochroa*); Brandt's cormorant (*Phalacrocorax penicillatus*), which uses its webbed feet to propel itself through the water to catch schooling fish; and the common

Figure 10.26 On only a half-dozen days a year is it clear enough to see the Farallones, about 20 miles west of the Golden Gate Bridge. (Alcatraz is in the foreground.). (Image © Richard Langs, 2009. Used under license from Shutterstock, Inc.)

murre (*Uria aalge*), which actually uses its wings to swim under water to depths of 150 meters (500 feet) and feeds on krill and fish.

The Farallones were established as a National Wildlife Refuge in 1972 and are managed by the US Fish and Wildlife Service. Although they are quite close to the mainland, the public is not allowed on the islands in order to preserve and protect the area for wildlife. In addition to birdlife, numerous species of seals and sea lions have rookeries on the islands, and whales, sharks, and numerous other marine species are found in the surrounding waters. Año Nuevo Island, also near San Francisco, has become a designated preserve as well and is owned and operated by the California State Park system. Like the Farallons, it is not open to the public in order to protect the plant and animal life both on the island and in the surrounding waters. Año Nuevo Island is also an important breeding ground for numerous species of seal and sea lions, and it supports nesting colonies of seabirds.

The Channel Islands, off the coast of Southern California, contain many rare and endemic species of plants and animals that have evolved in isolation from the mainland (Figure 10.27), including the island fox (*Urocyon littoralis*), the island spotted skunk (*Spilogale gracilis amphiala*), and the Santa Cruz Island bush mallow (*Malacothamnus fasciculatus* var. *nesioticus*). Of the eight islands, the four northern islands—San Miguel, Santa Rosa, Santa Cruz, and Anacapa—are actually an extension of the Santa Monica Mountains, whereas the four southern islands—Santa Catalina, Santa Barbara, San Nicolas, and San Clemente—were once connected to the Peninsular Ranges but were sheared off from the mainland by faulting 30 million years ago. The four Northern Channel Islands plus Santa Barbara Island are within the boundaries of Channel Island National Park as well as the Channel Islands National Marine Sanctuary. Both were established in 1980 to protect the natural and cultural resources on the islands and in the surrounding waters. For example, San Miguel Island is the only place in the world where five species of sea lions and

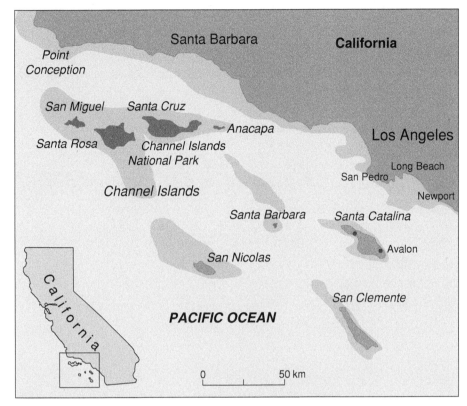

Figure 10.27 Map of the Channel Islands; the four northern islands plus Santa Barbara Island make up the Channel Islands National Park and National Marine Sanctuary.

plant and animal species is something that needs to be cherished and protected not only because it is a major aspect of the state's economy but because it is an iconic symbol of California itself. Californians strive to protect this beautiful coastal region from the effects of urbanization, habitat loss, and pollution because of its intrinsic value, because the coastal ocean region is worth saving, independent of its dollar value. We strive to save it so that sea otters can float in the kelp beds and eat abalone, so that clapper rails can nest in the wetlands, so that sea lions can breed on the Channel Islands, and so that Californians can walk on a beach, go on a whale watching trip, or surf in the waves.

seals maintain rookeries. It is estimated to support the largest concentration of seals and sea lions in the world. More than 60 species of seabirds use the Channel Islands and surrounding waters for nesting and feeding. The islands are also rich in cultural history; archaeologists estimate that the Chumash lived on various Channel Islands as long 8,000 to 30,000 years ago. The Chumash built plank canoes called tomols, which they used to navigate to other islands and to the mainland for hunting, fishing, and trading.

Summary

This chapter looks at the coastal ocean region of California; the variable coastline with its rocky shores, sandy beaches, dunes, and wetlands; and the nearshore and offshore ocean waters with their diverse natural communities and vast array of plant life, invertebrates, amphibians, fish, birds, and marine mammals. This beautiful resource of ocean and coastline and its wealth of

Selected Bibliography

California Coastal Commission. (1987). *California Coastal Resource Guide.* Berkeley and Los Angeles: University of California Press.

Dailey, M. D., Leish, D. J., and Anderson J. W. (Eds.). (1994). *Ecology of the Southern California Bight: A Synthesis and Interpretation.* Berkeley and Los Angeles: University of California Press.

Griggs, G., Patsch, K., and Savoy, L. (2005). *Living with the Changing California Coast.* Berkeley and Los Angeles: University of California Press.

Kildow, J., and Colgan, C. S. (2005). *California's Ocean Economy.* Prepared by the National Ocean Economics Program.

Leet, W. S., Dewees, C. M., Klingbeil, R., and Larson, E. J. (Eds.). (2001). *California's Living Marine Resources: A Status Report.* Sacramento, CA: California Department of Fish and Game.

Panel on the Southern California Bight of the Committee on a Systems Assessment of Marine Environmental Monitoring and Marine Board Commission on Engineering and Technical Systems National Research Council. (1990). *Monitoring Southern California's Coastal Waters.*. Washington, DC: National Academy Press.

Rawls, J., and Bean, W. (2003). *California: An Interpretive History.* New York: McGraw-Hill.

Schoenherr, A. (1995). *A Natural History of California.* Berkeley and Los Angeles: University of California Press.

Sustaining California and Beyond

"When we tug at a single thing in nature, we find it attached to the rest of the world."

–John Muir

Introduction

With a little over 4 percent of the country's land area, California hosts 12 percent of the country's population. Although the land size stays the same, the population has continuously increased (Chapter 9). It is projected that the population will continue to grow for the next few decades. With the current resources, will the state be able to support the needs of the current generation without compromising the environment for future generations? What is the status of the state in terms of available resources, practices, and possibilities to sustain the population growth and to protect the environment? These and related questions are explored in this chapter from both human and physical geographical perspectives. Unlike the previous chapters, the topics explored in this chapter are more exploratory, uncertain, and at times controversial. Because these topics are related to the future of all the people in California and beyond, however, it is important to look into these issues.

Needs of a Growing Population

Food

Currently, of the approximately 400,000 square kilometers (156,000 square miles) of land area in California, about 28 percent is farmland, which accounts for most of the state's *arable land*. With the increasing population in the state, the arable land has been decreasing continuously. It has been estimated that for each person added to the population approximately one acre of land is needed to meet the shelter and transportation infrastructure needs, and most of this comes from arable land. Data collected from a US agriculture census show that more than 480,000 hectares (1.2 million acres) of California farmland were lost in the five-year period from 1997 to 2002. Currently, the state sufficiently meets the food supply needs of its population, and agricultural products produced in the state are still being exported. Most of the foods imported from other states or countries are to meet consumer preferences rather than basic food supply needs. The question is how long the self-sufficiency in food supply will last with the increase in population and the decrease in arable land and water resources. If it becomes necessary to import foods to support the basic needs of the population, many new issues and challenges will surface.

Shelter

Land, construction materials, water, and energy are the fundamental natural resources that support the shelter needs of a population. Californians have had a larger share of these resources compared to other places in the country or the world with similar population sizes. Californians used to distinguish themselves from the rest of the major urban areas in the nation with low-rise, low-density residential areas and single-family homes. Although this still hold true to a certain degree, these claims may not last very long. Available land for new development is decreasing as competition with agriculture and environmental protection needs intensifies. To reflect the changes, multiunit high-rise and high-density residential housing have been developed in recent years. Although these developments require less land, they still add pressure on the demand for water and other resources and may intensify the environmental impact if the housing is not carefully planned.

Water

Water is key to a successful food supply system as discussed in Chapter 9. Water is also key to urban development, especially in the southern part of the state where annual precipitation falls short of supporting population needs. As discussed in Chapter 8, a large portion of the water supply in major urban areas in Southern California and the Bay Area are imported from other regions. Such a practice has resulted in serious environmental problems in the water source regions, as seen in the Mono Basin and the Owens Valley, where dust storms and other environmental problems intensified after the water was diverted to other regions. With increasing frequency of drought conditions in the state, water resources will play an increasing role in limiting the population growth.

Energy

Energy is at the center of the movement toward a sustainable state. The role of energy is twofold. First, energy is needed to produce food, transport water, provide mobility, build comfortable shelters, and support industry developments. By contrast, producing and consuming energy are the most important causes of environmental pollution and other problems, such as the release of greenhouse gases linked to global warming.

Annually, California consumes approximately 30 trillion kilowatts (10,000 trillion BTU) of energy, with about half of the energy in petroleum form and half split between electricity and natural gas. Figure 11.1 shows the different types of energy sources for the state. It is clear that California cannot generate enough energy to support its needs. The state produces about 70 percent of its electricity, less than 40 percent of its crude oil, and only 14 percent of its natural gas. All combined, about half of the energy consumed in the state is imported from other states or other countries. California imports more electricity than any other state in the country.

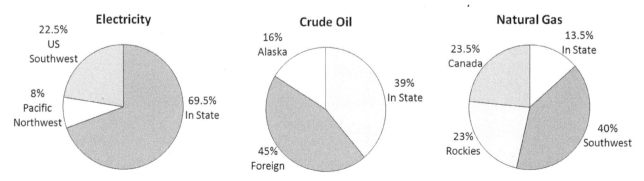

Figure 11.1 California energy sources.
(Data Source: California Energy Commission)

Figure 11.2 is a comparison between California and the rest of the country in energy sources used to generate electricity. For California, more than half of the electricity is generated by natural gas, a relatively clean energy source, although not renewable; whereas for the country, close to half of the electricity is generated by coal, which emits the most pollutants into the air compared to other energy sources for electricity. In this regard, California uses more clean sources to produce electricity compared to the country as a whole. In fact, California leads the nation in using non-hydropower *renewable energy* sources to generate electricity, including geothermal power, wind power, landfill gas, and solar power.

Because petroleum and natural gas are found only in sedimentary rocks, most in-state petroleum and natural gas production is located along the coastal areas and in the Central Valley, where sedimentary rocks dominate. In terms of production capability, California ranks third in the nation in refining capacity, and its refineries are among the most sophisticated in the world.

Despite the large amount of energy consumed, California leads the nation in low per capita energy consumption, partly because of its mild climate, which reduces heating and cooling needs. Residential and commercial energy consumption accounts for a little over one-third of the total energy consumption, which is lower than the national average. By contrast, California's energy consumption in transportation makes up the largest energy consumption sector (40 percent), higher than the national average. The higher proportion of transportation energy consumption in California is a result of several factors, including large land, relatively dispersed population, and long-distance commuting due to expensive housing. With most of the transportation energy consumption in the form of petroleum fuel, there is a great potential to reduce the environmental impact and foreign oil dependency by developing technology and strategies that will improve fuel efficiency, replace petroleum fuel with alternative clean fuels, and reduce transportation needs in the state. This topic is further discussed in the following sections.

Environmental Impact

Ecological Footprints

Ecological footprint is a measure developed in the 1990s and later adopted worldwide by scientists, environmental activists, and policy makers as one of the measures to gauge the sustainability of an environment. The measure calculates the area of land and the amount of resources needed to support the needs of a population and absorb the waste it generates under the prevailing technology. These footprints can be calculated from an individual level to the global level. On a global

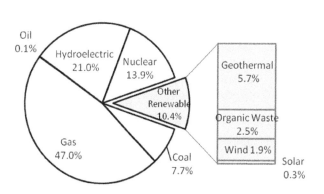

US Electricity By Source 2006
Source: Energy information administration Official Energy Statistics from the us government

California In-State Electricity by Source 2006
Source: California Energy Commission

Figure 11.2 California and US.electricity by sources.
(Data Source: California Energy Commission)

scale, the ecological footprint of the world population as a whole has increased steadily and has "overshot" the earth's *biocapacity*, according to some studies. Some estimate that the ecological footprint for a typical US resident is 24 global acres, whereas the footprint for a typical world resident is only 5.4 global acres.

The numbers should be used carefully so as not to oversimplify the complexities of the human–environment interaction. For example, the consumption of a product could have a completely different impact on the environment with different production, consumption, and disposal methods. Nevertheless, regardless of the definition or calculation methods, negative human impact on the environment is apparent: decreasing biodiversity is observed everywhere in the world; *land degradation* occurs in most farming areas; the overall world forest coverage is decreasing; increasing global warming due to human activities is a widely accepted fact; and severe air-water-soil pollution problems occur in many places. With a land area over 400,000 square kilometers (156,000 square miles), all the signs of environmental stress are seen in California.

With more than 38 million people living in the state and another 350 million domestic and international travelers visiting each year, the impact on the environment is clear. However, on a per capita basis, Californians are doing better than most of the residents in the United States. Collectively, California is one of the highest emitters of carbon dioxide among all the states in the country, but the state's per capita emission is among the lowest in the nation. This lower per capita consumption is partly the result of favorable climatic conditions that reduce the air conditioning and heating needs and partly the result of the high emission standards set by the state.

Land Degradation

Land degradation occurs when the healthy ecological processes of water, energy, and mass flow in the environment are interrupted as a result of the removal or disturbance of vegetation cover. In California, the land is more vulnerable to land degradation in the rigid mountain environment and the arid environment. Land degradation is also widespread in the agricultural regions, especially in the areas with large-scale farming operations and where livestock are raised in open range areas. The consequences of land degradation are complicated, and their effects reach far beyond the degraded land. Decreased agricultural productivity; increased flooding, landslides, dust storms, pollution, and sediment flow in streams and reservoirs; and loss of native flora and fauna are just a few of the visible consequences of land degradation.

Air Pollution, Water Pollution, and Solid Wastes

The California Environmental Protection Agency is the key organization that tracks the environmental pollution status of the state and makes policy recommendations to protect the environment and its inhabitants from harmful materials. In 2002, the first comprehensive report, *Environmental Protection Indicators for California*, was published and in subsequent years has been updated. The report documents the pollution status of various carcinogens in gas, liquid, and solid forms that go into the air, water, and land.

In general, as a result of the continuous efforts in fighting air pollution problems, the air pollution levels of most carcinogens in California have declined in past decades. The rate of decline varies, however, based on the region and the type of pollutant. In some cases, the level is still alarmingly high despite the decline. For example, Figure 11.3 shows that for one out of every three days in the South Coast area, the ozone level exceeded the state standard, although exposure to unhealthful levels has declined significantly in the same region. The ozone level has remained the same in the San Joaquin Valley, with an upward trend in recent years, whereas in the San Francisco Bay area, the level has been significantly lower than in other regions of the state.

Although transportation and energy consumption in the urban areas are major sources of pollution, agriculture is the main source of pollution in the rural areas. Following the same trend in the United States and the western world,

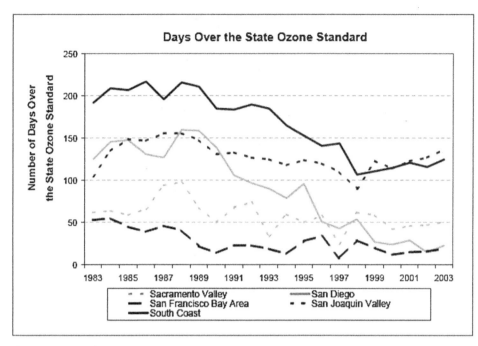

Figure 11.3 Number of days when ozone pollution levels exceeded the state health standard in the 20-year period from 1983 to 2003.

(Source: Environmental Protection Indicators for California, 2004 Update. California Environmental Protection Agency—California Resources Agency, January 2005)

farming practices in California have undergone revolutionary changes since the 1940s with the introduction of new technologies and management models. Machines, chemical fertilizers, and chemical pesticides were introduced; monoculture, big, and cooperative-owned farms replaced multi-crop, small, family-owned farms. These changes brought a significant increase in productivity and created a multibillion-dollar industry that became part of the core of the state's economy as discussed in Chapter 9. With the increased productivity and revenue, however, many problems, especially environmental problems, also surfaced and intensified. Chemicals used in the fertilizers and pesticides contaminated groundwater and generated air pollution in some areas; the chemicals remained in soil, and the monoculture practices decreased soil productivity over the time. Increased salinity caused by irrigation and chemical use has damaged the environment beyond where the salt resides. The Salton Sea, in the Imperial Valley farming area, was a classic example of a damaged environment that resulted from unsustainable farming practices. The "sea" is an accidental human-made lake created by the

accumulation of irrigation water in the Salton Sink, which used to be a dry land depression. For years, the increasing salinity and harmful chemicals from irrigation water created a troubled ecological environment with decreasing wildlife, diminishing biodiversity, and unhealthful living conditions. Restoration efforts over recent years are yielding signs of recovery; however, the salinity of the lake is still over 40 parts per million (35 PPM for the Pacific Ocean).

In 2007, an estimated 39.6 million tons of solid waste in California went into landfills. This is about 42 percent of the total waste generated (the other 58 percent is diverted). On average, each Californian generates more than 6.2 kilograms (14 pounds) of wastes each day; approximately 2.6 kilograms (5.8 pounds) goes into landfills. There are two major types of solid wastes: municipal and industrial. In California, construction, demolition, and agriculture are the major contributors to industrial solid wastes. Because industrial wastes are more predictable, guidelines have been established by different government agencies to manage the flow and diversion (through recycle and reuse) rates to minimize the wastes that go into landfills.

Most municipal solid wastes go into landfills. Figure 11.4 shows the closed and active solid waste disposal sites in California. To ensure environmental and public safety, modern landfills are strictly regulated, from site selection, construction, and operation management, to post-operation monitoring. They still take up large land areas, however, and pose potential threats to the environment.

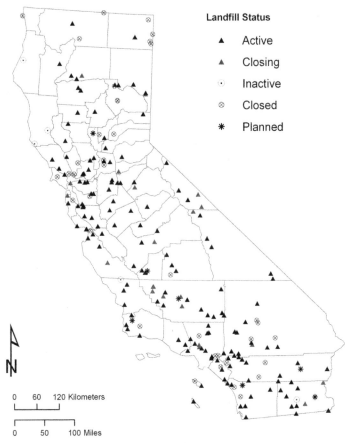

Data Source: California integrated waste management board. July 2009.

Figure 11.4 Past and current solid waste disposable sites.
(Source: Lin Wu, Data: California Integrated Waste Management Board, 2009)

Landfill Status

▲ Active

▲ Closing

⊙ Inactive

⊗ Closed

✳ Planned

Impact of Climate Change

The issue of climate change, especially global warming, has been brought to the center stage of world affairs in recent years. Climate change is considered the most threatening environmental issue of our time. From physical geography perspectives, California faces many of the direct threats of climate change: more frequent extreme weather events could cause widespread hazardous conditions; prolonged drought could severely affect the state's agriculture and water supply; and, with a long coastline, a rising sea level may intrude into many coastal communities and lowland areas.

On a positive note, the attention to the climate change in recent years has brought scientific research, technological advancement, and economic development opportunities to the state. California, for many years, has presented itself as a leader in fighting the climate change. Indeed, California has the toughest emission and environmental standards, and it is in the forefront of alternative energy research and practices.

Solutions and Challenges

New technologies, policies, strategies, and lifestyles in all areas need to be developed in order to sustain the state with its population and environmental needs. We now look into the current practices, experiments, and potentials in seeking solutions to sustain the state. We also explore the challenges and the downside of these solutions. It is not possible to touch on every important aspect of this topic; what's presented here is only a sample.

Renewable and Alternative Energy Sources

California has been a leading state in exploring solutions to the energy sources and energy pollution problems. California is considered to be "a model for the potential of alternative and renewable energy sources" for other states in the nation, as voiced by the US Department of Energy. The following sections present an overview of the alternative energy resources the state has already tapped into or is in the process of exploring. Although brief overviews of basic principles in science and technology behind each of the alternative energy sources may be necessary for the discussion, the focus is placed more on the geographic perspectives of each topic.

Solar Energy

Solar energy comes from the sun in the form of radiation. Because the energy is generated during the fusion process in the sun, there is no direct pollution on earth and the energy is renewable. Solar energy can be converted into electricity, or the heat from the sun can be used directly for cooking, heating water, and other uses that otherwise would need electricity or another power source. Because of the timing and locational restrictions, most of the direct heating uses of solar energy are limited to smaller scales, such as home water heaters. Because electricity generated from solar radiation can be stored and transported, more potential and possibilities exist for using solar power to generate electricity than for using it directly for heating purposes.

Although the sun emits radiation over the entire state 365 days during a year, the geographic distribution of the solar energy is uneven. The highest concentration of solar energy is in the southeastern corner of the state, particularly in the Mojave Desert (Map 10). This region experiences higher sun angles (more intense energy) and fewer cloudy or rainy days than elsewhere in the state. Because of this, California's large-scale commercial solar power plants are located here, though they are still limited in number. In recent years, the use of small-scale solar generators, such as rooftop solar generators, has increased significantly as the technology has become more available and affordable (Figure 11.5).

The collection or production of solar energy is not without challenges or negative impact on the environment. Current solar technology still makes solar energy more expensive than traditional forms of energy resources, and many of the technologies are still in the developmental stage. For example, cars powered directly by solar energy are still a concept and are nonexistent on the market (although one could argue that some electric cars that use solar-generated electricity could be considered solar-powered cars). Collecting solar energy takes a large space. With solar collectors such as photovoltaic cells (PVCs) replacing the natural groundcover, the natural atmospheric and biological processes are altered and could create a negative impact on the environment. The uneven geographic distribution of solar energy is also a challenge. Locations with high concentrations of solar energy are often long distances from the consumption locations and so require extensive infrastructure development for storing and transporting the electricity. These developments could add to the cost and environmental impact to the region. In addition, the production and disposal of PVCs and other solar devices also generate environmental wastes. Research and technology development

are still needed to maximize the resources and minimize the cost and impact.

Hydropower

Hydropower contributes about 19 percent of the total energy in California. About 80 percent of the hydropower is generated in the state; 20 percent is imported from neighboring states. The state has about 200 large reservoirs, and about half of them generate electricity. Hydropower generators use the potential energy carried by free-falling water to drive turbines that generate electricity. Because water is "moved" by the natural hydrologic cycle from lower elevations to higher elevations through evaporation and precipitation, hydropower is truly a renewable energy source.

Hydropower generation is highly restricted to locations with enough water in the stream and a significant elevation drop. These conditions are met only in the northern part of the state where rivers that originate from the Sierra Nevada and Klamath Mountains have enough water flow and elevation drop to generate hydropower. To tap into hydropower, a dam must be constructed to form a reservoir behind it. The water level difference in front of and behind the dam makes it possible to capture the potential energy in hydropower. In California, most hydropower operations are byproducts of the reservoirs, which were built primarily to manage and capture the water flow.

Hydropower is often considered to be a "clean" alternative to *fossil fuels* because the energy-converting process is mostly a physical process without releasing harmful chemicals into the environment; however, the energy is not derived without paying an environmental price. Building dams and reservoirs interrupts the natural flow of streams and alters the ecological balance of the watershed environment. The long-term impact of dam building on the environment was noticed just a few decades ago. In recent years, there has been a movement calling for the reduction of dams and reservoirs and for restoration of some watersheds to their original state. Some efforts were made to restore watersheds after old dams were removed. The restoration process is costly and sometimes not

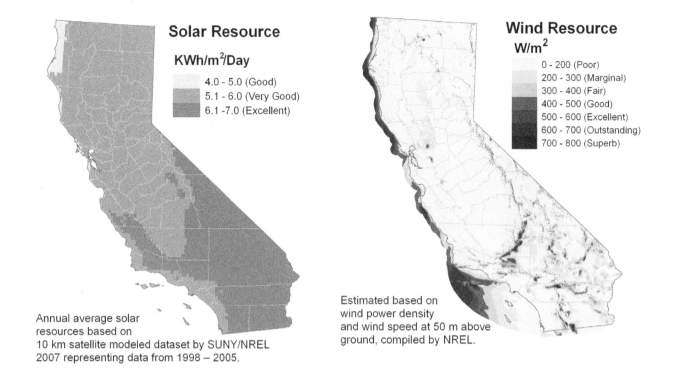

Solar Resource

KWh/m²/Day

- 4.0 - 5.0 (Good)
- 5.1 - 6.0 (Very Good)
- 6.1 -7.0 (Excellent)

Annual average solar
resources based on
10 km satellite modeled dataset by SUNY/NREL
2007 representing data from 1998 – 2005.

Wind Resource
W/m²

- 0 - 200 (Poor)
- 200 - 300 (Marginal)
- 300 - 400 (Fair)
- 400 - 500 (Good)
- 500 - 600 (Excellent)
- 600 - 700 (Outstanding)
- 700 - 800 (Superb)

Estimated based on
wind power density
and wind speed at 50 m above
ground, compiled by NREL.

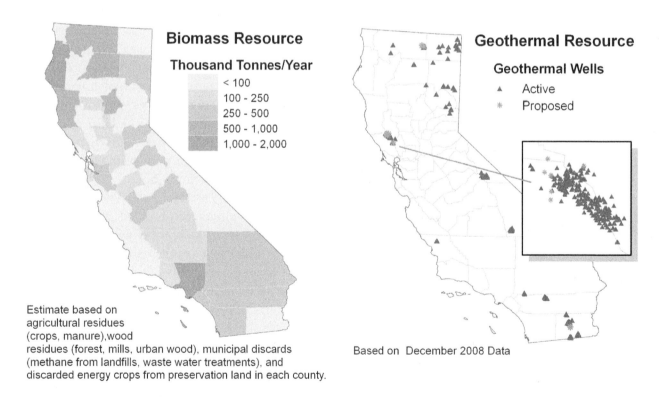

Biomass Resource

Thousand Tonnes/Year

- < 100
- 100 - 250
- 250 - 500
- 500 - 1,000
- 1,000 - 2,000

Estimate based on
agricultural residues
(crops, manure),wood
residues (forest, mills, urban wood), municipal discards
(methane from landfills, waste water treatments), and
discarded energy crops from preservation land in each county.

Geothermal Resource

Geothermal Wells

- ▲ Active
- ＊ Proposed

Based on December 2008 Data

Sources: Solar, wind, and biomass data are from National Renewable Energy Laboratory (NREL), Office of Energy
Efficiency and Renewable Energy, U.S. Department of Energy.   Geothermal data are from California Department of
Conservation, Division of Oil, Gas, and Geothermal Resources

Map 10 California renewable energy resources.

(Source: Lin Wu)

Figure 11.5 Solar panel on a residential area.
(Source: Lin Wu)

at all possible. In addition to these problems, most hydropower plants are located in remote areas, and the power generated must be transported long distances to the population centers. Because of all these restrictions, the potential of increasing hydropower resources within the state is limited.

Geothermal Power

The map of energy source distribution in California (Map 10) shows the geothermal resources in the state. Geothermal power accounts for part of the 11 percent of renewable energy resources for electricity generated in California. The potential of geothermal power in the state is not fully tapped yet. *Geothermal energy* is the heat beneath the earth's surface; when captured, it can be used to generate electricity. The energy can be captured by different technologies, depending on whether the source is water or steam and the temperature of the source. Dry steam power plants convert the energy from steam directly into electricity. Steam coming out of a production well drives the turbine connected to a generator, which in turn generates electricity. The steam is then injected back into the ground through an injection well. Dry steam power production is the oldest, yet still efficient, technology that is widely used. The Geysers, a dry steam power plant located along the border of Sonoma and Lake Counties, is the largest geothermal power plant in the world. The electricity generated by the power

plants in the Geysers produce 60 percent of the power of the North Coast area from north of San Francisco to the Oregon border.

Electricity can also be generated through a binary cycle as illustrated by Figure 11.6. The heat carried by the hot water or steam runs through an enclosed tube in close contact with a secondary enclosed system. The heat running through the first tube causes the fluid in the second tube to vaporize, which then drives the turbine connected to the generator. Because the fluids from inside the earth run through an enclosed system, air pollution is minimized. This operation is also effective with the geothermal temperature below 200°C (400°F), which describes a large portion of California's geothermal fields. In all these cases, underground rock structures that facilitate water or steam to recycle through the geothermal layers are essential.

Currently, there are about twenty geothermal power plants in California, and there is great potential for the development of more in the future. Most of the new development will use the binary cycle technology because of the advantages mentioned above. Although geothermal power is a relatively clean energy source, the process still releases chemicals into the air, especially hydrogen sulfide (H_2S) carried out by the underground heat. Special systems must be in place to reduce the emissions and to

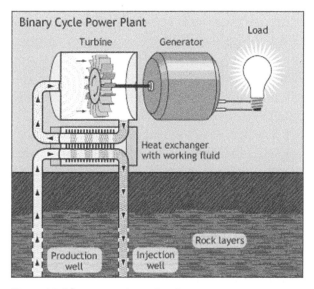

Figure 11.6 Binary cycle technology.
(Source: US Department of Energy, Energy Efficiency and Renewable Energy)

minimize the impact on air quality in the local area. In addition, most large geothermal fields are not close to population centers, and so the electricity generated must be transported.

Wind Power

A pioneer in wind power generation, California still leads the nation in wind power production. Wind power is another renewable energy resource and uses wind turbines to harness energy from strong winds high above the ground. Electricity is generated when the wind turns turbines connected to generators. Suitable locations for commercial-level wind power plants are limited mostly to mountain passes that have an annual average wind speed measured at 50 meters (183 feet) above ground that reaches or passes 6 or 7 meters per second (15 to 16 miles per hour). Most of these locations are in the Southern California mountains, including the mountains east of San Diego, San Gorgonio Pass near Palm Springs, the Tehachapi Mountains near Bakersfield, and areas in the Mojave Desert. In Northern California, there are a few sites in the Coast Ranges near San Francisco (Map 10).

Technology for wind power generators has improved significantly in the past decade, but it is still expensive to establish large–scale wind power plants. One limitation of power plants is the instability of the source. In most locations, wind speed follows a daily cycle and, for some, a seasonal cycle. Wind power production fluctuates with these cycles. With a forest of running wind turbines in an area, the disturbance to the local environment could be quite significant. Some studies have suggested high mortality rates of birds in these areas.

A recent development in wind power technology is the design of a vertical wind turbine (Figure 11.7b) that could run on a lower wind speed, need less space compared to a traditional wind turbine (Figure 11.7a) of comparable scale, and reduce disturbance to the environment. The new development in wind technology increases the potential of small–scale wind power generation in an urban environment and in residential use. Combined with small-scale solar technology development, the hybrid wind–solar power

generator could minimize the instability problem of both solar and wind power and reduce the need for power transfer lines.

Figure 11.7a A traditional wind turbine in Central Valley.
(Source: Lin Wu)

Figure 11.7b Vertical wind turbines on the roof top of the Adobe headquarters.
(Source: Used with permission of San Jose Mercury News Copyright © 2013. All rights reserved.)

Nuclear Power

Nuclear power is generated when unstable radioactive material undergoes nuclear fission—a process that generates enormous heat and radiation. Few radioactive materials exist in the earth's natural environment. The common radioactive material used in nuclear power generation is uranium-235 (U-235). Energy released from 1 gram (0.035 ounces) of U-235 when undergoing fission is 1 million watt-days—2.5 million times the energy generated by 1 gram of coal. In other words, it takes 2.5 metric tons of coal to generate the same energy as 1 gram of U-235.

However, electricity is not the only product generated by a nuclear power plant. The other product is radioactive waste material that could be a public health hazard if not properly transported and disposed of. Another public safety concern is the potential catastrophic consequences of leaking radioactive materials as a result of accidents, such as the Chernobyl disaster in 1986 and the Fukushima Daiichi nuclear power plant disaster during the March 2011 earthquake. The radioactive material used to generate electricity is the same material used to make nuclear weapons. Public policy makers are also concerned about the security of these materials. In addition, the cost of nuclear power is still higher than many of the other options.

Because of these concerns, nuclear power development in California declined in the 1990s despite the large amount of energy generation potential and the clean air in the generation process. In California, most nuclear power plants stopped operation in the last two decades, with the latest closure being the San Onofre's two units in 2013, because of safety concerns. Interests in nuclear power have been renewed in recent years, however, because the world is facing the consequences of global warming and the high cost of oil. These renewed interests are helped by development in technology that promises accident-free reactors and safe disposal of the radioactive waste materials. With a long history of nuclear power development in the state, this energy source holds great potential for the state in the future.

Other Energy Sources

With its long stretch of coastline, wave-generated power may hold potential for California as another renewable energy source. Ocean waves are created when winds move across the ocean surface. These waves hold a high amount of energy that could be captured to generate electricity. Although the power of a wave is easily observed at any rocky coastline when the wave pounds the rocks with its full force, capturing that wave energy is not an easy task. Around the world, different technologies have been developed that attempt to capture the wave energy near the shoreline or away from the shoreline. Currently, it is still not economically viable for the state to generate power from wave energy.

Another renewable energy source that has attracted attention in recent years is biomass. Through biological processes, plants and animals store energy in their tissues. After the organisms die, this energy can be converted into fuel for vehicles or to generate electricity. As is true for any other alternative energy sources, biomass has its limitations. It requires large land areas and water to grow biomass materials, and these resources are limited in the state; it requires harvesting and/or gathering and transporting biomass to processing centers; and, because it is burned to produce energy, it contributes to air pollution. Research has been conducted to explore high-efficiency or recycled bioenergy production, such as biofuel extracted from *algae* and the use of recycled food oil. However, experiment and practices are still very limited.

Better Cars and Fewer Cars

With vehicles consuming nearly half of California's energy, making more fuel-efficient cars, using cleaner fuels, and reducing the number of cars on the road will significantly reduce energy consumption and environmental impact while enhancing the quality of life for the more than 38 million Californians. Let's take a look at some of the developments in this area.

Electric and Hybrid Vehicles

The initial concept of electric vehicles dates back to the mid-1800s and early 1900s. Commercial development and mass production of these vehicles, however, didn't take off until the 1990s. Today electric and hybrid vehicles are the only noticeable alternative vehicles in the market.

Electric vehicles use electric motors to convert electricity, usually stored in battery packs, into mechanical power that runs the vehicles. The conversion process in the vehicle is emission free, but generators producing the electricity used to charge the batteries do emit pollutants. When electricity is generated in power plants, the emission is better controlled than power generation on individual vehicle levels. Collectively, the air pollution could be reduced even if the electricity that powers the electric vehicles is generated from fossil fuels. If renewable and cleaner energy sources are used to generate electricity, air pollution and greenhouse gas emissions could be significantly reduced.

Electric vehicles, however, have a limited driving range between each battery recharge. To overcome this limitation, hybrid vehicles, which run on both battery power and fuel power, were developed. Most car trips are within the range of battery power, but hybrid vehicles have the advantages of low emission and a long driving range when needed. As a result of the technological developments that make electric and hybrid vehicles affordable, incentive policies (e.g., the benefit of using carpool lanes), and soaring gasoline prices, electric and hybrid car purchases have increased drastically in the 2000s. These vehicles hold a great potential for the future.

Biodiesel

One alternative fuel for vehicles is biodiesel—liquid fuel derived from new and used vegetable and animal fats. With little modification, biodiesel can be used to power diesel engines. Although not completely emission free, biodiesel releases significantly less carbon dioxide and fewer other chemicals into the air, and it is a renewable energy source because vegetable and animal fats are ultimately produced by solar energy. The proportion of biodiesel among all the fuel sources is still small, but biodiesel production is increasing rapidly. Currently, California has more than a dozen biodiesel plants in production or under construction and

more than 60 biodiesel distribution sites; most of them are located around major metropolitan areas in Los Angeles and in the Bay Area. Like other alternative energy sources, biodiesel is not a fuel source without limitations. The amount of recycled animal and vegetable fats for producing biodiesel is limited, and large production of biodiesel calls for a large land and other input to produce vegetable or animal oil, which could take away valuable land and resources. Although it reduces the amount of carbon and other gas emissions, burning biodiesel increases nitrogen oxides (NO_x), one of the main ingredients of ground-level ozone. Research and development are needed to overcome these challenges to increase the use of biodiesel.

Hydrogen-Powered Cars

Although there is still a long way to go before hydrogen-powered cars become accessible to the general public, the opening of the first public hydrogen fuel station on the West Coast (Los Angeles area) in June 2008 signifies the determination of policy makers and car manufacturers to move forward with this new technology.

Hydrogen fuel technology is not new, but the use of hydrogen to power vehicles is a new development. In 2005, California was the first state to classify hydrogen as a transportation fuel. Hydrogen-powered cars use fuel cell technology to convert hydrogen into electricity to power cars. Fuel cell cars are, in fact, electric cars, except that the electricity is generated continuously in the fuel cells on board the vehicles instead of generated in power plants and stored in the batteries. Fuel cells generate electricity by mixing hydrogen with the oxygen in the air (Figure 11.8). The only emission from the vehicle when running on hydrogen power is water, which eliminates the greenhouse gases and other harmful chemicals from the car emission.

Hydrogen is a colorless, odorless gas that accounts for 75 percent of the entire mass of the universe. It is a renewable source. However, pure hydrogen does not exist in the natural environment; it is found only in combination with other elements, such as oxygen, carbon, and nitrogen. To use hydrogen as a power

Hydrogen generating technology options

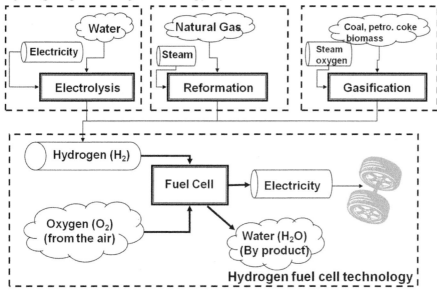

Figure 11.8 Oxygen fuel technology.
(Source: Lin Wu)

source, it must be separated from other elements, stored, and transported. Therefore, the main technological and economic challenges in developing hydrogen-powered vehicles are to develop economically viable and emission-free technology for producing, storing, and transporting hydrogen.

It should be noted that hydrogen, similar to electricity, is not a source of energy but a form of energy that can be stored and transported. Just as with electricity, it takes energy to produce hydrogen. When hydrogen is generated from renewable energy sources, however, hydrogen-powered cars can reach near-zero emission standards.

Alternative Transportation Systems

On average, public transportation produces 95 percent less carbon monoxide (CO), 90 percent fewer volatile organic compounds (VOC_s), and about half as much carbon dioxide (CO_2) and nitrogen oxide (NO_x) per passenger mile as compared to private vehicles. Developing a strong public transportation system is one key for the state to reduce transportation fuel demand and pollution. Based on Bureau of Transportation 2010 data, only 5.2 percent of commuters in California use public transportation, which is a 0.2 percent decrease from a decade ago, and

73.2 percent drive alone. Although these figures are along the same lines with the national figures, some states with large populations have a much higher percentage of commuters using public transportation. For example, during the same year (2010), 26.7 percent of New York commuters used public transportation.

Large population and access to advanced technology in California give the state some advantage in developing a successful public transportation system; however, it proved to be challenge because of the large land area, scattered population centers, and, perhaps most important, the attitude of Californians toward transportation choices. The progress in public transportation has been slow, but the potential is there.

To make public transportation systems more successful in the future, developments at state, regional, and local levels will have to be improved and integrated. An example of state-level improvement is the proposed fast rail system that will make the travel time between most of the population centers in the state two hours or less. This timeframe will increase possibilities for many of the commuters. A timeline for such a project, however, is still unclear. An example to improve regional scale transportation system is Northern California's BART improvement project, which would replace and modernize its aging fleet of 700 rail cars to make them more attractive, convenient, and comfortable for the passengers. Recent trends in local improvement have been to incorporate transit stops into local commercial centers and make improvements in pedestrian traffic patterns.

Another solution to getting cars off the streets is to reduce commuting time via better planning for urban and community development. By designing homes closer to workplaces and service centers, bringing jobs to population centers, improving walking- and biking-friendly infrastructures, and changing culture and life-styles, fewer commuting trips by cars will be needed. One example in the state is the city of Davis. A small town with a large university population, Davis is well known for its environment-friendly community design and atmosphere. It has a functioning public transportation system, and a reported 20 percent of the population commutes on bikes, which is 10 times more than the national average.

Because of the nature of a college town, it may be difficult for other places in the state to follow Davis's model; however, strategies such as improving bike routes and strategically placing transit stops could increase the number of people using alternative transportation.

Zero Waste

Increased environmental awareness and increased unavailability of waste-disposal sites have prompted efforts to recycle and reuse solid wastes and thus reduce the amount that goes into landfills. These efforts include new and enhanced

solid-waste management systems that collect and recycle green wastes and other recyclables in the urban areas, and mandatory construction and demolition diversion programs. These efforts have brought significant change in solid-waste diversion rates, as shown in Figure 11.9. In 1989, 90 percent of solid wastes in the state went into landfills; by 2006, the rate was reduced to 46 percent. As the saying goes, "Reduce, Reuse, and Recycle."

By contrast, because the amount of solid wastes generated each year has been steadily increasing in pace with the population and construction growth, the net amount of solid wastes that goes into landfills each year has not been reduced (Figure 11.9). It became clear that relying on diversion strategies alone could not solve the waste problems. Zero Waste, a concept related to waste management, was introduced and became increasingly adopted by communities worldwide, including about two dozen California cities and counties, such as Oakland, San Francisco, and San Juan Capistrano. Similar to many of the recycling programs developed in the 1980s and 1990s, the Zero Waste approach

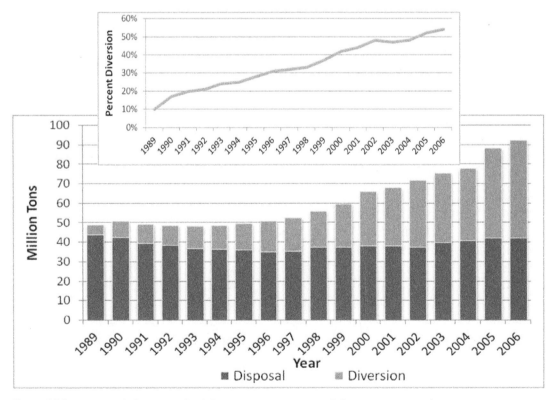

Figure 11.9 Historical changes of solid-waste generation and diversion rate in the state, 1989–2006.
(Source: California Integrated Waste Management Board)

also encourages maximizing the reuse and recycling of materials. The fundamental difference between the two is that Zero Waste also tackles products at the production level and approaches the problem as a cyclical system. It uses nature as a model. In the natural processes, there is no "waste"; materials go through different stages and states in cyclical processes. In practice, it may take years to reach the Zero Waste goals even at a community level; however, many cities and counties in the state have adopted it as a guiding principle to revolutionize solutions for waste problems.

Alternative Agricultural Practices

Modern farming practices have created serious pollution and land degradation problems. As part of the solution, there has been a movement toward sustainable farming, including organic farming and other practices. The main objective of sustainable farming is to reduce input while keeping the productivity and sustainability of the land. Here are a few key ingredients that make up the core of sustainable farming practices: (1) use multicrop rotation to increase soil productivity and quality, (2) diversify crops to create positive ecosystems that increase productivity while reducing water and other input; (3) base crop selection on site-specific conditions, such as climate, topography, water system, soil conditions, and land use history to optimize production while minimizing environmental impact.

With the awareness of environmental problems and food quality increases, organic farming - practices with crop rotation, multicropping systems, composting, green manure, biological pest control, and so on -, has

regained its place in agricultural industry. The introduction of the "organic" label in supermarkets in recent years has helped marketing the products from organic farms; however, these farms are still a very small portion of the agriculture industry in the state, and the number of organic growers is not increasing, although the farm-level sales values from the organic growers have increased, especially for vegetables and fruits (Figure 11.10).

In most cases, organic farming is more sustainable than nonorganic farming, but sustainable farming practices may or may not carry the organic label. In addition, organic farming is more labor-intensive and may not be an option for large-scale farms. In addition to organic farming, there are other options for improving sustainability in agricultural practices and food supply, including switching to renewable energy sources and promoting edible gardens in urban residential areas.

From time to time, industrialization and technology are often blamed for Earth's environmental problems; however, the goals of sustainable agriculture could not be reached without the help of technology. In fact, science and technology will play increasingly important roles in

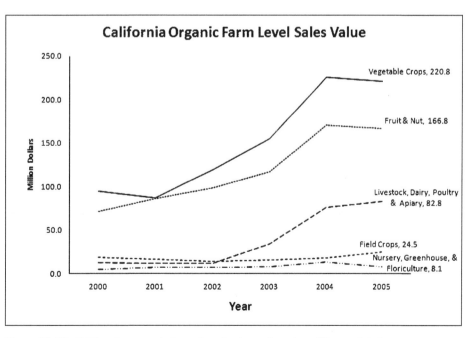

Figure 11.10 California organic farm-level sales values in millions of dollars.
(Source: Statistical Review of California's Organic Agriculture, 2000–2005, UC Agricultural Issues Center, University of California, Davis)

reaching sustainability goals. An example of this is the development of precision farming, which is farming with the assistance of remote sensing, global positioning systems, and computing technology to obtain information such as soil nutrients, wind speed, and potential-crop-yield from different locations in the field so that more precise decisions about water, nutrients, and other important aspects of farming can be made. What goes into the land is carefully calculated on the basis of the specific needs of each location in the field to reach the predicted crop yield. These technology and management models are still being developed and tested in the country, including several locations in California.

Sustainable Home

With the residential sector accounting for a significant portion of energy and water consumption and waste generation, it makes sense that a major effort to sustain the state should be placed at the home level. Fortunately, Californians have more possibilities for reducing the environmental footprint of their homes than do residents of most other states, because of the favorable climate and other environmental conditions.

Energy

Beyond using energy-saving lightbulbs and appliances, reducing the use of air conditions and heaters could make a significant difference in energy consumption in our homes. Application of some simple strategies could make homes more comfortable with less frequent (or no) use of air conditioners or heaters. For example, for new construction, facing the windows south and adding roof overhangs could increase solar heating during the winter and decrease it in the summer. Building multilevel homes against a slope, a possibility that exists in many California residential areas, could keep the temperature of a home more constant, resulting in less need for air conditioning and heating. For older homes, adding window awnings, draperies, and tints; installing wall insulation; and planting deciduous trees near sunny windows could serve the same purpose (Figure 11.11). Improving air circulation in a home could also reduce the use of air conditioners, as illustrated. With most of the residential areas in California located in relatively mild Mediterranean climate zones, application of sustainable home strategies has great potential to reduce energy needs.

Another California (especially Southern California) advantage is the abundant sunshine, which can be used to heat a home, heat water, or generate electricity. With improving solar technology, wind technology, and government-provided financial incentives, residential solar and wind device installations are increasing. The California Solar Initiative, launched in 2007 with a goal of installing 1,940 megawatts new solar systems in 10 years, offers solar incentives for energy users, including residential homeowners to install solar systems. Two years into the program, thousands of projects were completed with thousands of applications in the pipeline.

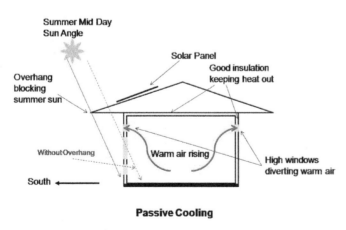

Passive Cooling

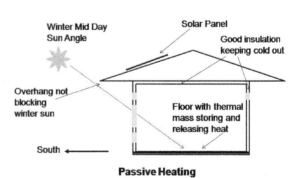

Passive Heating

Figure 11.11 Examples of an energy efficient house design.
(Source: Lin Wu)

Water

One of the largest consumers of water in a residential home is landscaping, especially in Southern California and in desert areas where natural rainfall is limited and most landscapes require irrigation. With increasingly dry conditions and a reduced water supply, mandatory water conservation is practiced in many places, which often requires reducing landscape water usage. The solution to this problem is more complicated than it appears. For example, landscape water consumption could be reduced or eliminated by replacing a landscaped area with pavement or artificial turf - a seemingly simple solution. Such a practice, however, would potentially increase the aridity of the area's microclimate by channeling more water directly into the ocean (instead of keeping it in the plants and soil) and by retaining more heat in the concrete or other pavement materials. This practice, if applied widely in a community, could increase the *urban heat island* effect and may result in health problems and increase energy consumption. Better solutions should be based on site-specific conditions and the use of low water consumption plants instead of eliminating permeable areas.

Combined with waste treatment strategies, *gray water*—water from showers, dishwashers, and laundry—could be treated on-site and be used for landscaping purposes. Such practices not only keep a healthy landscape that moderates the climate but also reduces the need for waste transfer and centralized treatment facility. Technology and practices in recycle gray water are used at institutional levels but are still limited on residential levels.

Waste

Despite the fact that many urban homes participate in recycling programs, a large amount of solid waste from homes still goes into landfills. For the Zero Waste approach to work, it requires many of us to change the way we shop, cook, eat, and so on around our homes. For example, a large portion of household waste that ends up in landfills is organic food waste, which could be composted and returned to the land as fertilizer, completing a natural cycle without needing the landfill. Such a process could be completed at the household level with a family-size composting system and backyard food production system, or it could be completed at the community, city, or regional level with food wastes collected separately from other types of recycled materials. Beyond the household level, however, any collection and processing system would add the environmental cost of collecting, transporting, and processing, which also generate "waste" along the ways.

Even if it is technically possible to reduce the waste to zero, how many of us in our community, state, country, and beyond are willing to change our lifestyles and make it work? Many of us would be willing to install an energy-saving or solar system to reduce nonrenewable energy use even if it would cost us more money; not too many of us are ready to give up the convenience of packaged foods or to devote time every day to carefully separating and recycling wastes, even if it would save us money. Many challenges are still ahead of us to reach Zero Waste at our homes.

Protect and Improve Biodiversity

California is fortunate to have many parks, open lands, and forested areas that become the sanctuaries of many plant and animal species. Different ecosystems have different degrees of environmental tolerance, and sometimes even moderate disturbances can trigger drastic changes in the environment. This is especially important in the urban and agricultural areas, where many of the native plants and animal species have been replaced. By applying some strategies such as promoting urban forests, increasing urban wetlands, adding rooftop gardens, and making wildlife protection part of the agriculture management practices, even in the urban and agricultural areas, biodiversity could still be protected and restored.

Summary

Sustainability is a complicated and far-reaching topic but an essential one that cannot be avoided in the context of geography. The chapter only touches the surface of various aspects related to the sustainability issues.

Perhaps one of the most important elements in the equation toward a sustainable California is education. With one of the most extensive education systems in the world, California produces hundreds of thousands of graduates each year, and a large number of them join the workforce in the state. It is crucial that the workforce in the future be well educated about the issues and solutions on sustainability in their fields, regardless of which fields they will be in. The success of any effort to improve sustainability requires integration and cooperation from all segments of society. In that regard, it is also necessary to educate the current workforce in sustainability with professional training and lifelong learning programs.

Well-trained scientists, engineers, and policy makers are also needed to move the sustainable process forward through research that will lead to the development of new technology and the formation of new policies and strategies.

Finally, the goals of sustainability cannot be reached without public acceptance, support, and eventually a widely accepted and practiced sustainable lifestyle. Education, formal or informal, is important to moving the state in this direction.

Selected Bibliography

California Energy Commission website *Information and statistics on energy sources and consumption in California.* (http://www.energy.ca.gov/).

California Environmental Protection Agency website *Information on environmental protection status, laws, etc. in California.* (http://www.calepa.ca.gov/).

California Integrated Waste Management Board website *Information on waste management status, policies, new technology, etc. in California.* (http://www.ciwmb.ca.gov/).

California Solar Initiative, California Public Utilities Commission website *Detailed information and data on California Solar Initiative.* (http://www.cpuc.ca.gov/puc/energy/solar/aboutsolar.htm).

Purvis, M. and A. Grainger., Eds. (2004). *Exploring Sustainable Development: Geographical Perspectives, Earthscan.* London, UK: Routledge.

US Department of Energy, Office of Energy Efficiency and Renewable Energy (EERE) website *Information and statistics on energy sources and consumption in the United States.* (http://www.eere.energy.gov/).

University of California Sustainable Agriculture Research and Education Program website *Information on sustainable agriculture status and development.* (http://www.sarep.ucdavis.edu/).

Wright, R. T. (2008). *Environmental Science: Toward A Sustainable Future.* Upper Saddle River, NJ: Prentice Hall.

California on the Threshold

*"**California** has become the first American state where there is no majority race, and we're doing just fine. If you look around the room, you can see a microcosm of what we can do in the world.... You should be hopeful on balance about the future. But it's like any future since the beginning of time—you're going to have to make it."*

–Bill Clinton

"As one went to Europe to see the living past, so one must visit Southern California to observe the future."

–Alison Lurie

Introduction

Reportedly, when the Chinese wish to see someone experience difficulties and problems, they extend the following "benediction": "May you live in interesting times!" As has been true throughout the history of the state, Californians certainly live in interesting times, in both positive and negative senses. In many ways a microcosm of the nation, California represents at the same time the best and the worst of a challenging future. In a broader sense, we can recognize in the state the embodiment of the rapid evolution of the larger global environment, social and physical. The fascinating reality is that many of the "problems" perceived by critics of California are merely manifestations of the inexorable changes a rapidly shrinking world is experiencing. Indeed, when the United Nations held a global summit on world development several years ago, the topics of discussion could well have been taken from a town hall meeting in almost any community in California: chronic unemployment, pervasive poverty, degradation of the physical environment, deterioration of the social order, and a widening gap between rich and poor. Yet, while we recognize these issues, we also concede the progress that has been made: advances in medicine, which have extended life expectancy; improvements in education, which have resulted in a higher literacy rate; and technological progress, which has brought overall increases in quality of life.

Perhaps what is most important is to identify the changes, understand the context, and then create a coherent plan for the future. Certainly, a major reality shift for California (and the world) has been the end of the Cold War, the advent of the war on terror, and the global economic crisis that started in the late 2000s. These things have brought with them significant changes in economics, politics, and social focus. The security and prosperity of California, the United States, and the world are highly dependent on the operations of the world economy, and foreign policy is being conducted with different presumptions from before. In an increasingly "borderless" world, trade is a key issue at the national and local levels, and the line between domestic and foreign policy is somewhat blurred. Trade, immigration, illegal narcotics, education, jobs, culture, health, environment, transportation, communication, entertainment, agriculture—all are at once both international and local in scope. In the area of immigration, California has become the most popular destination in the United States and thus has a crucial stake in the discussion of national policy on the issue. In the realm of illegal narcotics, Los Angeles is a (if not *the*) major entry port in the country. In the employment sector, the maquiladora program moved many jobs out of the state and into Mexico. The impact of immigration on education has been notable for the accompanying demand for multicultural and multilingual teaching. The health and medical establishments of California are now faced with significant problems such as tuberculosis, once deemed an all-but-eliminated disease. And continual population growth has meant increasing traffic congestion, strains to urban infrastructures, and environmental difficulties.

The Prospects

Many years ago, Curt Gentry wrote *The Last Days of the Late, Great State of California* (1968), a book in which he envisioned a catastrophic end to the state brought about by a devastating earthquake. More important, he used this natural disaster as a starting point to reflect on the problems the state had created for itself. Gentry questioned whether California would have destroyed itself even without a natural disaster. His conclusions were troubling. In essence, he portrayed a state in crisis, populated by dysfunctional residents constantly sowing the seeds of their own physical, cultural, and spiritual

destruction. His imaginary cataclysmic quake merely brought quicker closure to the process.

Several decades later, novelist Cynthia Kadohata wrote *In the Heart of the Valley of Love,* and continued the tradition of envisioning California in the future. Rather than attributing the decline of the Golden State to a natural disaster, Kadohata built on the theme of social-political-environmental decay. Taking Los Angeles as her symbolic representative, she described a situation of moral decay and corruption in government and society, of environmental pollution so bad that cancer and disfiguring skin diseases were the norm, and of class riots and daily violence so common as to evoke no surprise from citizen or official alike. Although lacking a catastrophic earthquake, Kadohata's fictional destruction of the state is no less troubling in its implications.

Of course, California has not yet been destroyed by human, social, or natural disaster. The thought of self-destruction, however, remains unsettling. Little progress has been made toward creating an efficient mass transit system in California. In a state where mobility and effective transportation are critical needs, the dominant mode is still the economically inefficient and energy wasteful automobile. Although some control has been imposed in the area of pollution, standards of air and water quality, noise regulation, and chemical waste disposal are still far from satisfactory. Indeed, revelations concerning hazardous chemical dumps in residential areas have created a whole new category of physical maladies suffered by unfortunate and unwary Californians.

Violence has become an accepted fact of life in large urban areas, and the number of guns being purchased reflects the level of confidence many residents have in current standards of police protection. Extremism of various types threatens the social fabric of the state. The persistence of groups like urban gangs, religious cults, terrorist groups, racial hatred groups, neo-Nazis, and the Ku Klux Klan raise doubts about the overall stability of California culture and society. The predominant philosophy, however, remains apathy. The majority of Californians persist in ignoring the pressing problems of the present and future. In their pursuit of pleasure, citizens of the Golden State tend to gloss over the cracks in the social, political, and environmental structures.

Certainly, many of the difficulties are a function of size. California has a yearly budget, a resource base, and a population greater than those of many countries. The complexity of maintaining an orderly society in a population that numbers 12 percent of the US population is often overwhelming. The challenge of coordinating interests in a geographical and political entity encompassing approximately 758,693 square miles is enormous. This diversity has frequently led to discussions about dividing the state. To date, these discussions have never moved past the proposal stage, although the legislature has debated and acted at counter purposes more than once, and it is still a remote possibility in the future.

In short, the prospects for California are mixed. Problems exist, and solutions are possible, but actions are often confused. The key to the future lies in the interaction between people and the natural environment. Whether they are in conflict or in harmony will determine the destiny of the state.

The People

Californians face many changes. The increase in population is both a growing problem and an exciting challenge, the solution to which can dramatically affect both the prosperity of the state and the lifestyle of the people. The changing ethnic character of California also will affect cultural styles, social stability, and political power (Tables 9.1–9.4). As with other states, the development of California will be influenced by the overall strains on the social fabric of the nation as a whole. The people, however, are the first resource of the state, and it is up to them to find the solutions.

Growth and Population

The first reality about people involves the number of them who call the state their home. Around 30 million persons resided in the state at the beginning of the 1990s; this figure rose to close to 34 million by the year 2000 and increased to over 38 million in 2012. Census statistics show no indication of a decline in the growth rate, and the upward trend of 2 percent per year (compared with the national average of 1.1 percent per year) means that by the year 2035 California's population could be approximately 64 million, almost double what it is today! The primary factor for this growth rate is the high immigration rate into California, both legal and illegal, compared with other states.

Although California's population density is not yet at a critical juncture, the distribution of people is a problem (Map 2). Concentration of population in the south is a continuing phenomenon that severely stretches existing resources. The south already has a water shortage potential. The south already has too many cars on the roads. The south already strains to provide employment for its population. The south already suffers a housing problem. Unfortunately, little relief is in sight. The north is in no better position to absorb excess population and at times expresses open reluctance to do so. This is a major issue, and so far, too little thought has been directed toward its solution.

Perhaps the reasons behind the gridlock in planning for population increases have something to do with changing economic patterns, shifting political attitudes, and tenuous ethnic balances. Whatever the reasons, the need to deal with continuing growth is a given. Depending on which authority is cited, over the next several decades the state's population is expected to surge by 10–20 million persons (around 40 percent from foreign immigration).

Of course, the significance of these population statistics is found in the reality of lifestyles of Californians. The way we respond to demographic change will help determine the quality of life for the future. Issues such as job availability, housing, water and energy supplies, waste management, education and social services, and human civility pose daunting challenges for Californians who seek to maintain the state's golden image. Certain issues, in particular, provide focus for consideration, including immigration, migration, and ethnicity, as they relate to population growth.

Immigration, Migration, and Ethnicity

One of the most difficult yet most exciting elements of modern California is the dramatic interplay of ethnic groups. Reflecting its position on the Pacific Rim, California is experiencing an astonishing readjustment in its ethnic mix, from an Anglo-dominated majority to a more international model. Driven by such factors as economic opportunity, geographic juxtaposition, freer trade as a result of the North American Free Trade Act (NAFTA), and political instability, new immigrants have been steadily moving (legally and illegally) into the state. Especially evident in urban settings is the multiethnic pattern of Buddhist temples and Korean churches, Little Saigons and Little Tokyos, Chinatowns and Hispanic barrios (representing most nations of Central America), and enclaves of Russians, Armenians, Iranians, Cambodians, Laotians, and many others.

This increase in new immigrants has generated negative reactions in many quarters. Some groups, feeling socially and economically threatened, have initiated anti-immigration activities in an effort to halt the tide. The passage of controversial Proposition 187 in the mid-1990s (ostensibly denying education and certain social services to undocumented aliens) demonstrated the intensity of the debate, as did California's unsuccessful but symbolic lawsuits against the federal government for reimbursement for and protection from the costs of such immigrants to the state. Of course, the actual impact of immigrants on California is by no means clear. Some vocal authorities allege that the immigrants are a drain on the welfare system and refuse to assimilate into the mainstream; other experts point to directly contradictory evidence suggesting that these new immigrants have a dynamic and positive effect on the socioeconomic fabric in California. What many fail to recognize is that this debate is hardly new, but rather reflects a national (and Californian) tradition of cycles of immigration and hostile reactions dating back two centuries. Californians have long struggled to resolve their ambivalent racial-ethnic attitudes toward one another.

Although the state has a long history of mixed races, it has also had a rather poor record of promoting social equality and harmony. The continuing tension between ethnic groups cannot be denied, and the 1992 Los Angeles race riots were neither unprecedented nor totally unexpected. As the state evolves in the new century dominated by regional and global changes, the citizens of California represent a cross section of reactions. Some white flight from the state has taken place, although this is occurring more among lower socioeconomic classes than among the better educated (and thus more employable). Some experts have also noted an interesting minor trend of "black flight" to the suburbs, as the newer immigrants preempt upwardly mobile black Californians. Thus, racial tension is not a whites-only phenomenon. Tension among black, Asian, Hispanic, and Middle Eastern populations is just as real. With California becoming the first Third World state in the United States, the measure of progress may be found in how modern Californians deal with this issue.

A brief summary of the crucial demographic data underscores the changes and challenges facing the state in the twenty-first century. First, California's population is still growing. The US Census Bureau projects that approximately 60 percent of the nation's total growth well into the new century will take place in a handful of states in the South and West, with California showing the largest gain. This would bring California's share of the nation's total population to 15 percent within the next 20 years.

Second, as with the nation as a whole, California's population continues to grow older. With current life expectancy in the United States estimated as high as 82 years (compared to 65 only half a century ago), the profile of California's population is maturing. Indeed, the US Census Bureau also projects that the 65-and-older population of the state will double over the next two decades. Contributing to the overall pattern, California's fertility rate will continue to decline.

Third, ethnic and migration patterns will significantly modify the state's makeup. The realities of national immigration statistics are particularly notable in California, with the state leading the nation in percentage of foreign-born immigrants. By far the largest groups are Asians and Hispanics, who by the early 2000s accounted for almost 85 percent of California's overall population growth (from both net migration and natural increase). It might be added that the past two decades saw a modest out-migration of whites and Native Americans. All of this reflects national adjustments, with predominant international "migration magnets" remaining the ports of entry (e.g., Los Angeles, San Francisco, San Diego) and internal "migration magnets" shifting to mostly noncoastal destinations (e.g., Phoenix, Las Vegas, Atlanta).

Finally, urbanization of the state will not abate. Already reflecting a 95 percent urban population, the increased immigration, natural increase, and internal readjustments promise to maintain large urban clusters, as well as boost secondary and tertiary population centers. Contrary to views of some pessimistic observers, the results will continue to be growth rather than stagnation and change rather than decline.

The Social Fabric

An additional, broad, people-oriented difficulty involves strains on the state's general social fabric. Various stresses of modern life are taking their toll on Californians. Sociologists have expressed concern over the fallout of "future shock" in America. As society moves faster and faster, people become more confused, alienated, and maladjusted. Unfortunately, California exhibits some of the more dramatic manifestations of this dysfunctional social order. Suicide and divorce rates both exceed the national averages. The state ranks in the top half in crime rates, and its expenditures on prisons and law enforcement lead the nation. California also ranks in the top five in homelessness, drug abuse, unemployment, and welfare recipients, and it is in the top ten for persons falling below the poverty line. It ranks number one in highest costs per patient day in hospitals. Its expenditures for the arts place it among the bottom five states, as does its percentage of eligible voters who bother to register. The media detail the escalating death rates from gang-related violence in virtually every metropolitan region. Add to these sobering statistics the disturbing images of the 1992 Los Angeles riots (following the verdict in the Rodney King beating trial) and the 1995 circus performance of the California legal system in the O. J. Simpson case, and the Golden State appears to have a rather brassy tarnish. Further damage to the perception of quality of life for California can be found in the too-common evidence of infrastructure decline, most dramatically chronicled by the bankruptcy of Orange County in the mid-1990s and the more recent bankruptcies of Stockton and San Bernardino.

These statistics are disturbing and raise important questions about the overall quality of life in the state. Certainly, California can ill afford to ignore the social, economic, and political imbalances that feed these problems. The actions taken to create a healthier and more stable society will help determine how many people wish to continue to live in the state.

The Pacific Rim–The Pacific Century

Many experts are suggesting that the global balance of power is dramatically shifting from Europe to the Pacific Rim. This shift is quite credible if we contemplate the approximately 50 countries with nearly half of the world's population that border the Pacific. Certainly for the United States, the shift has already occurred, with trans-Pacific trade now exceeding trade with Europe (a change that began in the mid-1980s). In an increasingly borderless world, trade has assumed a central role in international relations. The significance of this fact for California is immense.

Geographically, California sits squarely in the center of the new international economic coordinates. Many people see the immigration, economic, and political patterns emerging in the state as representing the future of the country. As a prime gateway to the Pacific Rim, California has already attracted much foreign capital. Real estate investment in Los Angeles is a prime example, with roughly 30 percent of choice downtown sites now owned by foreign corporations based in Pacific Rim countries. San Francisco real estate also reflects a remarkable pattern of Hong Kong Chinese ownership. In terms of product movement, thousands of cars move both ways between Californian and Asian ports daily; California farm products are major profit items in this market; and trans-Pacific communications (business and personal) have created an extensive linkage between California and virtually every site in the Pacific Rim.

Likewise, north–south ties encouraged by NAFTA are particularly evident in California. Economic partnerships between San Diego and Tijuana and between Calexico and Mexicali have helped create new jobs and markets on both sides of the border. The advantages and disadvantages of this arrangement are numerous. For Californians, it has often meant that jobs have been exported to Mexico, where American or multinational corporations can enjoy cheaper labor costs, more lax environmental restrictions, and preferred tax treatment from the host country. That arrangement has been a boost to the Mexican economy, with roughly 45 percent of that country's exports coming from maquiladoras. With increased global competition, however, even lower cost offshore sites (e.g., China, Malaysia, Central America) have reduced Mexico's share of jobs and dollars. Although in decline, the 3,000-plus maquiladora factories along the United States–Mexico border still employ more than one million workers. The question may be asked whether these jobs would have remained in California even without the maquiladoras. A related question may also be whether some of the improved income for Mexican workers along the border found its way back into California with higher cross-border consumer purchasing by Mexican workers.

Chain migration ties families throughout California to Mexico, El Salvador, Guatemala, and many other nations to the south. The value of trade flowing through California ports is astonishing; California airports also account for billions of dollars of Pacific trade. As noted previously, California in general (and Los Angeles in particular) is *the* gateway for swelling immigration from Asia and Latin America, and projections for the near future envision even more Pacific Rim immigrants, most of whom will be drawn to California.

As Asian consumer demand continues to rise because of both sheer numbers and increasingly sophisticated tastes, the potential for California businesses and products also rises. China, Korea, Taiwan, Singapore, Malaysia, and others represent growing opportunities for a rapidly changing California economy. Similarly, the increase in U.S. exports to Latin America has been dramatic, with some experts predicting that this market will soon surpass exports to Western Europe. Thus, if the historical bases of California's economy (aviation, agriculture, defense, oil, and apparel) have stabilized or declined, the emergent state economy will be increasingly buoyed by Asian and Hispanic entrepreneurial initiatives, driven in part by the fortuitous geographic location of the Golden State.

The Natural Environment

The natural environment is as critical to the state as its people. A fine balance must be struck to maintain maximum utility from the air, land, and water without creating irreparable damage. Nature has proved to be relatively tolerant of abuses perpetrated by humans, but the mistreatment has consequences. Recent developments in environmental control have had mixed results. Several acts are of particular interest.

Federal Regulations

The *National Environmental* Policy *Act of* 1969 is the basis for most modern environmental protection laws. It states that US national policy is to foster "Harmony between man and his environment." The act mandated the establishment of a Council on Environmental Quality, directed the president to present an annual state-of-the-environment message, and led to the use of environmental impact reports on federal activities that would affect the quality of the environment.

Of course, the act does not apply to individual, corporate, or state actions—a major shortcoming. During the act's existence, even enforcement of environmentally responsible actions by the federal government has been fraught with inconsistencies and compromises. At best, the act can be said to establish a guiding principle that should be followed. Other federal acts pertaining to environmental protection tend each to be addressed to one type of environmental problem and may involve regulation of states, private industries, and individuals. Thus, various acts attempt to regulate water quality, air quality, drinking water, endangered species, solid waste disposal, toxic substances, mining, coastal zones, insecticides, scenic rivers, and various other issues. Frequently, the federal government "assigns" authority to the individual states to enforce certain of the acts. However, in addition to the federal government, state governments have also engaged in efforts to regulate the environment. California's record of protecting the natural environment has followed a somewhat predictable pattern.

State Regulations

The California Environmental Quality Act of 1970 was patterned closely after the federal act. It committed the state to maintaining a quality environment for the people of California "now and in the future." Following this act, the state legislature passed laws that addressed specific problems. Over half of the state codes soon contained sections on environmental problems and controls. Laws were passed that dealt with general pollution, water quality, air quality, particulate pollutants, open space, parks and recreation areas, wildlife and wilderness areas, and nuclear power. These laws have attempted to set reasonable standards to protect the environment from excessive polluting or encroachment on natural areas.

California's efforts to regulate and clean up the air are well known throughout the nation. Recognizing that air pollution transcends boundaries, the state established regional air basins and placed responsibility at the local and regional levels for nonvehicular sources of pollution. It maintained primary statewide control over vehicular pollution, with

some authority shared with regional agencies. Examples of these regional agencies are the South Coast Air Quality Management District, the Bay Area Air Quality Management District, the Sacramento Metropolitan Air Quality Management District, the Mojave Desert Air Quality Management District, and the San Joaquin Valley Air Quality Management District.

Regulations at both state and regional levels attempt to deal with agricultural and other burning, toxic air contaminants, emission-emitting equipment, fuel vapor and fuel-burning emissions, and other sources of air pollution (Figure 12.1). With the strictest requirements in the nation, California has been accused by automakers and others of setting draconian standards for air quality. As with many environmental rules, however, the issue seems less a question of feasibility than one of reduced profit margins. Apparently, Californians will have to continue to ponder their priorities: decent air quality versus corporate profits.

Water quality has also posed serious problems for the state. Both the quantity and the safety of water for human and animal use are crucial issues. Various enactments under state authority prohibit dumping or otherwise allowing noxious objects in waters, establish standards for drinking water, control ground water contamination, regulate coastal marine environments, provide for toxic cleanup, and require regional plans to achieve compliance. Reflecting the historical economic and political influence of agribusiness, both the allocation of water for environmental purposes (fish and wildlife) and the enforcement of quality standards often have been delayed and sidetracked.

Figure 12.1 The Wilshire Corridor and the downtown Los Angeles skyline in the distance, encased in smog and fog.
(Image © Jose Gil, 2009. Used under license from Shutterstock, Inc.)

Another area of growing concern for Californians involves environmental hazards. In recent decades, it has become evident that plans for dealing with environmental hazards have been woefully inadequate. These hazards may be broken down into two major categories: *human-created* and *natural*.

In the realm of human-created hazards, one of the most evident is toxic waste sites. Since the mid-1990s, California has ranked second to only New Jersey in the number of toxic cleanup sites targeted for action by the Environmental Protection Agency. Spread all around the state, 98 sites are currently listed as high priorities requiring billions of dollars to correct problems created by mining companies, industries, oil producers, commercial landfill/dump operators, military bases, and agribusiness. Soil and groundwater contamination, wildlife and aquatic life destruction, and human illnesses have been directly attributed to several of these sites. Progress on correcting these catastrophes has been modest, but steady.

In the area of natural hazards, California has demonstrated a surprising lack of coherence in planning. In the early 1990s, the Oakland Hills firestorm demonstrated the danger of casual treatment of land use in an explosive urban–wildland interface. More than 3,000 dwellings and 25 lives were lost, with damage estimates approaching $2 billion. Twenty years later (2007), firestorms swept through chaparral-covered residential areas stretching from Malibu to the Mexican border, destroying million-dollar homes in Laguna Beach, Malibu, Thousand Oaks, Altadena, San Diego, and other communities. Yet, despite the huge property losses, the massive expenditures for firefighting, and the overall damage, homes were being rebuilt on the same sites within a year. The October 2007 California wildfires were some of the worst on record. This series of wildfires in Southern California burned 2,000 square kilometers (500,000 acres) from Santa Barbara County to the United States–Mexico border. Roughly 1,500 homes were destroyed, 9 people died, and 85 people were injured, including 61 firefighters. Governor Arnold Schwarzenegger declared a state of emergency for seven California counties, and President George W. Bush ordered federal aid to help the state and local response efforts. In total, 6,000-plus firefighters fought the series of wildfires, along with units of the United States Armed Forces, the National Guard, and approximately 3,000 prisoners. Repeats of the firestorm episodes occur with yearly regularity (2008 and 2009 in Southern California and 2012 in Northern California). Likewise, despite massive flooding along the Russian River in the north and flooding and mudslides in Malibu in the south, homes were quickly rebuilt in the same neighborhoods. Even after the 1989 Loma Prieta and the 1994 Los Angeles/Northridge earthquakes, life (and construction) began anew with little regard for the existence of the earthquake hazard zones. The logical question at this point is not whether California's public policy should address this problem, but when!

The necessity for planning to mitigate hazard losses is a fairly modern concern, not just for California but also for the nation as a whole. Several factors have brought about this change. Population growth has led to increased demands for land that would have been ignored in the past. Aesthetics also have evolved over time: Where prime "bottomland" may have been the ideal of the past, now rocky cliffsides, waterside locales, "rural" brushlands, and mountain slopes exert strong appeal for their scenic qualities. Similarly, economics may make less costly land in fault zones, desert regions, slump zones, or outwash plains acceptable residential choices. Thus, many of the hazard losses noted previously occurred precisely because people located their homes and businesses in inherently hazardous areas. Certainly, the desire to own and occupy one's own house and property is well established in the American psyche. The importance of this desire is reflected in the "new housing starts" statistics collected and reported by the federal government. Typically, a decline in these numbers is viewed as a negative in the national (or state) economy. Yet, though the desire to see these numbers increase is understandable, the reality of a declining base of safe, usable land is ever-present. The problem's scope is exacerbated by the preference of Americans to live in single-family, detached residences. Clearly, this preference places even greater pressures on static land resources, resulting in the increased use of potentially risky sites.

Set against the desire to build in questionable areas is the societal mandate to protect citizens from harm. When the government acts on behalf of the "public good," does this mean it should protect people from their own possible folly? That question has placed California's government (and those of other states as well) in the middle of a classic political debate: How much regulation should government exert over its own citizens? There are two basic choices.

First, government may choose (by benign neglect) to allow persons to build on the San Andreas Fault, in fire-prone Malibu, along the banks of the Russian River, in the Santa Ana River floodplain, or anywhere else. Thereafter, if losses occur, the government may (1) do nothing, (2) provide public assistance, or (3) encourage private/insurance assistance. If the government knows of the risk, however, is it then liable to the injured homeowner (as some court rulings have held)? And if the property owners know the risk, should other citizens be forced to spend their tax dollars to "bail out the stupid"?

Second, the government could attempt to prevent the losses in the first place. This would involve prohibitions against certain uses of land in hazardous areas. Of course, this notion immediately raises constitutional questions of private property rights, deprivation of property without compensation, and similar issues. Of course, the logic of preventing people from engaging in possibly harmful activities seems reasonable. As may be seen from the heated arguments over motorcycle helmet or seatbelt laws or driving-and-texting prohibitions, not everyone readily accepts such "parental" wisdom. Challenges to land use regulations are spirited and instantaneous, so California government agencies have hesitated to establish even reasonable restrictions on building in hazardous zones. The irony is that even where these same agencies do not restrict land use for fear of legal challenge, they may face subsequent lawsuits for failing

to prevent homeowners from building in the hazardous zone. It is a seemingly circular problem for which neither the courts nor legislatures at state or federal levels have been willing to provide a responsible policy. Until such responsibility is exercised, losses such as in the Oakland Hills, Malibu, and Laguna Hills will recur with monotonous regularity. Can such preplanning ever be accomplished? The answer is a qualified yes. A very modest effort has been made to establish reasonable land use rules in the state along its vast coastline.

The California Coastal Zone Conservation Act of 1972 and the Coastal Act of 1976 established control over development of the state's coastline. These acts set up a Coastal Commission responsible for all development within the coastal zone. They have been relatively effective in preserving natural coastal environments from further erosion and have placed severe limitations on high-density development and modifications of the natural landscape.

Various specific regional planning programs have also been added to state law. The Tahoe Regional Planning Compact saw five California counties and the state of Nevada join together to create a regional development plan that would transcend state jurisdictions. The Tahoe Regional Planning Agency has met with mixed success in its efforts to provide a comprehensive approach to the area. Similarly, the San Francisco Bay Plan was set up to protect the delicate balance of the bay and to coordinate all development bordering its waters. As a result of its efforts, overall water quality has steadily improved, and the bay is gradually becoming an attractive recreation and wildlife environment again.

The trend for future environmental legislation is encouraging. Refinements of both federal and state laws have occurred on a regular basis. California has recently dealt with tobacco smoke, noise, engine emissions, hazardous wastes, beverage containers, and general recycling mandates. Federal legislation has likewise updated and refined the procedures established to protect the environment. In the area of legislation, the people occasionally have shown a willingness to pay the price for improvement.

The Water Problem

The persistent problem of water allocation in the state has not been resolved. Few other issues can raise emotional and political hackles as effectively as this one, and Californians recognize that at least in *their* state, water is power. California has been able to achieve its enviable position because of the riches made possible by extensive irrigation. Agricultural and residential properties appreciate in value precisely because water can be brought to otherwise barren regions.

Despite a comprehensive water management system, however, a crisis is present. As various and regular droughts illustrate, even one dry year has far-reaching consequences throughout the state. In vast portions of California, failure of water supplies not only impinges on normal agricultural and household uses but also exacerbates a perennial fire danger. The brushfire season is one of growing concern as more people build in former brush areas. Many Californians have learned, to their dismay that disaster can strike at home.

The water future is truly a question mark because of several factors. Water from the Colorado River declined drastically in the mid-1980s as a result of the Central Arizona Project, but new sources have yet to be found. Long-term solutions have not been determined or implemented as Californians continue to argue about sources, rationing, and other issues. While drought threatens one year and floods the next, residents seem unable to agree on viable alternatives. Unless the state can find a satisfactory approach soon, the water crisis will haunt Californians for many years to come.

Several water-related issues in particular warrant attention, each of which has an important role in the state's future. First, increased population and growth means that existing supplies must be better managed. It seems obvious that land use planning and water supplies are closely linked. Indeed, some legislative proposals have specifically suggested that one should not occur unless the other is part of the process. As early as the mid-1990s, the Department of Water Resources projected that even in non-drought years, supplies would be inadequate to meet the growing demand. This fact suggests a built-in limit to future growth unless some viable answers are found. Another growth issue relates to urban versus agricultural uses. As residential uses expand, the amount of water available for agriculture must necessarily decline. Mindful of this trend, many farm operations are installing drip irrigation systems, as well as converting from crops such as alfalfa, corn, and cereals to less water-intensive crops such as vines and fruit trees. This trend has also inserted the water supply issue into the debate about farmland preservation, agricultural profits, and unchecked urban sprawl. The overall unlikelihood of new interbasin transfer projects has significant implications for ongoing water management. Undoubtedly, better use and cleanup of major existing aquifers will be necessary, as will better methods of capturing runoff for recharging groundwater sources.

The second issue is the pollution of water resources, particularly by salts from irrigation, fertilizers, and pesticides, which compromises the health of humans, wildlife, and the land itself. The problem occurs when semiarid land is brought into production through irrigation. As the water, fertilizers, and pesticides are applied to the soil, plants absorb the water but leave much of the chemical salts in the soil. With inadequate additional water to leach the salts from the ground, a steady buildup results. Depending on the underlying soil layers, the salts eventually percolate into and pollute the groundwater or build up to the point that the surface plants cannot survive the soil's salt content. Furthermore, if runoff occurs, the salts can find their way into surface waters, pose a hazard to birds and aquatic life, and compromise the quality of this surface water. Traditional solutions to this problem have included evaporation ponds, diversion drains, and removal of land from production. Unfortunately, each solution has its shortcomings and, to date, conflicting goals have prevented effective resolution. What is certain is that unless serious attention is paid to this problem, a decline in both environmental quality and in farm production will follow.

The third issue involves the distribution of the state's existing water supplies. Specifically, the Sacramento–San Joaquin Delta is the pivotal supply point for most of California's drinking water. This fact has engendered much debate over the years concerning pumping and diverting water from the Delta to Southern California and to San Francisco. The oft-proposed Peripheral Canal Project to divert water around the Delta has led to some of the bitterest political and environmental battles in the state. Clearly, a balance must be struck between the needs of the Delta for sufficient water to maintain its delicate ecosystem and the needs of the state for water to supply its growing population. Unfortunately, planning for the future has been irregular and inconsistent, and no responsible management plan for the Delta has emerged. Until that happens, the situation promises only to get worse.

The final issue is the need for water recycling and conservation. With projected shortfalls, as noted previously, improving efficiency becomes crucial. Periodic droughts have brought with them increased awareness of the need for such efforts, with various water agencies using recycled water for irrigating crops or for recharging local aquifers. Likewise, public education programs have attempted to convert domestic users into nonwasteful consumers, with low-flow toilets and showers and drought-resistant landscaping. Assuming that the projections of continual population growth are accurate, conservation and recycling efforts will become a fact of California life.

The Problem of Movement

The enduring difficulty of handling movement of persons and goods around this huge state will intensify in the future. Californians have made some efforts toward resolving this problem, but a coordinated and committed approach is needed. Population pressures, air pollution concerns, energy scarcity, and other problems mandate development of viable alternatives and expansion of experimental techniques that will provide greater transportation efficiency.

Most discussions categorize California's transportation question into the three areas: land, sea, and air. Because of the state's varied topography, different problems have arisen and different usages have developed for these forms.

In terms of transportation by water, California has become a leading international trade focus, particularly for the nations of the Pacific Rim. An extensive shoreline with many excellent ports has encouraged ocean transportation, and the ports of Los Angeles, San Francisco, Oakland, Long Beach, San Diego, Richmond, Sacramento, Stockton, and others serve the varied needs of this trade. The rise of container ships has also spurred this area of transportation in California. Although travel by water does not account for a major portion of people movement, it does constitute a significant mode of industrial mobility.

The revolution in air travel has certainly affected California in a major way. Access to the state is immediate and easy for travelers from throughout the country and the world. Major airlines serve most large cities in the state, with San Francisco and Los Angeles connected by air with almost any other place on the globe. As far as intrastate transportation is concerned, the extreme size of California is no longer as formidable an obstacle for the casual traveler or business executive who needs or desires to move quickly from one end of the state to the other. Regional airlines now can provide connections almost anywhere in the state. To a lesser degree, the airlines now also provide rapid cargo movement for the state. Although volume may be lower than land or ocean transport, air cargo does serve as an immediate and efficient means of prompt delivery.

Surface or land transportation is by far the dominant form in California. Here, the choice is movement by rail or by road, with the road the primary choice in recent years. As noted in Chapter 3, the railroad played a key role in the development of California. In addition, the growth of San Francisco and Los Angeles was facilitated by the San Francisco Key Route Electric Railway and Henry Huntington's Pacific Electric Line in Southern California. Railroads, however, have largely surrendered their role as passenger carriers and now concentrate on the transport of goods. The freight-hauling function of railroads has been augmented by piggybacking, container cargo, and other techniques for maximizing the efficiency of rail shipping. Passenger carriage is not viewed as a profitable venture, and most railroads prefer to leave this aspect of rail transport to Amtrak or short-distance commuter lines. The once-popular passenger train is now largely a romantic tale from the past, although a few notable passenger lines persist.

California's road system is another story: this is the heart of transportation in the Golden State. As energy-inefficient and costly as it may be, the private automobile is perhaps the predominant icon of the state. From the early beginnings of El Camino Real connecting the 21 missions, the California road system has grown into a phenomenal network of arteries, freeways, expressways, and roads. Financed by a state gasoline tax, the highway system has become a thing of wonder, reflecting the fact that California is the heaviest user of automobiles of any state in the country.

To provide some organization and structure, the California Master Highway Plan was adopted in 1959. This plan envisioned an eventual pattern whereby approximately 60 percent of the state's total road travel would move on the freeways and expressways. This system of rapid thoroughfares was targeted for completion in the early 1980s as a coordinated means of dealing with the state's transportation needs. Unfortunately, the system was already overused and inadequate long before the 1980s arrived. California's heavy urban commuter traffic placed a severe strain on the system, and the few newly constructed freeways have been outdated even before being put into use. The additional fact of California's position as the leading trucking state placed an impossible burden on the highway system and demonstrated that alternatives were sorely needed.

Experiments have been undertaken in an attempt to solve some of the commuter pressure. One of the most notable efforts is the BART (Bay Area Rapid Transit) system.

Recognizing the peculiar and unique topographical problems of the area, with its peninsulas, growing population, and limited space, planners sought a system whereby commuters could be moved more rapidly in and out of the region's urban work centers. Traffic snarls on the various bridges in the Bay Area emphasized the need to get individual motorists out of their cars and into a mass transit system. What eventually emerged was an automated electric diorail train/subway system that connected downtown Oakland and San Francisco with outlying areas such as Concord, Berkeley, Hayward, Richmond, Fremont, and Daly City. The system was built to provide rapid service and dealt with some geographic problems by constructing a subway tunnel under the Bay's floor and through the Berkeley Hills. Although BART has been plagued with difficulties and was almost inadequate from its date of completion, it nonetheless demonstrated that such a system could operate and would be patronized by the public. To date, it is most likely California's most successful effort in mass transit.

Another effort to solve the problem of moving people about the Bay Area is found in the CalTrain commuter lines, which extend light-rail connections from San Francisco southward to San Jose and beyond. A similar internal line (the Muni Metro) helps move commuters within San Francisco itself. Given the traffic congestion in the region, along with the uniquely configured Bay and shoreline traffic patterns, mass transit seems particularly useful in this region. Like its neighbors to the south, however, the Bay Area's transportation mode is still dominated by the automobile.

The case of Southern California is even more problematic in terms of traffic congestion and gridlock on the one hand and viable solutions on the other. Various studies have confirmed that commuter traffic in California is the worst in the nation, with the greatest problems occurring in the Los Angeles–Orange County region. Indeed, in recent years, the Federal Highway Administration identified the ten busiest highway interchanges in the nation and found that the top nine were in California. Five of these were in Los Angeles County, and three in Orange County; the other was in Alameda County. The statistics behind these facts are staggering. In Southern

California, peak commuter hour speeds drop to around 20 miles per hour; most commuters drive alone (around 73 percent by most estimates); there are more than 31 million motor vehicle trips per weekday; and the number of cars on the road just keeps increasing.

Compared with the Bay Area, how has the greater Los Angeles area responded? Los Angeles has had much less success in the realm of mass transit. The Southern California Rapid Transit District historically has relied on a bus system to serve public transportation needs. It has been exceptionally difficult, however, to entice Southern California drivers out of private automobiles. Indeed, private automobiles still account for the overwhelming bulk of commuter traffic, as rush hour freeway traffic shows. Some experiments have proven partially successful, such as Commuter-Computer, Park-n-Ride, bus and car pool lanes, and special lanes on metered ramps. In certain heavily traveled corridors, Caltrans has even added toll lanes to speed affluent commuters along.

Political pressure has resulted in commuter train runs (Amtrak) between Los Angeles and San Diego, Los Angeles and Orange County, and Los Angeles and Riverside/San Bernardino. Perhaps more noticeably, the 1990s saw the initiation of commuter-oriented Metrolink trains to connect outlying areas with the downtown core. Even more dramatically, the county transportation agency undertook a comprehensive effort that resulted in the planning and partial construction of multiple light-rail routes connecting various areas in the Los Angeles basin. Planners envisioned 400 miles of light-rail, subway, and commuter lines and similar connections by 2010, labeled the Blue Line, the Green Line, the Red Line, and so on. Since 2010, plans for further expansions have slowly moved forward. Portions already completed are considered qualified successes, particularly by frustrated commuters who are now able to escape the crowded freeways. Nonetheless, it remains a major challenge to pry drivers out of their cars. Even various incentives to carpool, use public transit, or find alternative travel arrangements have not been overwhelmingly successful.

What does the future hold in the area of transportation for Californians? Indications are that some Californians are beginning to recognize the need for alternatives. San Diego's "Tijuana Trolley" is an interesting representative effort. These bright red electric trolley cars cover a 16-mile route between downtown San Diego and the Mexican border, carrying in excess of 10,000 people daily. This light-rail system was conceived and funded at the local and state level and has filled a real need in the area. Depending on its continuing success, extension of the system is contemplated for the future.

Sacramento remains heavily reliant on automobile transport, but efforts to expand a light-rail system have also been pursued. As the Sacramento metropolitan area continues to grow, such a system may well be crucial if gridlock is to be avoided (Figure 12.2).

Meanwhile, the state highway system apparently has reached critical mass. Few additional projects are planned, and maintenance of existing freeways and expressways has proved a substantial chore. Some modifications are in process, including a proposed system of rail and bus corridors along center medians of existing freeways. Other modifications have included

Figure 12.2 A bullet train on a high-speed rail similar to the one proposed in California. The estimated time to travel the 663 kilometers (412 miles) from Los Angeles to Sacramento is 2 hours 17 minutes. The planned route will run over 1,300 kilometers (800 miles) connecting major population centers from San Diego to Sacramento and the Bay Area.
(Source: Image © Shi Yali, 2009. Used under license from Shutterstock, Inc.)

video camera monitoring of heavily traveled routes, computerized signboards for traffic information, metered ramps, and other electronically controlled devices.

Basically, however, these are attempts to keep a transit system of private automobiles in operation and so represent short-term answers only. In 2009, California urban roads were ranked the worst in the nation for lack of repair and maintenance, demonstrating that short-term thinking seldom resolves long-term problems. It remains for long-range planners to develop more comprehensive solutions. With the continuing air pollution problem, the increasingly limited energy sources, and the population pressures in the state, alternatives to the internal combustion engine and private auto are needed.

Some discussion has revolved around electric or solar-powered vehicles and hydrogen-powered cars. Some planners have even envisioned totally automated freeways. More interest has been generated in modified residential-work patterns. "Old" abandoned systems of mass transit are being reexamined with an eye to learning and profiting from lessons of the past, such as the Los Angeles Pacific Electric Red Cars (trolley) or Angel's Flight cable system of downtown L.A.'s Bunker Hill area. The one certainty is that creativity, innovation, and dedication are needed in this vital area of California life.

The Future

What can California expect from the future? So far, the interplay between people and the environment has been quite uneven. On the positive side, Californians are generally healthier, younger, better educated, more affluent, and more advanced scientifically and technologically than people anywhere else in the nation. On the negative side, they often seem confused about how to use and control the technology that has made their lives better. Furthermore they often are myopic when it comes to recognizing the impact of their actions on the environment.

One obvious area of interaction between people and the environment is population size. With demographers projecting a population continuing to grow into the foreseeable future, Californians still cannot fully deal with the impact their current population is having on the environment. Given a doubling of the current population, how will California respond? How will land use policies be set? How will the already critical water problem be resolved? How will strained energy sources serve future needs? How will the economy of the state absorb additional workers? Most important, how will quality of life in the Golden State be affected? Solutions to these open-ended questions will depend on the responses of thoughtful and concerned Californians.

Certain selected issues bear attention as a final commentary on California's future. Although not meant to be a comprehensive list, the following points are worth brief attention. First, can the economy of California sustain its rank of eighth largest GNP in the world? Although the economic downturn of the 1980s, the early 1990s, and 2008 suggested a dislocation of prosperity, subsequent events have suggested change rather than decline. Certainly, some companies have relocated in recent years to other states. Certainly, job decline has been substantial in the aerospace, defense, computer, and manufacturing industries. The influence of the World Trade Organization and NAFTA, however, has placed California in the forefront of economic growth. In fact, job creation by smaller, more adaptable, knowledge-intensive companies has been substantial. Areas such as telecommunications, medical enterprises, and foreign trade are growing segments of the new California economy. Much of this growth has been buoyed by a state government that oversees and officially supports offices of tourism, business development, technology, foreign investment, filmmaking, and world trade. All of this activity continues to be fueled by easy access to universities, ports, and airports; by high-tech firms; and by ambitious new residents.

Second, can California continue to meet the challenge of educating its citizenry? As a gateway to the nation, California plays a particularly crucial role in providing appropriate education to established and new residents alike. To answer the multitude of problems facing the K–12 (precollegiate) systems, the state has turned its attention to better funding, reduced class sizes, and alternate methods of training teachers. The very diversity of California's population creates unusual challenges in the classroom, and satisfactory solutions have been difficult to find. Higher education in the state has found itself facing a crisis brought about by increasing enrollments, declining state revenues, and rapidly changing technologies. Planning efforts at both the University of California and the California State University systems have attempted to address the problems by rethinking delivery systems, increasing private sector support, and challenging the faculty to find better ways of educating students. The state's community college system has likewise tried to re-engineer itself to cope with the disconnect between demand and funding. Given the critical importance of education to the state's prosperity, these will continue to be central issues well into the future.

Third, can California continue its prominent role in developing technology and ideology related to sustainable development and living? For half a century, California has been the leader in setting high vehicle emission standards to reduce pollution from tail pipes. Could the state lead the nation (and the world) by not only setting the high standards but also developing clean and affordable fuel technology, along with other solutions, to make it technologically and economically viable to ultimately meet the zero-emission goals? California has pioneered alternative energy resources development in the past by experimenting with solar, geothermal, wind, and nuclear power technology at different scales. Will California tap into the experiences and move forward in developing long-term sustainable, environmentally friendly, safe, and economically viable alternative energy resources? With a slightly smaller ecological footprint than the national average, Californians still leave behind a huge imprint on the environment regardless of recent advancements in environmentally friendly technology, policy, and practices. Will Californians, more educated and more accessible to traditional wisdom from different cultures on sustainable living, emerge as leaders in closing the gap between US and other world citizens on sustainable living, from both practical and ideological perspectives? Answers to these questions will shape the future of the state.

Finally, can California manage to understand and establish priorities for its own well-being? For example, is it more important to build additional prisons or to address the issues of education, health, welfare, and the environment? Can the state avoid being buried in its own garbage by adopting a serious waste management plan? Can the ponderous, inefficient, and poorly respected court and legal system ever reform itself? These serious questions require all Californians to rise to the challenge of keeping the state a dynamic and positive force in the global environment.

Summary

As California continues its development and growth, many changes face its citizens. As in the past, the economic profile of the state reflects a dynamic response to new technologies, new structures, and new social configurations. The state's economy has historically undergone cyclical ups and downs but has managed to position itself as an economic leader. The pattern of expansion of population and wealth does not seem to be changing for the future, although the demographic makeup may be undergoing more dramatic shifts.

The major issues and challenges for the state's future are not localized alone. As with California's past history, the state may well be leading the nation in its experiments to address the issues of environment, resources, and population management.

Selected Bibliography

California Department of Finance website (www.dof.ca.gov).

California State Government website (www.ca.gov).

Gentry, C. (1968). *The Last Days of the Late Great State of California*. New York: G. P. Putnam's Sons.

Hundley, N. (2001). *The Great Thirst: Californians and Water: A History.* Berkeley and Los Angeles: University of California Press.

Hyslop, R. S. (1990). *Hazards and Land Use Planning: The Unresolved Dilemma.* Ann Arbor, MI: University Microfilms.

Kadohata, C. (1992). *In the Heart of the Valley of Love.* New York: Viking Press.

Lantis, D. W., Steiner, R., Karinen, A. E. (1989). *California, the Pacific Connection.* Chico, CA: Creekside Press.

Miller, C., and Hyslop, R. (2000). *California: The Geography of Diversity.* Palo Alto, CA: Mayfield Publishing.

Reisner, M. (1987). *Cadillac Desert: The American West and Its Disappearing Water.* New York: Penguin Books.

Trover, E. L. (State Ed.), and Swindler, W. F. (Series Ed.). (1972). *Chronology and Documentary Handbook of the State of California.* Dobbs Ferry, NY: Oceana Publications, Inc.

US Census Bureau website (www.census.gov).

Western Water Education Foundation. *Western Water.* Various editions.

Winchester, S. (1991). *Pacific Rising: The Emergence of a New World Culture.* New York: Touchstone.

Absolute humidity – The amount of water vapor in the air measured by weight, volume, or vapor pressure.

Adiabatic process – A physical process in a system (e.g., air) resulting in temperature change without any energy exchange between the system and its surrounding environment, causing cooling in rising air and warming in descending air.

Agglomeration – In geography, a reference to clustering or aggregating similar activities or urban environments.

Algae – (Latin for "seaweed") A large and diverse group of simple, typically autotrophic organisms, ranging from unicellular to multicellular forms, such as the giant kelps. Most are photosynthetic and "simple" because they lack the many distinct cell and organ types found in land plants. The largest and most complex marine forms are called seaweeds.

Alluvial Fan – A fan-shaped landform formed when sediments carried by streams are deposited at or near a foothill region.

Alluvium – Materials deposited by streams and consisting of coarse deposits, such as sand and gravel, and finer-grained deposits, such as clay and silt.

Alpine – Living or growing on mountains above the tree line.

Alta California – A former Spanish colony of New Spain, in the modern-day states of California, Nevada, and N. Arizona.

Angiosperms – A plant whose ovules are enclosed in an ovary; a flowering plant.

Appropriative water rights – The most common use-based water rights in the United States; most commonly found in the western states where water is scarcest.

Aquifers – Underground beds or layers of permeable rock, sediment, or soil that yields water.

Aquitards – Fine-grained layers, such as clay and silt that do not store water.

Arable land – Land that is, or can be, used for growing crops.

Arêtes – Sharp ridges formed during glaciation.

Asthenosphere – The "soft" layer of the earth underneath the solid lithosphere.

Au naturel – Naked or in a "natural state."

Baby boomers – The cohort of persons born between the years of 1946 and 1965.

Baleen – A filter-feeder system inside the mouths of baleen whales.

Barrio – Area or neighborhood in a city occupied largely by Spanish-speaking residents.

Bathymetry – The study of the "beds" or "floors" of water bodies, including the ocean, rivers, streams, and lakes.

Bays – Bodies of water partially enclosed by land but with a wide mouth, affording access to the ocean.

Beach – A landform along the shoreline of an ocean, sea, lake, or river consisting of loose particles, which are often composed of sand, gravel, pebbles, and occasionally mollusk shells or coral.

Bio capacity – The capacity of an area to provide resources and absorb wastes.

Biome – A major regional or global biotic community, such as a grassland or desert, characterized chiefly by the dominant forms of plant life and the prevailing climate.

Breakwaters and jetties – Structures constructed on coasts to protect from the effects of weather, waves, and longshore drift, thus slowing coastline erosion.

Caballero – Horseman or rider.

California Coastal Commission/California Coastal Act – The mission of the Coastal Commission is to "Protect, conserve, restore, and enhance environmental and human-based resources of the California coast and ocean for environmentally sustainable and prudent use by current and future generations." The California Coastal Commission was established by voter initiative in 1972 (Proposition 20) and later made permanent by the California legislature through adoption of the California Coastal Act of 1976.

California Cooperative Snow Surveys Program – Began in 1929 as a water management and regulation tool; about 50 agencies pool their efforts in collecting snow data at nearly 270 snow courses in the mountains; snowpack information, along with precipitation and river runoff data, are used by the Department of Water Resources to develop forecasts of expected snowmelt runoff and total annual water runoff on the major rivers.

California Department of Water Resources – Responsible for managing and protecting California's water; works with other agencies to benefit the state's people, and works to protect, restore and enhance the natural and human environments: www.water.ca.gov/

California Doctrine and California Water Code – States that the state of California owns all of the water in the state (California Constitution, California Water Code [CWC]); in California, rights are usufructuary and pertain to the use of the water, not actual ownership of it (California Constitution, CWC). The State Water Resources Control Board is the state agency in charge of administering and allocating water rights.

California Farm Bureau – A nongovernmental, nonprofit, voluntary membership California corporation whose purpose is to protect and promote agricultural interests throughout the state of California.

Cenozoic – A geologic era from 65 million years ago to the present.

Central Valley Project (CVP) – (1933) Federal water management project in California under the supervision of the United States Bureau of Reclamation; provides irrigation and municipal water to much of the Central Valley.

Channel Islands National Marine Sanctuary (CINMS) – (1980) A reserve area off the coast of California; established in 1980, it encompasses the waters that surround Anacapa, Santa Cruz, Santa Rosa, San Miguel and Santa Barbara Islands (five of the eight Channel Islands of California).

Channelization (Channelized) – Any activity that moves, straightens, shortens, diverts, or fills a stream channel, including the widening, narrowing, straightening, or lining of a stream channel (with concrete, gravel or culverts) that alters the amount and speed of the water flowing through the channel.

Chumash Indians – Native American people who historically inhabited the central and southern coastal regions of California, in portions of present day San Luis Obispo, Santa Barbara, Ventura, and Los Angeles Counties; they also occupied three of the Channel Islands (Santa Cruz, Santa Rosa, and San Miguel).

Cirque – Bowl-shaped depression formed during glaciation.

Cismontane – Situated on this side of the mountains; in California, the west side of the Sierra Nevada.

Clean Water Act – The primary federal law in the United States governing water pollution; passed in 1972, the act established the goals of eliminating releases of high amounts of toxic substances into water, eliminating additional water pollution by 1985, and ensuring that surface waters would meet standards necessary for human sports and recreation by 1983.

Coast redwoods (*Sequoia sempervirens*) – Evergreen trees growing in a narrow band along the Pacific Coast from southern Oregon to central California. The state tree of California, it lives 2,000 years or more and is the tallest living tree on earth, reaching up to 110 meters (360 feet) in height. Currently categorized as an endangered species due to commercial logging that began in the 1850s, an estimated 95 percent of the original old-growth redwood forest has been cut down.

Coastal wetlands – Salt-water and freshwater wetlands located within coastal watersheds that drain into an ocean or gulf.

Coastline – A line that forms the boundary between the land and an ocean or lake.

Colorado River Aqueduct – A water conveyance in Southern California built between 1933 and 1941 and operated by the Metropolitan Water District of Southern California (MWD); impounds water from the Colorado River and is one of the primary sources of drinking water for Southern California.

Condensation – The phase change from water vapor to liquid water.

Conifers – Any of various mostly needle-leaved or scale-leaved, chiefly evergreen, cone-bearing gymnospermous trees or shrubs, such as pines, spruces, and firs.

Continental margin – The zone of the ocean floor that separates the thin oceanic crust from thick continental crust. Together, the continental shelf, continental slope, and continental rise are called the continental margin; continental margins constitute about 28 percent of the oceanic area.

Continental shelf – The extended perimeter of each continent, much of which was exposed during glacial periods but is now submerged under the oceans.

Conurbation – A large, high density urban sprawl created by the expansion and blending of adjoining suburbs and cities.

Copepods – Groups of small crustaceans found in the sea and nearly every freshwater habitat. Some species are planktonic (drifting in seawaters); some are benthic (living on the ocean floor).

Cultural personality – A concept promoted by many authorities that suggest that a specific area or region has its own unique personality; often termed "personality of place."

Cyclonic storm – Also called a wave cyclone or midlatitude cyclone, a storm associated with a low pressure system originating in the midlatitude ocean and traveling from west to east over the ocean and land, bringing precipitation along its path.

Delta – A triangle-shaped landform formed when sediments carried by streams are deposited along a coastal area where the streams enter the ocean.

Dendritic Drainage – A treelike stream pattern that forms in areas with high amounts of precipitation and less confined geological structures.

Dendrochronology – The dating and study of annual rings in trees; the study of climate changes and past events by comparing the successive annual growth rings of trees or old timber.

Depositional coast – A coast that gains sediment over time.

Depression years – Decade of extreme economic turmoil generally lasting from around 1929 to 1939.

Doctrine of correlative water rights – A legal doctrine limiting the rights of landowners to a common source of groundwater (such as an aquifer) to a reasonable share, typically based on the amount of land owned by each on the surface above. Under California law, the owners of overlying land own the subsurface water as tenants in common, and each is allowed a reasonable amount for his/her own use.

Drought – A long period of abnormally low precipitation, especially one that adversely affects growing or living conditions.

Dunes – Hills of sand built either by wind or water flow, occurring in different forms and sizes.

Ecological footprint – One of the measures of sustainability that calculates the area of land needed to provide resources to support the needs of a population, and to absorb the waste it generates under the current technologies in use.

Edaphic – Of or relating to soil, especially as it affects living organisms.

Effluent – Typically an outflow from a sewer or sewage system; also a discharge of liquid waste, as from a factory or nuclear plant.

El Niño – A band of anomalously warm ocean water that occasionally develops off the western coast of South America and can cause climatic changes not only across the Pacific Ocean but also worldwide.

Endangered Species – A species of organisms facing a very high risk of extinction. There are currently 3,079 animals and 2,655 plants classified as endangered worldwide, compared with 1998 levels of 1,102 and 1,197, respectively.

Endemism – Prevalence in, or peculiarity to, a particular locality or region.

Environmental lapse rate – The rate at which temperature decreases with the increase of elevation.

Epicenter – The surface location above an earthquake focus.

Erosional coast – A coast that loses sediment over time.

Estuaries – The parts of the wide lower course of rivers where their currents are met by the tides; arms of the sea that extend inland to meet the mouth of a river.

Evaporation – The phase change from liquid water to water vapor.

Evapotranspiration – The process of transferring moisture from the earth to the atmosphere by evaporation of water and transpiration from plants.

Extant – Still in existence; not destroyed, lost, or extinct.

Extinction – The fact of being extinct or the process of becoming extinct; the state of being no longer in existence; the condition of a species that has ended or died out: the extinction of a species of fish.

Fault – A fracture or a zone of fractures where the rocks are broken under pressure and are accompanied by vertical or horizontal displacement of rock layers on one or both sides of the fracture.

Fauna – Animals, especially the animals of a particular region or time period, considered as a group.

Federal Bureau of Reclamation – A federal agency under the US Department of the Interior that oversees water resource management, specifically as it applies to the oversight and operation of the diversion, delivery, and storage projects that it has built throughout the western United States for irrigation, water supply, and hydroelectric power.

Firestorm – A rapidly spreading and large fire that typically destroys vast swathes of land and structures.

Flora – Plants considered as a group, especially the plants of a particular country, region, or time period.

Floristic province – A geographic area with a relatively uniform composition of plant species; adjacent floristic provinces do not usually have a sharp boundary, but rather a soft one, a transitional area in which many species from both regions overlap.

Folding – The bending of rock layers under compression, which results in a rock layer called a fold.

Footloose industries – Economic activities that are not tied to a particular place or supply of raw materials, but rather can be located anywhere.

Forty-niners – People that went to California in 1849 during the gold rush.

Fossil fuel – An energy source, such as coal, petroleum oil, or natural gas, that formed over millions of years from the remains of plants and animals underneath the earth's surface.

Friars – Brothers, as in brothers in a religious order, such as Franciscans.

Future shock – Psychological and physiological disturbance caused by an inability to cope with rapidly changing society and technology.

Gastronomic – Pertaining to food and cuisine.

Gentrification – The restoration and refurbishing of deteriorated urban properties by middle-class buyers, typically resulting in displacement of lower-income people.

Geographic determinism (environmental determinism) – The theory that suggests geographic (or environmental) factors are the major determining agents of a group or culture.

Geothermal energy – Energy derived from the heat in the interior of the earth.

Greenhouse gases – Gases in the atmosphere that absorb and emit radiation within the thermal range, thus leading to global warming.

Gross domestic product – The market value of all officially recognized final goods and services produced within a region in a given period of time

Gross state product – A measurement of the economic output of a state.

Groundwater – The water located below the ground.

Groundwater basin – An area underlain by permeable materials capable of storing and furnishing a significant supply of groundwater.

Gymnosperms – Plants, such as cycads or conifers, whose seeds are not enclosed within an ovary.

Gyre – In oceanography a large system of rotating ocean currents driven by large wind movements.

Halophytic – A plant that grows in waters of high salinity, such as in saline semi deserts, mangrove swamps, marshes and sloughs, and seashores.

Headlands – A point of land, usually high and often with a sheer drop that extends out into a body of water; often used as a synonym for **promontory**.

Herbaceous – Plants or plant parts that are fleshy as opposed to woody.

Hollywood – A city in Southern California, though the term is more often used to refer to the movie industry in California.

Horace Greeley – Founder of the Liberal Republican Party, and the editor of a leading newspaper of his time (the 1800s).

Horns – Sharp peaks formed during glaciation.

Hydrologic cycle – The movement of water throughout the hydrosphere.

Hydrologic regions – Major drainage areas that share common characteristics of precipitation and runoff, and are also individual water-planning areas.

Hydrology – (*hydro* means "water") Refers to the science that encompasses the occurrence, distribution, movement, and properties of the earth's waters and their relationship with the environment within each phase of the hydrologic cycle.

Hydrosphere – The watery "envelope" that contains all the water vapor in the atmosphere, and the liquid and solid water on and under the land surfaces and in streams, rivers, lakes, and oceans.

Identity clusters – The grouping together of people of similar ideals, values, tastes, and self- images.

Igneous rock – Rock formed of molten materials from inside the earth under cooling process above (extrusive igneous rock) or underneath (intrusive igneous rock) the earth's surface.

Indigenous – Of or belonging to the people native to a particular area.

Interior drainage – A drainage area where streams do not have outlets to the ocean.

Isohyets – Lines of equal precipitation on a map.

Isotherm – Lines of equal temperature on a map.

John Muir – Referred to as Father of the National Parks; a Scottish-born American naturalist, author, and early advocate of preservation of wilderness in the United States. His activism helped to preserve the Yosemite Valley, Sequoia National Park, and other wilderness areas; founded the Sierra Club, which is now one of the most important conservation organizations in the United States.

Kelp (forests) – Large seaweeds (algae) belonging to the brown algae; grows in underwater "forests" in shallow, nutrient-rich ocean water.

Known for their high growth rate; the genera *Macrocystis* and *Nereocystis* can grow as fast as half a meter a day, ultimately reaching 30 to 80 meters (100 to 260 feet).

Krill – Small crustaceans found in all the world's oceans. The name *krill* comes from the Norwegian word *krill*, meaning "young fry of fish." Krill are a critical component of the California Current ecosystem.

Laissez faire – A policy of noninterference in the economics and lives of others.

Land degradation – Deterioration in the quality of land, its topsoil, vegetation, and/or water resources, caused by excessive or inappropriate human impacts.

Leeward – The downwind side of a mountain, typically the eastern slopes in California.

Lithosphere – The solid rock layer of the earth's surface, composed of the earth's crust and upper mantle, divided into lithospheric plates.

Littoral Drift – The process whereby beach material is gradually shifted laterally as a result of waves meeting the shore at an oblique angle.

Los Angeles Aqueduct – A water conveyance system built and operated by the Los Angeles Department of Water and Power; designed and built under the supervision of the department's Chief Engineer William Mulholland; delivers water from the Owens River in the Eastern Sierra Nevada Mountains to Los Angeles, California.

Los Angeles Department of Water and Power (LADWP) – The largest municipal utility in the United States, serving over four million residents. Founded in 1902 to supply water to residents and businesses in Los Angeles and surrounding communities; in 1917, it started to deliver electricity. LADWP can deliver 7,200 megawatts of electricity and, in each year, 760 billion liters (200 billion US gallons) of water.

Maize – Spanish for corn.

Maquiladora – Typically an American-owned factory run on the Mexican side of the border to take advantage of cheaper labor and looser labor and environmental regulations.

Marine Mammal Protection Act (MMPA) – (1972) The first act of Congress to call for an ecosystem approach to natural resource management and conservation; signed into law in 1972 by

President Richard Nixon; MMPA prohibits the taking of marine mammals and enacts a moratorium on the import, export, and sale of any marine mammal, along with any marine mammal part or product within the United States

Marine terrace (coastal terrace) – A flat surface at the base of a cliff formed by erosion by waves.

Maritime transportation – Watercraft carrying people (passengers) or goods (cargo). Sea transport has been the largest carrier of freight throughout recorded history.

Market segmentation – A division of a population into identifiable groupings, especially for marketing purposes.

Megalopolis – A very large or extensive metropolitan area, often consisting of multiple adjoining cities.

Mercalli scale – A way to measure earthquake intensity, the shaking and impact of an earthquake.

Mesic – Of, characterized by, or adapted to a moderately moist habitat.

Mesozoic – A geologic era from 240 to 65 million years ago.

Metamorphic rock – Rock formed of pre-existing rock under extreme heat and pressure.

Migration magnets – Sometimes also called poles of attraction, this refers to places/locales that exert unusual attraction for different reasons, among them jobs, established communities of like people, compelling physical features, etc.

Million acre-feet of water (maf) – One acre-foot is the amount of water that covers one acre to a depth of one foot.

Missions – As defined in California history, Spanish Catholic religious outposts.

Mono Lake Committee – A nonprofit committee that works to protect Mono Lake and the surrounding basin from habitat destruction as a result of LADWP water withdrawals for the Los Angeles aqueduct.

Moraine – A mound, ridge, or mass of rocks, gravel, sand, clay, etc. carried and deposited directly by a glacier, along its side, at its lower end, or beneath the ice.

Multicropping – In agriculture, the practice of growing two or more crops on the same land during the same growing season.

National Audubon Society – (1905) A nonprofit, environmental organization dedicated to conservation; uses science, education, and grassroots advocacy to advance its conservation mission; named in honor of John James Audubon, an ornithologist and naturalist who painted, cataloged, and described the birds of North America.

National wildlife refuge – A designation for certain protected areas of the United States managed by the United States Fish and Wildlife Service; a system of public lands and waters set aside to conserve America's fish, wildlife, and plants.

Native plants – Plants that have developed, occur naturally, or existed for many years in an area.

Naturalistic school – A writing style characterized by realism and a focus on environmental influences, often very negative ones.

Nearshore waters – The fertile shallower waters overlying the continental shelf (to a depth of 100 meters [330 feet]); an area of strong upwelling and kelp forests.

Nomadic – Moving around the landscape, typically engaged in hunting, gathering, and fishing.

Nonnative Plants – Originating in a part of the world other than where they are currently growing.

Nonpoint source pollution – The cumulative effect of a region's residents going about their everyday activities, both urban and rural, flowing into a water body, and containing a variety of pollutants, such fertilizers, pesticides, and heavy metals.

Nova Albion – The name given to California by Sir Francis Drake of England.

Offshore ocean waters – Less productive waters, where the continental shelf drops off to the deep sea floor (extending 320 kilometers [200 miles] offshore).

Ollas – Earthen or ceramic jars or pots used to hold water or for cooking.

Orographic precipitation – Precipitation on the windward side of a mountain caused by the uplifting of air along the mountain slope, typically the western slopes in California.

Pacific flyway – A major north–south route of travel for migratory birds in America, extending from Alaska to Patagonia. Every year, migratory birds travel some or all of this distance both in spring and in fall, following food sources, heading to breeding grounds, or travelling to overwintering sites.

Pacific Rim – In very general terms, refers to the land that borders the Pacific Ocean, including the Americas, Oceana, and Asia.

Padres – Literally "fathers"; used to refer to Catholic priests.

Pangaea – A super continent believed to exist 20 million years ago, named by Alfred Wegener who proposed continental drift theory in the 1920s, which led to the development of plate tectonics theory decades later.

Paternalistic – Behavior that may be benevolent but which limits other persons' choices in the belief that the decision maker knows best.

Paul Bunyan – A mythological, giant lumberjack of American folk culture, credited (tongue-in-cheek) with creating the Great Lakes, the Grand Canyon, Mount Hood, and other North American landmarks.

Pelagic – Of, relating to, or living in open oceans or seas rather than waters adjacent to land or inland waters.

Per capita – The average per person; a term often used in a statistical context.

Percolation (percolates) – The downward movement of water from the surface through soil and rock material to the water table.

Perennials – Having a life cycle lasting more than two years.

Permeability – The measure of how easily something (like water) can flow through a substance (like soil and rock materials).

Phytoplankton – Minute, free-floating aquatic plants. (See *algae*.)

Plant community – An assemblage of plant populations sharing a common environment and interacting with each other, with animal populations, and with the physical environment.

Plate tectonics – Theory about earth's landform building process based on moving lithospheric plates.

Pleistocene – A geologic epoch (period) within the Cenozoic era from 1.8 million years ago to approximately 10,000–11,000 years ago.

Point Conception – A headland along the coast of California, located in southwestern Santa Barbara County; the point where the Santa Barbara Channel meets the Pacific Ocean, and the border between the mostly north–south trending portion of coast to the north and the east–west trending part of the coast to the south; the natural division between Southern and Central California.

Point source pollution – Pollution that comes from a single source, such as a factory or wastewater treatment plant.

Precipitation – The transfer of water from the atmosphere to the surface as rain, snow, hail, or sleet; water in the form of rain, snow, hall, etc. falling from the sky and reaching the ground.

Primary sewage treatment – The simplest form of sewage treatment, solid sewage (sludge) is settled out and while grease and oils rise to the surface and are skimmed off.

Prohibition – A popular reference to the 18th Amendment, which banned production, transportation, sale (and presumably consumption) of alcoholic beverages between the years 1920 to 1933.

Psychographic profiling – The development of categories of consumer lifestyles, values, opinions, and preferences using demographic data.

Public trust doctrine – The principle that certain resources are preserved for public use and that the government is required to maintain them for the public's reasonable use.

Pueblos – In California, the name for the civilian communities (villages or towns).

Raker Act – (1913) An act of Congress that permitted building of the O'Shaughnessy Dam and flooding of Hetch Hetchy Valley in Yosemite National Park, California

Rancheros – The ranchers or cattlemen of California.

Relative humidity – The amount of water vapor, the gaseous form of water, in the air relative to the maximum amount of water vapor the air can hold, which varies by temperature.

Relict species – An organism or species of an earlier time surviving in an environment that has undergone considerable change.

Renewable energy – Energy from continually replenished resources, such as solar radiation, wind, ocean waves, and geothermal heat.

Reservoir – A natural or artificial lake, storage pond, or impoundment from a dam that is used to store water.

Restore Hetch Hetchy – A grassroots nonprofit organization seeking to restore the Hetch Hetchy Valley in Yosemite National Park to its original condition

Reurbanize – To rebuild, repair, upgrade, and restore urban areas to draw the population (business and residential) back to the original heart of the city.

Richter scale – A way to measure of an earthquake's magnitude, the energy released during an earthquake.

Riparian water rights – Under the riparian principle, all landowners whose property is adjoining to a body of water have the right to make reasonable use of it. If there is not enough water to satisfy all users, allotments are generally fixed in proportion to frontage on the water source. These rights cannot be sold or transferred other than with the adjoining land, and water cannot be transferred out of the watershed.

Rookeries – A colony of breeding animals, including birds and marine mammals.

Runoff – The surface water flowing to, or in, stream and rivers to lakes and oceans.

Salt marshes and mudflats – Areas with shallow, quiet waters, such as bays, lagoons, and estuaries, where water is salty and still enough for suspended particles to settle to the bottom; found within the intertidal zone, and thus submerged and exposed approximately twice daily. In the past these areas were thought to be unhealthy and economically unimportant and were often dredged and developed into agricultural land.

Sea cliffs – Also called **abrasion coasts**, a form of coast where the action of marine waves has formed steep cliffs.

Sea dog – An English sailor, oftentimes a pirate.

Secondary sewage treatment – Designed to substantially degrade the biological content of the sewage using aerobic biological processes; the next level up from primary treatment.

Secularization – In the context of California, the process of removing the influence of the Catholic Church and its missions and turning the church land over to individuals who had been loyal to the "new" Mexican government.

Sedimentary rock – Rock formed of consolidated sediments under pressure and chemical process.

Sequoia (*Sequoiadendron giganteum*) – Tree that occurs naturally only in groves on the western slopes of the Sierra Nevada of California. The world's largest single tree by volume, growing to an average height of 84 meters (275 feet) and 6–9 meters (20–30 feet) in diameter, and living on average 3,200 years. Sequoia bark is fibrous and thick (79 centimeters [31 inches]), providing it with significant protection from both fires and insect infestations.

Shangri-La – An imaginary, exotic, and romantic place.

Silicon Valley – The southern region of the Bay Area in California, home to many of the world's largest technology corporations and small start-up companies. The term originally came from the region's silicon chip innovators and manufacturers, but now it is used to refer to high-tech industry in California.

Snowpack – Layers of snow that accumulate in geographic regions and high altitudes where the climate includes cold weather for extended periods during the year; an important water resource that feed streams and rivers through snowmelt.

Snow surveying – A technique for providing an inventory of the total amount of snow covering a drainage basin or a given region.

Solar radiation – Radiation from the sun, also called shortwave radiation, a form of electromagnetic radiation with wavelengths ranging 0.4–0.7 micrometers.

Southern California Bight – A bight is a wide open-mouthed bay; the Southern California Bight extends from Point Conception past the southern border of California into Baja, Mexico, and includes the Channel Islands.

Sovereignty – Power to govern or rule a region.

Species diversity – The effective number of different species that are represented in an area.

Squatters' rights – The possession and use (and sometimes ownership) of property by occupying it in spite of the legal owner's rights.

State Water Project (SWP) – A state water management project in California under the supervision of the California Department of Water Resources; the world's largest publicly built and operated water and power development and conveyance system; provides drinking water for more than 23 million people.

State Water Resources Control Board – The State Water Board's mission is to preserve, enhance and restore the quality of California's water resources, and ensure their proper allocation and efficient use for the benefit of present and future generations.

Stewardship – In geography, a concept that embodies the principle that natural resources should be managed for the good of the planet.

Stomata – The tiny pores in a plant leaf that serve as the site for gas exchange.

Stratus clouds – A low cloud layer that often uniformly covers the sky and may be accompanied by light rain or drizzle.

Subalpine – Of, relating to, inhabiting, or growing in mountainous regions just below the timberline.

Sublimation – Changing from solid water directly to a gas form of water.

Taxonomy – The classification of organisms in an ordered system that indicates natural relationships.

Temperate – Moderate or mild in temperature.

Temperature – The measure of the average motion of atoms or molecules of a substance, indicating the degree of hotness or coolness.

Temperature inversion – Temperature increase with elevation at or near ground level.

Tepary beans – Annual twining plant that produces edible beans, typically brown or white.

Terrestrial radiation – Radiation from the earth's surface or atmosphere, also called longwave radiation; a form of electromagnetic radiation with wavelength averages of 10 micrometers.

Tertiary sewage treatment – Tertiary treatment is the most advanced level of wastewater treatment and removes stubborn contaminants that secondary treatment was not able to clean up. Wastewater effluent becomes even cleaner in this treatment process through the use of stronger and more advanced treatment systems.

Third world – In current usage, countries of the world that tend to be poorer or developing nations.

Tinsel town – A term often used to refer to Hollywood. (*See* Hollywood.)

Trade winds – The prevailing pattern of easterly surface winds found in the tropics, the trade winds blow predominantly from the northeast in the Northern Hemisphere and from the southeast in the Southern Hemisphere.

Transmontane – Being or situated beyond the mountains; in California, the east side of the Sierra Nevada.

Transpiration – The transfer of water to the atmosphere from plants (through the stomata of leaves).

Transport – The movement of water throughout the atmosphere.

Travelogue – A presentation, written or otherwise, about travels to different places.

Tree line – The edge of the habitat at which trees are capable of growing. In California, tree line is determined by elevation; above the tree line trees cannot tolerate the environmental conditions (usually cold temperatures or lack of moisture).

Troika – From the Russian language, now used to refer to a group of three.

Trolley – A trolley car or streetcar.

Troposphere – The atmospheric layer from ground level to an altitude of 6–16 kilometers (4–10 miles).

US Army Corps of Engineers – A federal agency under the Department of Defense; the world's largest public engineering, design, and construction management agency. Although generally associated with dams, canals and flood protection in the United States, USACE is involved in a wide range of public works throughout the world. The Corps of Engineers provides outdoor recreation opportunities to the public, and provides 24 percent of US hydropower capacity.

US Fish and Wildlife Service – A federal government agency within the United States Department of the Interior dedicated to the management of fish, wildlife, and natural habitats.

US Geological Survey – A scientific agency of the United States government; scientists of the USGS study the landscape of the United States, its natural resources, and the natural hazards that threaten it.

Upwelling – An oceanographic phenomenon that involves wind-driven motion of dense, cooler, and usually nutrient-rich water toward the ocean surface, replacing the warmer, usually nutrient-depleted surface water.

Urban heat island – Higher temperatures in an urban area caused by more heat retained due to the replacement of natural surcease with concrete surfaces.

Urban realms – A geographic model that suggests cities are made up of small self-sufficient areas known as "realms," each with independent focal points.

Urban sprawl – Unplanned and uncontrolled spread of urban landscape (residential and commercial) into areas surrounding urban cores, often farmland, green space, and open space.

Vascular plants – Plants that have the vascular tissues xylem and phloem, including all seed-bearing plants (gymnosperms and angiosperms), ferns, and allies.

Vernal pools – Temporary pools of water that provide habitat for distinctive plants and animals.

Vertical zonation – The vertical vegetation distribution pattern on a mountainside as a result of temperature and/or precipitation change with changing elevation.

Viticulture – The cultivation of grapevines, frequently for wine making.

Watershed – The land area drained by a particular group of rivers, streams, or creeks.

Wind – Horizontal air movement caused by pressure differences.

Wright Irrigation District Act of 1887 – A state law of California passed in 1887 that allowed farming regions to form irrigation districts and get water to where it was needed; this enabled the diverting of waters from the Merced, San Joaquin and Kings Rivers in California's Central Valley.

Yankees – A variable phrase used to describe someone from New England or, outside the United States, any American.

YBP (years before present) – A time scale used mainly in geology and other scientific disciplines to specify when events in the past occurred. Because the "present" time changes, standard practice is to use 1 January 1950 as the origin of the age scale, reflecting the fact that radiocarbon dating became practicable in the 1950s.

Zooplankton – Plankton that consists of animals, including the corals, rotifers, sea anemones, and jellyfish.

Index

A

Abies magnifica. See Red fir forest
Ablation, 28
Aboriginal lifestyle, 43
Aboriginal tribes of California, 44–45, 45f
Active recreation, category of, 112
Active volcanoes of world, 18f
Adam, John, 140
Adiabatic process, 72
Aeolian landforms, 15, 29
Aesculus californica. See White buckeye
African American population, 110
Age ghettos, 107–108
Agribusiness, 62
Agricultural industry, 59
Agriculture, 56, 96–98, 109, 175, 189, 189t
  in California, 8, 164
  to Chumash Indians, 165
  farms and farmers, 190
  field crops, 191–192
  gross cash receipts, 191f
  livestock and poultry, 191, 191f, 191t
  natural resources, 190
  opportunities and challenges, 193–194
  products and distribution, 190
  role, rural environment, 137–138
  sector, 180
  vegetables, fruits, and nuts, 192–193
  water resources, 175–176, 176f
Air pollution, 220–222, 241, 241f
Air relative humidity, 72
Air travel, revolution in, 245
Alaska earthquake, 23
Alluvial fans, 29
  at Copper Canyon Fan, 30f
Alluvium, 159
Alternative agricultural practices, 231–232
Alternative energy resources development, 249
Alternative energy sources
  geothermal power, 225–226
  hydropower, 223–225
  nuclear power, 227
  solar energy, 222–223
  wind power, 226
Alternative transportation systems, 229–230
American immigration, 4, 8

American period, water usage in, 165–166
American rural ethic, 142
Ammophila arenaria. See European beachgrass
Amusement park, Santa Cruz Boardwalk
    Beach, 105f
Angel's Flight cable system, 248
Angiosperm, 116, 117t
Anthropogenic impacts on oak woodlands, 126
Anza Borrego Desert State Park, desert
    climate in, 83f
Appropriative rights, 166–167
Aqueducts
  construction information, 173f
  map of, 170f, 283f
Aquifers, 153, 153f, 159
Aquitards, 159
Argonauts, 91–92
Artemisia californica. See California sagebrush
Artemisia tridentata. See Big Basin sagebrush
Aster chilensis. See California aster
Asthenosphere, 18
Automobile, 9, 9f
  transport, 247

B

Baby boom generation, 139
Badlands, 35
Bakersfield
  agriculture, 96–97, 96f
  temperature, 69
Balaenoptera musculus. See Blue whale
Balboa Park, 103, 103f
Baleen whales, 208, 208f
BART system. *See* Bay Area Rapid Transit system
Basaltic rocks, 18
Basin and range landforms, 16
Basin and range topography, 22
Basque population, 97
Bay Area in CalTrain commuter lines, 246
Bay Area Rapid Transit (BART) system, 9, 246
Bay Bridge
  during 1989 earthquake, 25
  Loma Prieta earthquake on, 26f
Bedroom communities, 141
Beef industry, 53–54

Bidwell, John, 4
Big Basin sagebrush (*Artemisia tridentata*),
    121, 130
Binary cycle technology, 225, 225f
Biocapacity, 220
Biodiesel, 228
Biodiversity, protecting and improving, 232
Biomes, 119–120, 120f, 182f
    chaparral and coastal sage scrub, 128–129
    coniferous forest, 121–124, 122f
    desert scrublands and woodlands, 129–131
    grasslands and marshes, 126–127, 127f
    oak woodlands, 124–125
    and plant communities, 119
Blackburn, Thomas C., 53
"Black gold" rush, 57–58
Blind thrust fault earthquake, 26
Block faults, 21, 22f
Blue oak (*Quercus douglasii*), 118, 124, 125
Blue whale, 208
Bodie, 52f, 54
Boomerang effect, 138
Buckbrush (*Ceanothus cuneatus*), 129
Budget, 237
Budget-related issues in California, 148–149
Bullet train, 247f
Burlington Northern Santa Fe train in California, 55
Bush, George W., 242

C

Cabrillo, Juan Rodriguez, 46
Calatrava, Santiago, 187
Calico
    mining, 92f
    tourist consumption, 92
California aster (*Aster chilensis*), 116
California Coastal Zone Conservation
    Act of 1972, 243
California Department of Water Resources, 154
California Doctrine, 167–168
California Environmental Protection
    Agency (CEPA), 84
California Environmental Quality
    Act of 1970, 241
California fescue (*Festuca californica*), 116
California flora
    geographic divisions, 118–119
    taxonomic divisions, 116–118
California Indian population, 52–53
California Irrigation District Act (1917), 168–169
California Master Highway Plan, 246

Californian floristic provinces, 119–120, 281f
California sagebrush (*Artemisia californica*), 116
California Solar Initiative, 232
California State Department of Finance, 146
California Water Code, 167–168
CalTrain commuter lines, Bay Area in, 246
Campuses, 196–197
Capitol Records Building, 183, 184f
Castilleja pruinosa. See Indian paintbrush
Cattle production, forms of, 54
Cattle ranching, 49
Ceanothus cuneatus. See Buckbrush
Cedar Fair's Knott's Berry Farm, 105
Central Arizona Project, 244
Central Coast Region, 163
Central Pacific Railroad Company, 54
Central place theory, 140
Central Valley, 39–40, 161–162, 164
    agriculture, 96–98, 176f
    climatic conditions, 65
    landforms in, 41f
    and sedimentary rocks, 21
Central Valley Project (CVP), 157, 170–171
Central Valley Project Improvement Act, 171
Century of Dishonor (Jackson), 52
CEPA. *See* California Environmental
    Protection Agency
Cercidium spp. See Palo verde
Cercis occidentalis. See Western red bud
Chain migration, 240
Chandler, Raymond, 5
Channel Islands, 30f
Chaparral, 83
    vegetation, 128–129
Chapman, Charles Edward, 2
Charleston, temperature in, 70, 70f
Chernobyl disaster, 227
Chumash Indians, 200
Circular motion, 18
Cirque, 28
Cities, California, 146–147
Citrus fruits, 192–193
Civil Liberties Act of 1988, 60
Civil War, 54
Clean Water Act (1977), 175
Cleveland National Forest, 77
Climate, 7
    change, 84–85
    characteristics of, 65, 85
    classification, 81, 82f, 278f
    desert, 83–84
        in Anza Borrego Desert State Park, 83f

economic geography, 179
highland, 84
Mediterranean, 81–82
regional variations, 81–84
Climographs for representative stations, 68f
Clouds, fog and, 74–77
Coastal Act of 1976, 243
Coastal area
mild, comfortable climate in, 65
in Santa Barbara, 30f
Coastal landforms
beaches and dunes, 211–213
islands, 214–215
marine terraces and headlands, 211, 211f
sea cliffs, 211, 211f
wetlands, 213–214
Coastal markets *vs.* inland, 195, 196f
Coastal ocean region
Chumash, 200–201, 201f
coastal landforms
beaches and dunes, 211–213
islands, 214–215
marine terraces and headlands, 211, 211f
sea cliffs, 211, 211f
wetlands, 213–214
environmental protection, 201–202
marine mammals, 208–209, 210t
nearshore waters
El Niño, 205
kelp forests, 204–205
Northern California, rugged coastline of,
199, 200f
ocean economy, 201
offshore waters, 207–208
Pacific Ocean and California current
coastal upwelling, 203, 203f
continental shelf, 204
North Pacific gyre, 202, 202f
SCB
municipal wastewater, 206–207
stormwater runoff, 206, 206f
Southern California, wide sandy
beaches of, 119, 200f
Coastal sage scrub biomes, 128–129
Coastal tribes, 44
Coast landforms, 15, 29–31
at Point Reyes National Seashore, 17f
wave motions and processes, 31f
Coast live oak (*Quercus agrifolia*), 125, 126
Coast Ranges, 33–34
coast landforms in, 15
at Crescent City, 34f

landforms in, 33f
and sedimentary rocks, 21, 22
Coast redwood (*Sequoia sempervirens*),
117, 118f
distribution, 122
giant sequoia and, 121t
Cocos Plate, 21
Cold War, 61, 62
Colonization, 46
Colorado Desert, crescent-shaped sand
dunes in, 29
Colorado River Aqueduct (CRA), 171
Colorado River Region, 163
Colorado River tribes, 45, 53
Colorado River Valley, 164
Colorado River, water from, 244
Columbia tourist consumption, 92
Commercial energy consumption, 219
Commercial oil well, 57
Community college districts, 147
Community services
districts, 147
in rural areas, 139
Commuter train, 247
Condensation, 152, 153
Coniferous forest biome, 121–124, 122f
Continental crust, 18
Contra Costa County communities, 95
Convectional precipitation, 72–74
Convergent plate boundary, 19
Copper Canyon Fan, alluvial fans at, 30f
Cotton, 191, 192f
Counties, 145–146
CRA. *See* Colorado River Aqueduct
Creosote bush (*Larrea tridentata*), 130
Crossbreeding, 54
Culture, 48
diversity of, 87
geographic oddities, 88–104
agriculture, 96–98
beach scene, 101–102
gold country, 91–92
isolation/uniqueness, 88
logging Paul Bunyan style, 90–91
neon glitter, 98–101
perceptual regions, 88–90, 89f
resort mecca and commercial hub, 103–104
San Francisco Bay Area,
sophistication, 94–96
wine country, 92–94
Cupressus macrocarpa. See Monterey cypress
CVP. *See* Central Valley Project

D

Dana Point, beach areas of, 102
Dana, Richard Henry, 5
Dawn redwood (*Metasequoia glyptostroboides*), 122
Death Valley
    alluvial fan at, 30f
    by block faulting, 34–35
    computer rendering of, 17f
    flood, 74f
    temperature, 70
de Galvez, Jose, 47
Delicatessens, 101
Dendritic drainage pattern, 29
Dendrochronology, 155
Department of Water Resources, 244
de Portolà, Gaspar, 2–3
Depositional processes
    coast landforms, 30
    connecting erosional and, 27
    glacial landforms, 28
Desert climate, 83–85
    in Anza Borrego Desert State Park, 83f
Desert scrublands and woodlands biome, 129–131
Disneyland, 185, 187f
Divergent plate boundary, 19
Diversity
    of Bay Area, 94
    of culture, 87
Divorce rates, 239
Doctrine of correlative water rights, 168
"Domestic use" theory, 166
Dominant species, 125
Downtown Los Angeles, temperature, 69
Drake, Francis, 47, 88
Drip irrigation systems, 244
Droughts, 154–157, 156t, 157f
Dry farming, 56–57
Dry steam power plants, 225

E

Earthquakes, 17, 22–27
    contributors to, 19
    convergent plate boundary, 19
    destruction, 23
    displacement caused by, 24
    divergent plate boundary, 19
    energy released during, 22
    epicenter of, 22
    fire due to, 24, 25f
    focus of, 22
    with highest magnitude, 24
    intensity by Mercalli Scale, 23, 23f
    magnitude of, 22–23
    mortality due to, 24, 26
    in Northridge, 25–27, 26f
    prediction and future, 27
    property loss due to, 24
    in remote area, 23
    San Fernando, 23
    San Francisco, 24–25
    structure of, 22f
    transform plate boundary, 19
Eastern Highlands, 34
Eaton, Fred, 171, 172
Ecological footprints, 219–220
Economic geography
    agriculture, 189, 189t
        farms and farmers, 190
        field crops, 191–192
        livestock and poultry, 191, 191f, 191t
        natural resources, 190
        opportunities and challenges, 193–194
        products and distribution, 190
        vegetables, fruits, and nuts, 192–193
    climate, 179
    entertainment industry, 182–184
    housing market, 194–195, 194f, 195f
        coastal *vs.* inland markets, 195, 196f
    location, 179
    natural resources, 179
    population, 180–182, 180f, 181t, 182t
    Silicon Valley, 196–197, 197f
    tourism, 184–189
Economics
    agriculture's role, 137–138
    growth, military-related, 62
    impacts, changing, 134
    readjustments, 62
    slump, 63
Economy, 1, 62, 248
Education, 248–249
    in rural regions, 139
Electricity, 225
Electric vehicles, 227–228
Elementary school districts, 147
Elephant seals, 208, 209f
El Niño, 205
El Niño southern oscillation (*ENSO*), 158
Employment
    challenge in rural areas, 139
    in defense-related industries, 62
    sector, 236

Enclave villages, 144
Encounter restaurant at LAX, 107f
Endemic species, 117–118
Energy, 218–219, 232
    efficient house design, 232, 232f
    sources, 218f
    usage, 10–11
Enhydra lutris. See Sea otters
ENSO. See El Niño southern oscillation
Entertainment industry, 182–184
Entrepreneurs, 57
Environmental lapse rate, 69, 84
Environmental pollution, 236
Environmental Protection Indicators for
    California, 220
Epicenter of earthquakes, 22
Eriophyllum confertiflorum. See Golden yarrow
Erosional processes
    coast landforms, 30
    and depositional processes, connecting, 27
    glacial landforms, 28
Eschrichtius robustus. See Gray whale
Ethnic groups of California, 109–110
Ethnicity, 238–239
European beachgrass, 212
European exploration
    mission settlement patterns, 47–48
    Spanish dominance, 46–47
European immigrants, 46
European settlement patterns, 43
European-style farming, 45
Evaporation, 152, 153
Evidence of human presence, 44
External forces and processes, 27–31
Extrusive igneous rocks, 22

F

Fan palm, 130–131, 130f
Farallon Plate, 19
    shifted westward, 21
    subduction of, 21
Farm, 190
    products, 240
    workers, unionization of, 98
Farm Belt, 96–98
Farmers, 190
Fast-food establishments, 106
Fault distribution in California, 20f
Federal Clean Water Act, 207
Federal Regulations, 240–241
Feedlot cattle, 191

Festuca californica. See California fescue
Field crops, 191–192
Films, 58–59
Fire
    adaptation to, 129
    due to earthquake, 24
Fishing, 180
Floodplains, 29, 29f
Floods, 157–158
    farming, 44
Flora
    geographic divisions, 118–119
    taxonomic divisions, 116–118
Floristic provinces, 119
    Californian, 119–120, 120t, 181f
    Great Basin, 121
    Sonoran, 121, 130
    Vancouverian, 120–121
Fluvial landforms, 15, 28–29
Focus of earthquakes, 22
Fog, 75f
    and clouds, 74–77
Food, 217–218
    production system, 233
Foothill pine (*Pinus sabiniana*), 125
Foreign immigrants, 62, 239
Foreign incursions
    beef industry, rise of, 53–54
    decline of California Indian population, 52–53
    gold rush, 50–52, 51f
    mountain men, sailors, pioneers, and heroes, 50
Fort Ross, 47
Fruits, 192–193
Fuel cell cars, 228
Funeral, California, 108–109
Fur traders, 50

G

Galactic metropolis, 143
Garreau, Joel, 144
Gentrification, 142–143
Geographic determinism theory, 43
Geographic divisions of California flora, 118–119
Geographic location, 11–13, 11f, 275f
Geological history, 19–22
Geomorphic regions, 16f, 162f, 277f
    Central Valley, 39–40
    Coast Ranges, 33–34
    division, 31
    Great Basin and Southeast Deserts, 34–35
    Klamath Mountains, 38

Geomorphic regions (*Cont.*)
    Modoc Plateau and Southern Cascades, 38–39
    Peninsular Ranges, 31–32
    Sierra Nevada, 35–38
    Transverse Ranges, 32–33
Geopolitical landscape, 146f
    cities, 146–147
    counties, 145–146
    school districts, 147
    special districts, 147
Geothermal energy, 225
Geothermal power, 225–226
Geysers, 225
Ghost towns, 91–92
Giant kelp. *See Macrocystis pyrifera*
Giant sequoia (*Sequoiadendron giganteum*),
    115, 116f, 117, 122
    and coast redwood, 121t
Glacier landforms, 15, 28
    in Sierra Nevada, 16f, 84
Global positioning systems, 232
Gold
    production, 52
    rush, 50–52, 54
Golden Gate Bridge, 74, 94, 96
Golden yarrow (*Eriophyllum confertiflorum*), 116
Goods movement, problem of, 245–248
Gottmann, Jean, 143
Grain silo, 137f
Granite, 18, 22
Grapes, 193
Grapes of Wrath (Steinbeck), 6, 8, 59
Grass family (*Poaceae*), 116
Grasslands biome, 126–127, 127f
Gray water, 233
Gray whale, 208
Great Basin
    floristic provinces, 121
    and Southeast Deserts, 34–35
Great Basin bristlecone pine (*Pinus longaeva*),
    117, 118f
Great Valley. *See* Central Valley
Greeley, Horace, 7
Gridlock, 148
Gross state product (GSP), 201
Groundwater, 152, 153, 159, 161, 161f
    basin, 159, 161f
    contamination, soil and, 242
    law, 168
    percolating, 168
GSP. *See* Gross state product
Gymnosperms, 117

H

Harbor seals. *See Phoca vitulina*
Harrington, John P., 53
Health and medical establishments of California, 236
Hetch Hetchy Valley Reseroir, 162
Hetch Hetchy water delivery system, 174, 175f
Highland climate, 84, 85
High Sierra, geological origin, 36
High-tech industry, 196
    in foothill region, 92
Hispanics
    foreign-born immigrants, 239
    population in California, 1
Hog wallows. *See* Vernal pools
Hollywood, 58–59, 59f, 182–184
Housing bubble, 195
Housing market, 194–195, 194f, 195f
    coastal *vs.* inland markets, 195, 196f
Hsi lai Buddhist Temple in Hacienda Heights, 101f
Human-created hazards, 242
Humboldt State University in Arcata, 91
Humpback whale. *See Megaptera novaeangliae*
Huntington Beach, 102
    offshore oil rigs at, 57, 57f
Hurricanes, 80–81
Hutchinson, W. H., 46
Hybrid vehicles, 227–228
Hybrid wind–solar power generator, 226
Hydraulic mining, 165, 166f
Hydrogen fuel technology, 228
Hydrogen-powered cars, 228–229
Hydrologic cycle water resources, 151–153,
    152f, 153f, 155f
Hydrologic regions, 156f, 161–163
Hydropower, 223–225

I

Ice sheets, 28
Immigration, 1, 8, 44, 238–239
    African American to California, 110
    large-scale, 87
Imperial sand dunes, 30f
Indian paintbrush (*Castilleja pruinosa*), 117
Individual tribal groups, 44
Industrial-manufacturing enterprise, 48
Industry
    agriculture. *See* Agriculture
    entertainment, 182–184
    high-tech, 196
    tourism, 184–189

Inland markets, coastal *vs.,* 195, 196f
Intensity of earthquake, 23
Interior drainage basins, 34
Internal forces, 27
In the Heart of the Valley of Love (Kadohata), 236
Intrusive igneous rock, 22
Irrigating crops, recycled water for, 245
Irrigation colonies, 56–57
Irrigation farming
  basic concept of, 56
  techniques in, 56
Irving, Washington, 5
Islands, 214–215
Isolation, 88
Isotherm maps, 66–67

J

Japanese Americans, 60
Jobs challenge in rural areas, 139
Joint school district, 147
Joshua tree (*Yucca brevifolia*), 115, 129f
Juan de Fuca Plate, 21
June gloom, 77

K

Kadohata, Cynthia, 236
Kino, Eusebio Francisco, 88
Klamath Mountains, 38
  complex stream systems in, 17f
  landforms in, 39f
    fluvial, 15
  metamorphic rocks, 22
  volcanic rocks at, 19
Köppen climate classification, 81
Kroeber, Alfred, 53

L

LADWP. *See* Los Angeles Department of Water
  and Power
Lagenorhynchus obliquidens. See Pacific
  white-sided dolphin
Land
  breezes, 79–80, 80f
  building processes, 17, 23
  commission, 54
  cultivation, 56
  degradation, 220
  transportation, 245
  water and. *See* Water, and land

Land Act of 1851, 54
Landforms
  aeolian processes and, 29
  in Central Valley, 41f
  coastal processes and, 29–31
  in Coast Ranges, 33f
  definition, 15
  fluvial process and, 28–29
  glacial process and, 28
  in Klamath Mountains, 39f
  in Northeast Region, 39f
  in Sierra Nevada, 37f
  in Southeast Deserts, 35f
  in Transverse Ranges and Peninsular
    Ranges, 32f
Larrea tridentata. See Creosote bush
Last Days of the Late, Great State of California,
  The (Gentry), 236
Latin American population, 101
Latitude, temperature controls, 66–67, 69f
Leisure World, 107–108
Lessinger, Jack, 144–145
Lewis, Sinclair, 5
Lifestyles
  change in California, 8–9
  choices of, 133
Light-rail system, 247
Limited resources in rural areas, 137
Literary visions of California, 5–6
Lithospheric plates, 18
Littoral drift, 212
Livestock, 191, 191f, 191t
Living places, 143–145
Local government subdivisions, geopolitical
  landscape of, 145–147
Lodgepole pine (*Pinus contorta*), 124
Loma Prieta earthquake, 6, 96
  on Bay Bridge, 26f
London, Jack, 5
Los Angeles
  annual precipitation variations in, 71, 71f
  location, 21
  neon glitter, 98–101
  real estate investment in, 239–240
  riots, 239
  skyline of, 94, 101f
  traffic in, 246–247
Los Angeles Aqueduct, 171–172
  Mono Lake extension of, 172–174
Los Angeles Basin, 16
  folding and faulting system, 31–32
  photochemical smog in, 75, 76f

Los Angeles City Water Company, 171
Los Angeles Department of Water and
    Power (LADWP), 172
Lowest-elevation montane forest, 124
Low population density in rural areas, 136
Lumber, role of, 90

M

Macrocystis pyrifera, 204
Magma, 19
Magnitude of earthquake, 22–23
Mammoth Crater, 39
    at Modoc Plateau, 40f
Manhattan Beach, 102
Manzanar Detention Camp, 60f
Marine Mammal Protection Act (MMPA), 209
Marshall, James, 51
Marshes biome, 126–127, 127f
May gray, 77
McTeague (Norris), 5
McWilliams, Carey, 7
Meandering streams, 29
Median housing, 194, 194f
Medicine, advances in, 235
Mediterranean climate, 81–82, 85
    temperature of, 82
Megalopolis, 143–144
Megaptera novaeangliae, 208, 208f
Mendocino Highlands, 34
Mercalli, Giuseppe, 23
Mercalli scale, 23
    earthquakes intensity by, 23f
Mesquites (*Prosopis spp.*), 117
Metamorphic rocks, 22
Metasequoia glyptostroboides. See
    Dawn redwood
Metrolink trains, 247
Metropolitan water district (MWD), 171
Mexican–American War, 50
Mexican food, 106
Mexican laborers and blacks, 60
Mexican period, water usage, 164–165
Mexican restaurant, 110
Mexican revolution, 49–50
Mexican rule in California, 3
Micropolitan, 144
Midwestern Graffiti, 7–8
Migration, 238–239
    growth, 63
Military-related economic growth, 62

Mimulus rattani, 117
Mission Bay, 103
Mission San Juan Capistrano, 3f
Mission settlement patterns, European
    exploration in California, 47–48
MMPA. *See* Marine Mammal Protection Act
Mobile home parks
    age ghettos, 107
    in San Jose, 107f
Modoc Plateau
    extrusive igneous rocks, 22
    lava beds, 39f
    lava tube at, 40f
    Mammoth Crater at, 40f
    and Southern Cascades, 38–39
Modoc tribes, 52
Mojave tribes, 44
Mono Lake, 172–174, 174f
Montane forest community, 122
Monterey cypress (*Cupressus macrocarpa*),
    117, 118, 118f
Moraines, 28
Mormon settlement in San Bernardino, 56
Mortality, 24, 26
Mother Lode region, 38, 50, 52, 91, 92
    routes into, 51
Mountain, winds in, 80, 80f
Mount Lassen, 39
    volcanic landforms in, 15
Mount Shasta, 16f, 38–39
    volcanic landforms in, 15
Mount St. Helens, 39
Movie industry, 59
Mulholland, William, 171
Municipal corporations, 146
Museums, 100
MWD. *See* Metropolitan water district

N

NAFTA. *See* North American Free Trade Act; North
    American Free Trade Agreement
Napa Valley vineyard, 93
Nassella pulchra. See Needle grass
National Environmental Policy Act of 1969, 240
National parks, 184, 185
    annual recreational visits to, 187t
    distribution of, 186f
Native Americans
    culture, 46
    out-migration of, 239

period, water usage, 164
population, 44
Natural disaster, 236
Natural environment, 240
Federal Regulations, 240–241
State Regulations, 241–243
Natural gas, 219
Natural resources
agriculture, 190
economic geography, 179
Needle grass (*Nassella pulchra*), 116
Neon glitter, Southern California and Los Angeles, 98–101, 99f, 101f
Newport Beach, 103f
Nodal region of California, 89
Non-hydropower renewable energy sources, 219
Nonnative grasses, 126
Nonorganic farming, 231
Norris, Frank, 5, 55
North American Free Trade Act (NAFTA), 238
North American Free Trade Agreement (NAFTA), 104
North America Plate, 18, 19
movements, 21, 22
Transverse Ranges from, 22, 32
North coastal forest community, 122
North Coast Region, 162–163
Northern oak woodlands, 124–125
North Lahontan Region, 163
Northridge earthquake, 23, 25–27, 26f
Nuclear power, 227
Nuts, 192–193

O

Oakland–East Bay area, 94
Oakland Hills, firestorm demonstration, 242
Oak woodlands biome, 124–125, 125f
anthropogenic impacts on, 126
Obama, Barack, 148
Ocean currents, temperature controls, 70
Oceanic crusts, 18
Ocean transportation, 245
Octopus, The (Norris), 5, 54
Offshore oil rigs at Huntington Beach, 57, 57f
Offshore waters, 207–208
Oregon oak (*Quercus garryana*), 125
Organic farming, 231
Orographic precipitation, 72
Owens Valley, 60, 60f, 164, 171–172
Oxygen fuel technology, 228, 229f

P

Pacific Century, 239–240
Pacific Electric Red Cars, 248
Pacific Plate, 18, 19, 22, 32, 33
plate movements, 21
shifted westward, 21
sliding along San Andreas Fault, 21
Pacific Rim, 239–240
Pacific white-sided dolphin, 208, 209f
Palm Springs, 111
Palo verde (*Cercidium spp.*), 117
Pangaea, 18
Part-time services in rural areas, 136
Paul Bunyan style, logging, 90–91
Pearl Harbor
attack on, 59
Japanese bombing of, 59
Peninsular Range, 16, 31–32
L.A. Basin on, 17f
landforms in, 32f
Penturbia, 144–145, 145f
People of California, 237
growth and population, 237
immigration, migration, and ethnicity, 238–239
social fabric, 239
Perceptual regions of California, 88–90, 89f
Perennial grasses, 126
Permeability, 153
Permit system, 167
Petco Park, 112f
Petroleum, 219
Phoca vitulina, 204, 208, 209f
Photochemical smog, 75–77
in Los Angeles Basin, 75, 76f
Photovoltaic cells (PVCs), 223
Physeter macrocephalus. See Sperm whales
Physical landscapes in California, 16–17
Pinus contorta. See Lodgepole pine
Pinus longaeva. See Great Basin bristlecone pine
Pinus sabiniana. See Foothill pine
Pinyon-juniper woodlands community, 130
Plants
biomes, 119
communities, 119
human-caused reductions in, 127t
introduction of, 131, 131f
management methods, 45–46
Plate boundary zone, 22
Plate tectonics, 18–19, 18f, 21
Central Valley, 40
dynamics along, 19f

Pleistocene, 36
Poaceae. See Grass family
Point Reyes National Seashore, coast
    landforms at, 17f
Political subdivisions, 145
Pollution of water resources, 244–245
Population, 1, 6, 7, 50
    challenges, 148
    distribution, 12f, 135f, 276f
    growth, 61, 63, 148, 237, 242
    movement, problem of, 245–248
    trends in rural California, 138–139
Ports, permit control of, 147
Poultry, 191, 191f, 191t
Precipitation, 70–71, 71f, 152, 153, 279f
    controls of, 72–74
      orographic process, 72
      pressure belts and storm tracks, 72
    convectional, 72–74
    orographic, 72
Pressure belts
    global wind and, 73f
    precipitation controls, 72
Prisons, 239, 242, 249
Property loss, earthquake, 24, 129
Prosopis spp. See Mesquites
Public health, 46
Public policy makers, 227
Public transportation system, 229
Pueblo de Los Angeles, 48, 48f
PVCs. *See* Photovoltaic cells

Q

Quality of life, 63
Quercus agrifolia. See Coast live oak
Quercus douglasii. See Blue oak
Quercus garryana. See Oregon oak
Quercus lobata. See Valley oak

R

Racial attitudes, 60
Racial diversity in rural regions, 140
Radiation temperature inversion, 75
Radioactive material, 227
Railroads, 54–56, 245–246
Rain shadow, 72, 84
Raker Act (1913), 174
Ranchos, 49, 50, 54, 102, 165
Real estate investment in Los Angeles, 239–240

Reasonable use theory, 166
Recreation, 111
    in rural regions, 140
Recycled water for irrigating crops, 245
Red fir forest (*Abies magnifica*), 124
Redwood forests in northwest California,
    90, 90f
Redwoods (*Sequoia sempervirens*), 115
Relative humidity, 72, 74, 77
Relict species, 117–118
Renewable energy
    resources, 224f, 284f
    source, 227. *See also* Alternative energy
      sources
Residential energy consumption, 219
Resort community, 111
Restaurants, 87, 106–107
Retrofitted neighborhood, 144f
Rice, 191–192, 193f
Richter scale, 22–23
Rim Fire, 85
Riparian water rights, 166, 167
Roads, 8–9, 138f, 246
Rock
    oldest, 19
    structure and distribution, 22
    volcanic, 19
Roosevelt, Franklin, 59
Routes into California, 50
Runoff, 153
    statewide, 158f
Rural California, 135–137, 136f, 141
    agriculture's role in, 137–138
    challenges and problems in, 139–140
    economic, 137
    fire stations, 136f
    population trends, 138–139
    roads in, 138f
    small size and scale in, 136
Rural health care in rural regions, 140

S

Sacramento River Basin, 40
    flood plain at, 29f
Sacramento River Region, 161
Sacramento–San Joaquin Delta, 245
Sagebrush scrub community, 130
Sailors, 50
Salad bowl culture, 109–111
Salinas Valley, 34
Salton Sea, 163

San Andreas Fault, 19
   aerial view of, 21f
   earthquake, 24, 25
      prediction, 27
   northern section movement, 24
   Pacific Plate sliding along, 21
   southern section movement, 24
   transform plate boundary, 21
   and Transverse Ranges, 22, 33
San Clemente, 102
San Diego County
   at Camp Pendleton, 102
   issue for, 104
   resort mecca and commercial hub, 103–104
   temperature, 70, 70f
San Fernando earthquake, 23, 33
San Francisco
   earthquake, 24–25, 25f
   real estate, 240
   World Series broadcast in, 6
San Francisco Bay, 34
   Coastal fog in, 75
   Plan, 243
   sophistication, 94–96
San Francisco Bay Region, 163
Sanitation, 46
San Jose–Santa Clara Valley, 62
San Jose–South Bay area, 94
San Pablo Bay, 34
Santa Ana winds, 77–78, 78f
   pressure and wind patterns during, 78f
Santa Barbara
   coastal area in, 30f
   exclusive, wealthy security of, 102
   oil spill, 57–58, 57f
   sundown wind of, 78–79, 79f
Santa Barbara Mission, 48, 48f, 165
Santa Cruz Boardwalk Beach amusement
      park, 105f
Santa Cruz monkey flower, 117
Santa Monica Mountains
   earthquake in, 26
   sedimentary rocks in, 32
Santa Rosa Island, Southern California
      tradition on, 44
Santa Ynez Valley, Solvang in, 102f
Sardine fisheries, 207–208
Saroyan, William, 6
SCB. See Southern California Bight
Scenic Mount Lassen in northeastern
      California, 91f
School districts, 147

Schwarzenegger, Arnold, 242
Sea breezes, 79–80, 80f
Seafood, 106
Seal Beach, 102
Sea lions, 208
Sea otters, 202, 203f
Sea stacks, 30
Sea Wolf, The (London), 5
Secondary school districts, 147
Secularization, 165
Security and prosperity of California, 236
Sedimentary rocks, 21, 22
   in Santa Monica Mountains, 32
Self-destruction, California, 236
Sequoiadendron giganteum. See Giant sequoia
Sequoia sempervirens. See Coast redwood;
      Redwoods
Serra, Junipero, 48
Shamanism, 46
Shelter, 218
Sierra Nevada, 33, 35–38, 91
   Eastern, 38f
   glacial landforms in, 15, 16f
   glaciers in, 84
   intrusive igneous rock, 22
   landforms in, 37f
   precipitation in, 70
   temperature, 69, 69f
   volcanic rocks at, 19
   Yosemite Valley in, 28, 28f
Sierra snowpack, 158f
Silicon Valley, 62, 196–197, 197f
   computer industry, 95
Sinclair, Upton, 5
Ski resorts, 65
Skull Cave, 40f
Skyline of Los Angeles, 94, 101f
Small-scale solar technology development, 226
Small–scale wind power generation, potential of, 226
Smith, Jedediah, 4
Smog, photochemical, 75–77
Snowfall, time series of, 160f
Snowpack, 158–159, 158f, 160f
   time series of, 160f
Snow surveying process, 159
Social fabric, 239
SOD. See Sudden Oak Death
Soil contamination, 242
Solar energy, 222–223
Solar radiation, 67
Solid wastes, 220–222
   management systems, 230

Solvang, 187, 188f
  in Santa Ynez Valley, 102f
Sonoma County, 47
Sonoran floristic provinces, 121, 130
Soule, J. B. L., 7
South Coast Air Basin, 6
South Coast Region, 163
Southeast Asia population, 101
Southeast Deserts
  Great Basin and, 34–35
  landforms in, 35f
Southern California
  beach scene, 102
  neon glitter, 98–101
  tradition on Santa Rosa Island, 44
Southern California Bight (SCB)
  municipal treatment plants, sewage, 206–207
  stormwater runoff, 206, 206f
Southern Cascades
  Modoc Plateau and, 38–39
  stratovolcano of, 16f
Southern coastal playgrounds, beach-and-seashore
     scene of, 101–103
Southern oak woodlands, 125
Southern San Joaquin Valley, irrigation, 169f
South Lahontan Region, 163
Space, 148
Spain, American colonies of, 49
Spaniards, California and, 2–4
Spanish period, water usage, 164–165
Special districts, 147
Species diversity, 116t
Spectator sports, 111–112
Sperm whales, 208
Sprawl, 148
Starbucks, 107
State parks, 185, 186f
State Regulations, 241–243
State Water Project (SWP), 157, 169
Steinbeck, John, 6, 8
Stereotype, 101
Storm tracks, precipitation controls, 72, 73f
Stovepipe wells, 129f
Stratovolcano of Southern Cascades, 16f
Subalpine forests, 123, 124
Sublimation, 28
Subterranean stream, 168
Suburban areas, 141–142, 142f
Suburbanization, impact of, 97
Sudden Oak Death (SOD), 126
Suicide rates in California, 239
Summer dry patterns, characteristics of, 82

Sundial Bridge, 188f
Sundowner. *See* Sundown wind
Sundown wind of Santa Barbara, 78–79, 79f
Surface transportation, 245
Sustainability, 233, 234
Sutter Buttes, 40
Sutter, John, 51
SWP. *See* State Water Project
Synagogues, 101

T
Tahoe Regional Planning Agency, 243
Taxonomic divisions of flora, 116–118
Technology development, 249
Temperature, 65–66
  controls, 66–70
    latitude, 66–67, 69f
    ocean currents, 70
    topography, 67–69
  in desert valleys, 83
  distribution, 66f, 67f, 280f
  inversion, 69, 74
    radiation, 75
    type, 75
  of Mediterranean climate, 82
Terminal moraines, 28
Terre Haute Express (Soule), 7
Texas, tornadoes in, 80–81
Third Wave, The (Toffler), 134
Tidelands Act of 1955, 57
Timber production, 90
Topography, temperature controls, 67–69
Tornadoes, devastating hurricanes and, 80–81
Tourism, 184–189
Tract homes in San Diego, 61, 61f
Trade centers, permit control of, 47
Traffic congestion, 148, 148f, 246
Transcontinental railroad, 88
Transform fault. *See* Transform plate
     boundary
Transform plate boundary, 19, 21
Transpiration, 152, 153
Transportation, 54–56, 152, 153
  alternative systems, 229–230
  energy consumption, 219
  ocean, 245
  problem of movement, 245–248
Transverse Ranges, 15–16, 32–33
  landforms in, 32f
  from North America Plate, 22, 32
  San Andreas Fault and, 22, 33

Trappers, 50
Treaty of Guadalupe Hidalgo, 54
Tree, vertical cross section of elevation, 123–124
Tribal peoples, American impact on, 52
Tulare Lake, 24, 162
Tule fog, 75
Tuolumne River water delivery system, 174, 175f
Turlock Irrigation District, 168
Two Years before the Mast (Dana), 5

U

Unemployment, 59
Unified school district, 147
Uniform region, 89
Unionization of farm workers, 98
Uranium-235 (U-235), 227
Urban areas, 136, 142–143
Urban heat island effect, 233
Urbanization, 63, 97, 239
Urban sprawl in San Diego, 61, 61f
Urban villages, 143–144
US Geological Survey (1920), 169

V

Valley
   deepening and widening, 17
   deep V-shaped, 29
   hanging, 28
   U-shaped, 28
   winds in, 80, 80f
Valley oak (*Quercus lobata*), 124, 125f
Vancouver, George, 4
Vancouverian floristic provinces, 120–121
Vascular plants, taxonomic groups of, 116, 117t
Vegetables, 192–193
Vegetation, 115
   adaptation to fire, 129
   biomes of. *See* Biomes
   destruction, 131
   floristic province. *See* Floristic provinces
Vernal pools, 127, 127f
Vertical wind turbines, 226
Vertical zonation, 36
Vidal de la Blache, Paul, 89
Violence, 236
Visalia, farm service community, 97
Volcanoes
   active, 18f
   landforms, 15
   rocks, 19

W

Wambaugh, Joseph, 6
Waste, sustainable home, 232
Water, 218, 232
   distribution system, development of, 168–169
      CRA, 171
      CVP, 170–171
      historic lake levels, 174
      Los Angeles Aqueduct, 171–174
      SWP, 169
      Tuolumne River and Hetch Hetchy, 174, 175f
   and land, 69
   percolation of, 153
   pollution, 220–222
   problem, 244–245
   quality, 241
   supply systems, 65
Water resources
   agriculture, 175–176, 176f
   description, 151
   future, 177
   history, 163–164
      American period, 165–166
      appropriative rights, 166–167
      doctrine and water code, 167–168
      evolution of water rights, 166
      groundwater law, 168
      Native American period, 164
      riparian rights, 166
      Spanish and Mexican period, 164–165
   hydrologic cycle, 151–153, 152f, 153f, 155f
   natural environment, 174–175
   pollution, 244–245
   state hydrology, 154, 154t
      droughts, 154–157, 156t, 157f
      floods, 157–158
      groundwater, 159, 161, 161f
      hydrologic regions, 156f, 161–163
      snowpack, 158–159, 158f, 160f
   urban, 176–177
   usage
      impacts and issues, 174–177
      rules, 167–168
Watershed, 161, 161f
Westerly winds, 77
Western red bud (*Cercis occidentalis*), 117
Wetlands, 213–214
White buckeye (*Aesculus californica*), 125
White-collar workers, 62
White flight, 98
Whitman, Walt, 5

Wiggins, Kate Douglas, 5
Wildfires, 242
Wind
  farm, 11f
  in mountains and valleys, 80, 80f
  power, 226
  Santa Ana, 77–78, 78f
  sea and land breezes, 79–80, 80f
  sundown, 78–79
  westerly, 77
Wine Country, 92–94
Winter rain patterns, characteristics of, 82
World War II, 59–61

Wright Act of 1887, 56
Wright Irrigation District Act (1887), 168

Y

Yosemite Valley, 28, 28f
Yucca brevifolia. See Joshua tree
Yuman tribes, 44

Z

Zanja madre, 165
Zelinsky, Wilbur, 89
Zero Waste, 230–231, 233

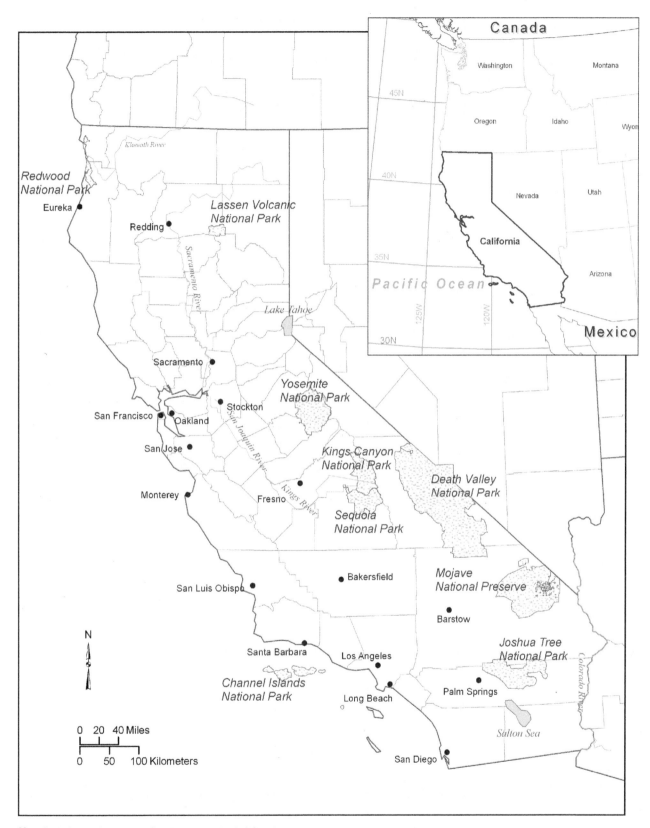

Map 1 Selected Geographic Features in California.

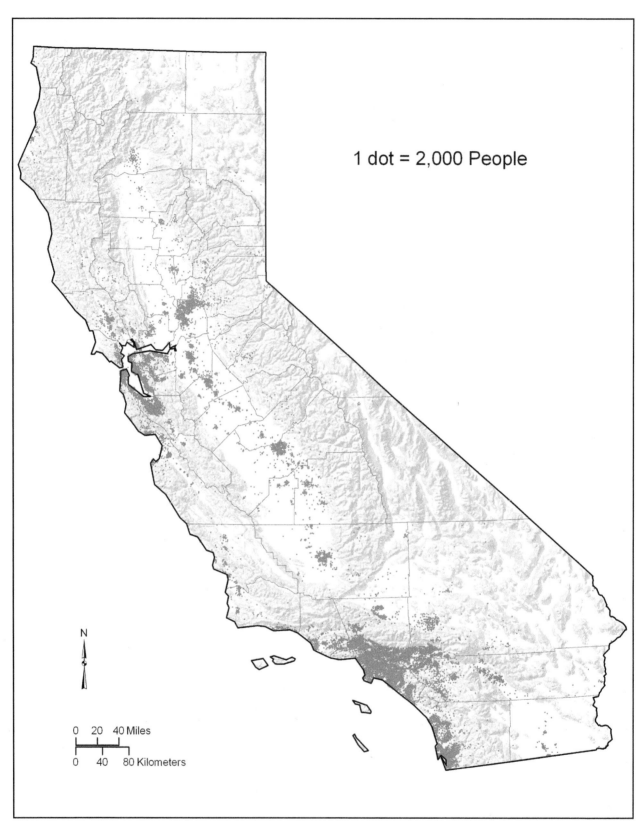

Map 2 California Population Distribution.

(Sources: US Geological Survey, ESRI Data and Map, US Census Bureau, Lin Wu)

1 dot = 2,000 People

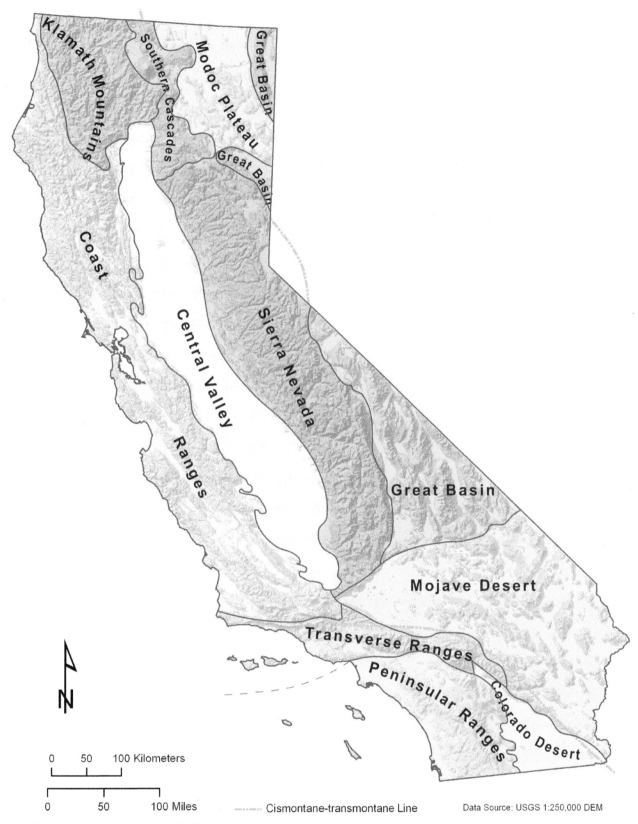

Klamath Mountains

Southern Cascades

Modoc Plateau

Great Basin

Great Basin

Coast

Central Valley

Sierra Nevada

Ranges

Great Basin

Mojave Desert

Transverse Ranges

Peninsular Ranges

Colorado Desert

N

0   50   100 Kilometers

0   50   100 Miles

Cismontane-transmontane Line

Data Source: USGS 1:250,000 DEM

Map 3 Geomorphic Regions of California.

(Source: Lin Wu)

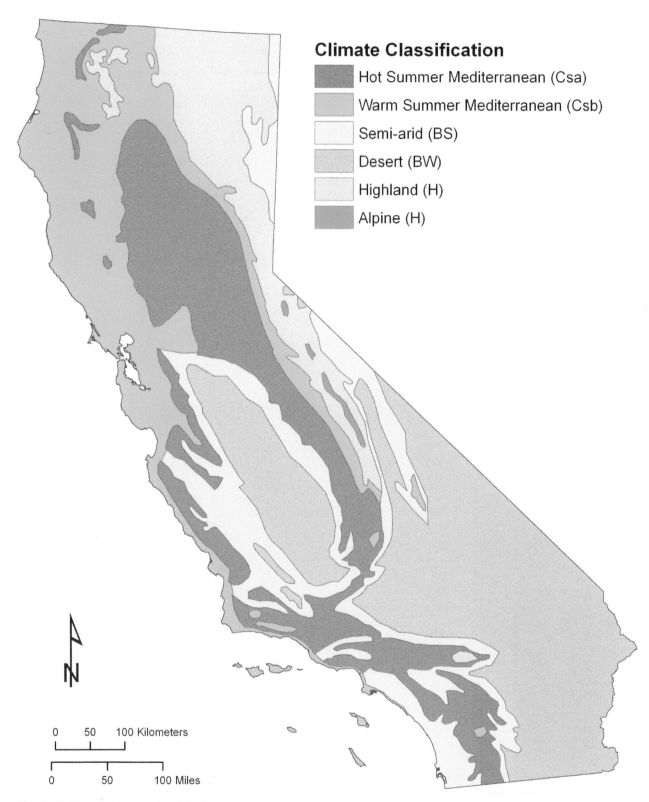

Climate Classification

- Hot Summer Mediterranean (Csa)
- Warm Summer Mediterranean (Csb)
- Semi-arid (BS)
- Desert (BW)
- Highland (H)
- Alpine (H)

0   50   100 Kilometers

0   50   100 Miles

Map 4 California Climate Classification.
(Source: Lin Wu)

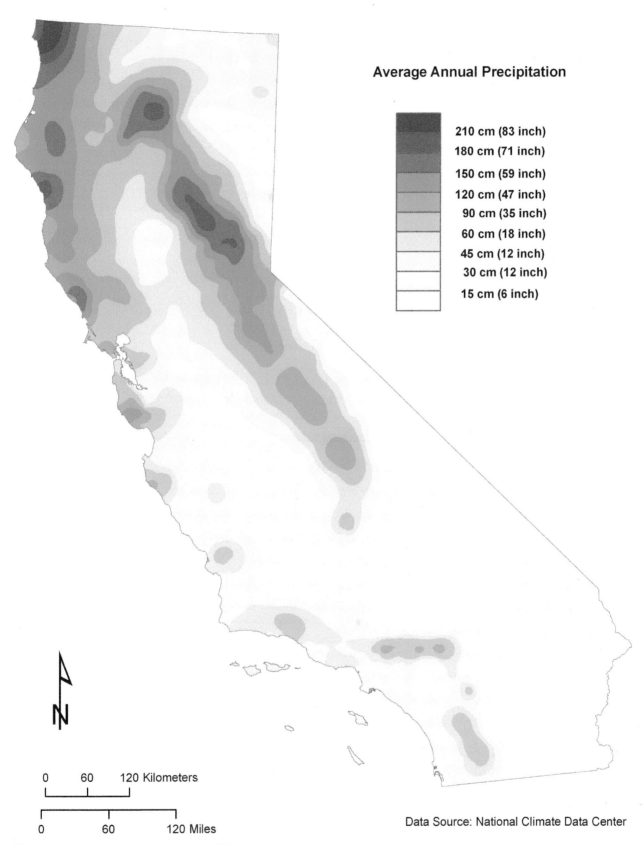

Average Annual Precipitation

210 cm (83 inch)
180 cm (71 inch)
150 cm (59 inch)
120 cm (47 inch)
90 cm (35 inch)
60 cm (18 inch)
45 cm (12 inch)
30 cm (12 inch)
15 cm (6 inch)

N

0   60   120 Kilometers

0   60   120 Miles

Data Source: National Climate Data Center

Map 5 Precipitation Distribution in California.
(Source: Lin Wu)

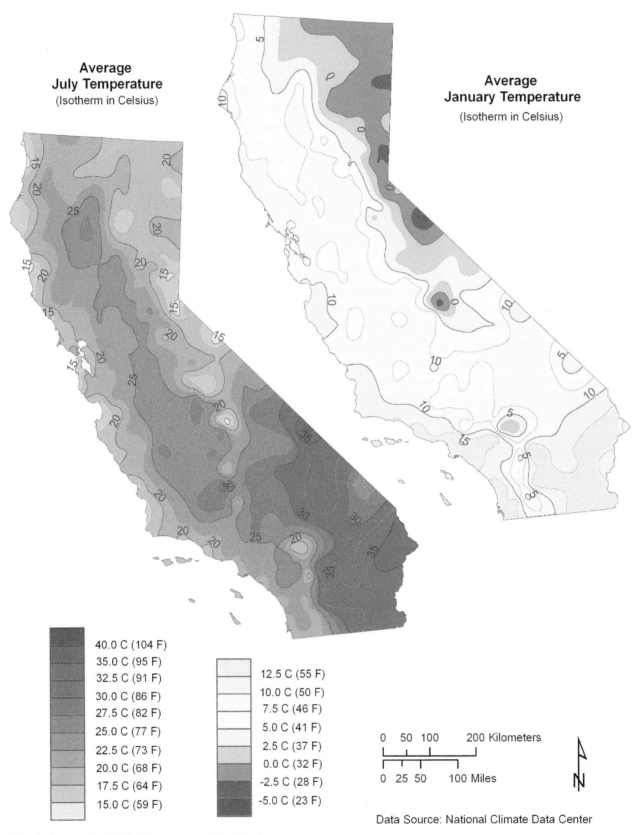

Average July Temperature
(Isotherm in Celsius)

Average January Temperature
(Isotherm in Celsius)

40.0 C (104 F)	
35.0 C (95 F)	
32.5 C (91 F)	12.5 C (55 F)
30.0 C (86 F)	10.0 C (50 F)
27.5 C (82 F)	7.5 C (46 F)
25.0 C (77 F)	5.0 C (41 F)
22.5 C (73 F)	2.5 C (37 F)
20.0 C (68 F)	0.0 C (32 F)
17.5 C (64 F)	-2.5 C (28 F)
15.0 C (59 F)	-5.0 C (23 F)

0   50   100   200 Kilometers

0   25   50   100 Miles

N

Data Source: National Climate Data Center

Map 6 January and July Temperature Distribution.
(Source: Lin Wu)

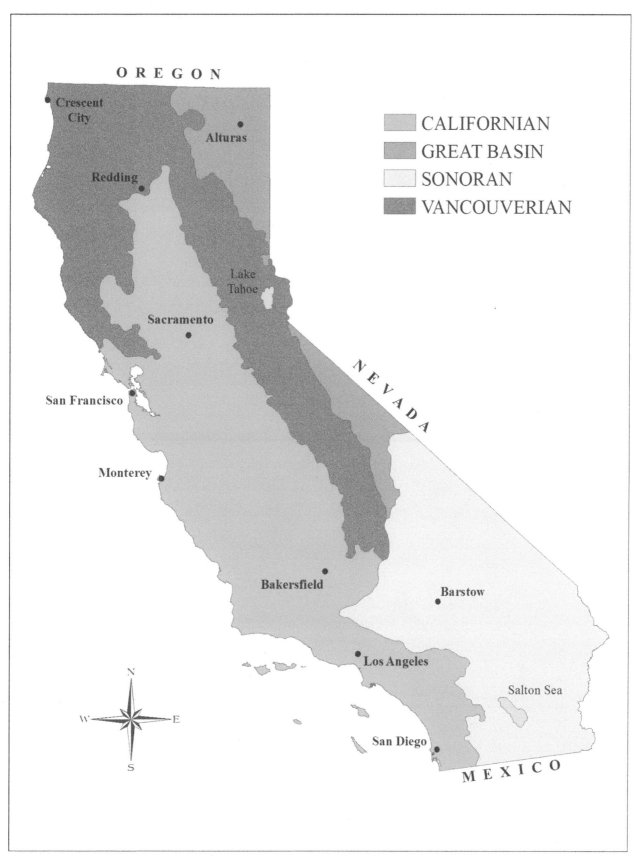

Map 7 Floristic Provinces of California.

(Source: Bryan Wilfley)

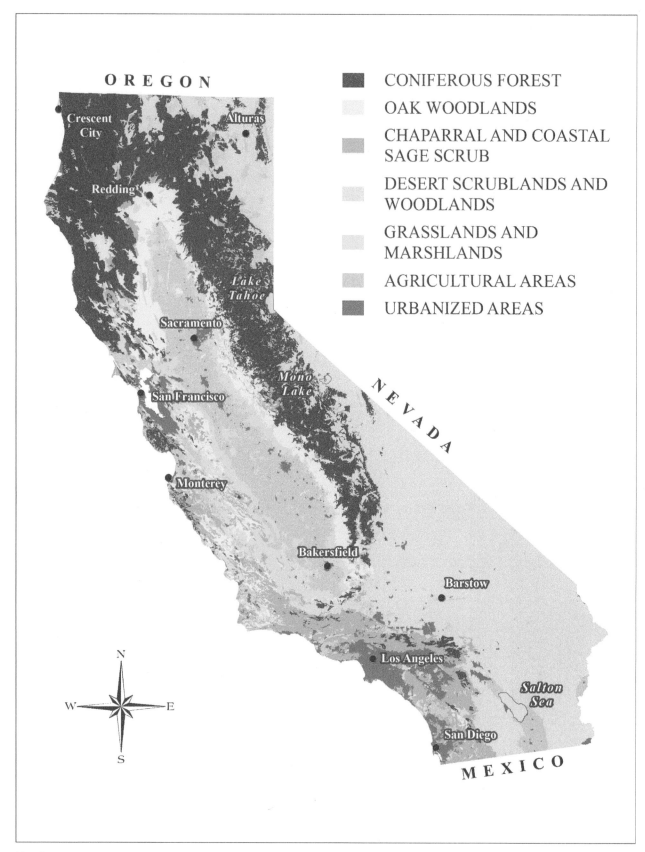

Map 8 Distribution of the Five Primary Biomes of California Vegetation.
(Source: Ari and Bryan Wilfley)

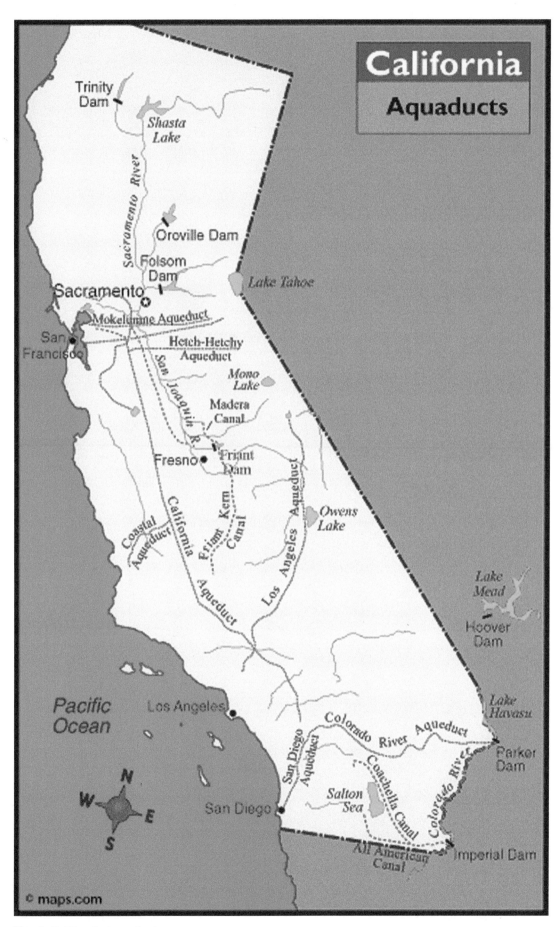

Map 9 California Aqueducts.

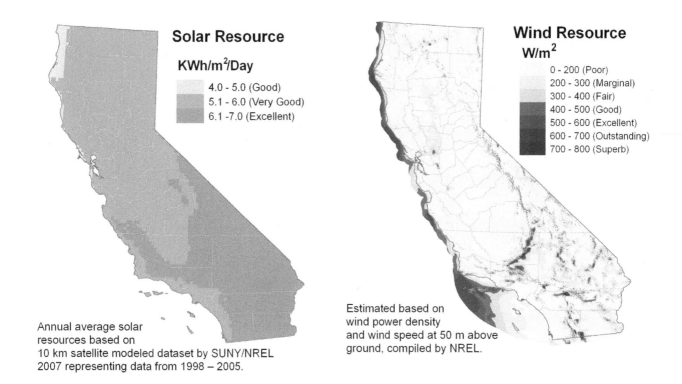

Solar Resource

KWh/m^2/Day

- 4.0 - 5.0 (Good)
- 5.1 - 6.0 (Very Good)
- 6.1 - 7.0 (Excellent)

Annual average solar
resources based on
10 km satellite modeled dataset by SUNY/NREL
2007 representing data from 1998 – 2005.

Wind Resource

W/m^2

- 0 - 200 (Poor)
- 200 - 300 (Marginal)
- 300 - 400 (Fair)
- 400 - 500 (Good)
- 500 - 600 (Excellent)
- 600 - 700 (Outstanding)
- 700 - 800 (Superb)

Estimated based on
wind power density
and wind speed at 50 m above
ground, compiled by NREL.

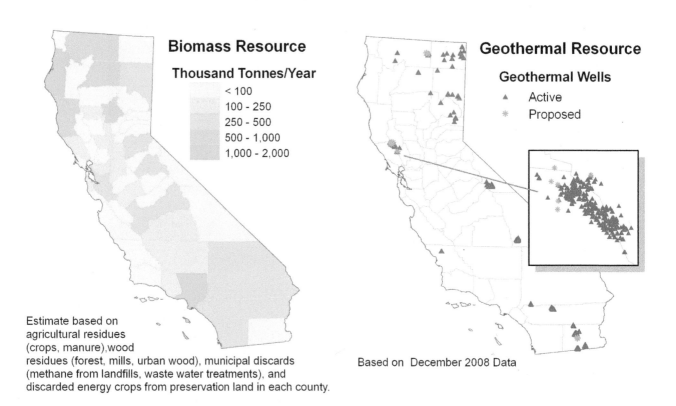

Biomass Resource

Thousand Tonnes/Year

- < 100
- 100 - 250
- 250 - 500
- 500 - 1,000
- 1,000 - 2,000

Estimate based on
agricultural residues
(crops, manure),wood
residues (forest, mills, urban wood), municipal discards
(methane from landfills, waste water treatments), and
discarded energy crops from preservation land in each county.

Geothermal Resource

Geothermal Wells

- ▲ Active
- ✳ Proposed

Based on December 2008 Data

Sources: Solar, wind, and biomass data are from National Renewable Energy Laboratory (NREL), Office of Energy
Efficiency and Renewable Energy, U.S. Department of Energy. Geothermal data are from California Department of
Conservation, Division of Oil, Gas, and Geothermal Resources

Map 10 California Renewable Energy Resources.

(Source: Lin Wu)

Figure 2.1b Mount Shasta, stratovolcano of the Southern Cascades.
(Photo: Lin Wu)

Figure 2.9 One of the many waterfalls in Yosemite Valley.
(Photo: Lin Wu)

Figure 3.2 Santa Barbara Mission—a jewel in the chain of California missions.
(Image © RonGreer.Com, 2009. Used under license from Shutterstock, Inc.)

Figure 4.18 A scene of desert climate in the Anza Borrego Desert State Park.

(Photo: Lin Wu)

Figure 5.10 Skyline of Los Angeles, a major metropolitan center of the United States.

(Image © Konstantin Sutyagin, 2009. Used under license from Shutterstock, Inc.)

Figure 4.12 A visible temperature inversion/smog layer against mountains surrounding the Los Angeles Basin viewed from Dodger Stadium during a baseball game.

(Photo: Samuel Yang)

Figure 6.5 Example of a coniferous forest biome. (Image © javarman, 2009. Used under license from Shutterstock, Inc.)

Figure 6.1 The famous giant sequoia (*Sequoiadendron giganteum*), found only on the western slopes of the Sierra Nevada.

(Photo: Kirsten Zecher)

Figure 6.10 Vernal pools in the Central Valley.

(Image © rayvee, 2009. Used under license from Shutterstock, Inc.)

Figure 7.2 Rural fire stations like this (Running Springs, CA) are typically staffed by volunteers and/or part-time "paid-call" firefighters.
(Photo: Richard Hyslop)

Figure 7.5 Lake Tahoe is a prime example of a rural locale with first-rate scenic amenities.
(Image © Mariusz S. Jurgielewicz, 2009. Used under license from Shutterstock, Inc.)

Figure 7.7 A representative suburban neighborhood in eastern Los Angeles County.
(Photo: Richard Hyslop)

Figure 8.8 Lingering snowmelt, July 2006. Virginia Lakes at 10,000′.
(Photo: Kirsten Zecher)

Figure 8.16 Present-day Mono Lake.
(Photo: Lin Wu)

Figure 9.8 A grape grower and his grandson at work in the Napa Valley.
(Photo: Kris Surber)

Figure 9.5 Mickey's neighborhood in Disneyland. With approximately 15 million visitors annually, the park is ranked the second most visited amusement park in the world.
(Photo: Lin Wu)

Figure 10.11 Kelp forest with starfish, bright orange garibaldi, swaying kelp, and brown sea fans.
(Image © Kelpfish, 2009. Used under license from Shutterstock, Inc.)

Figure 9.7 Sundial Bridge at Turtle Bay in Redding.
(Photo: Lin Wu)

Figure 6.8 Hills of Tehachapi, example of oak woodlands biome.
(Image © Richard Thornton, 2009. Used under license from Shutterstock, Inc.)